The Adequacy of the Fossil Record

The Adequacy of the Fossil Record

Edited by

STEPHEN K. DONOVAN

University of the West Indies, Jamaica

and

CHRISTOPHER R. C. PAUL

University of Liverpool, UK

JOHN WILEY & SONS

Chichester • New York • Weinheim • Brisbane • Singapore • Toronto

Other Wiley Editorial Offices

John Wiley & Sons, Inc., 605 Third Avenue,
New York, NY 10158–0012, USA

WILEY-UCH Verlag GmbH, Pappelallee 3,
D-69469 Weinheim, Germany

Jacaranda Wiley Ltd, 33 Park Road, Milton,
Queensland 4064, Australia

John Wiley & Sons (Asia) Pte Ltd, 2 Clementi Loop #02–01,
Jin Xing Distripark, Singapore 129809

John Wiley & Sons (Canada) Ltd, 22 Worcester Road,
Rexdale, Ontario M9W 1L1, Canada

Library of Congress Cataloging-in-Publication Data

The adequacy of the fossil record / edited by Stephen K. Donovan and Christopher R.C. Paul.
p.   cm.
Includes bibliographical references and indexes.
ISBN 0 471 96988 5 (cloth)
1. Paleontology, Stratigraphic. 2. Fossils. I. Donovan, S. K. II. Paul, C. R. C. (Christopher R. C.)
QE711.2.A33 1998
560'.17–dc21   98–10110
                                                                                            CIP

British Library Cataloguing in Publication Data

A catalogue record for this book is available from the British Library

ISBN 0 471 96988 5

Typeset in 10/12pt Palatino from the editor's disks by Vision Typesetting, Manchester
Printed and bound in Great Britian by Biddles Ltd, Guildford and King's Lynn
This book is printed on acid-free paper responsibly manufactured from sustainable forestation, for which at
least two trees are planted for each one used for paper production.

Contents

List of Contributors

Michael J. Benton, Department of Geology, University of Bristol, Bristol BS8 1RJ, UK

Martin A. Buzas, Department of Paleobiology, National Museum of Natural History, Smithsonian Institution, Washington, DC 20560, USA

Alan H. Cheetham, Department of Paleobiology, National Museum of Natural History, Smithsonian Institution, Washington, DC 20560, USA

Stephen J. Culver, Department of Palaeontology, The Natural History Museum, Cromwell Road, London SW7 5BD, UK

H. Allen Curran, Department of Geology, Smith College, Northampton, Massachusetts 01063, USA

Stephen K. Donovan, Department of Geography and Geology, University of the West Indies, Mona, Kingston 7, Jamaica

Benjamin J. Greenstein, Department of Geology, Cornell College, 600 First Street West, Mount Vernon, Iowa 52314–1098, USA

Elizabeth M. Harper, Department of Earth Sciences, University of Cambridge, Downing Street, Cambridge CB2 3EQ, UK

Jeremy B.C. Jackson, Center for Tropical Paleoecology and Archaeology, Smithsonian Tropical Research Institute, Box 2072, Balboa, Republic of Panama

Carl F. Koch, Program in Geological Sciences, Old Dominion University, Norfolk, Virginia 23529–0498, USA

Charles R. Marshall, Department of Earth and Space Sciences, Molecular Biology Institute, and Institute of Geophysics and Planetary Physics, University College of Los Angeles, Los Angeles, California 90095–1567, USA

David M. Martill, Palaeobiology Research Group, Department of Geology, University of Portsmouth, Burnaby Building, Burnaby Road, Portsmouth PO1 3QL, UK

John M. Pandolfi, Center for Tropical Paleoecology and Archaeology, Smithsonian Tropical Research Institute, Box 2072, Balboa, Republic of Panama

Christopher R.C. Paul, Department of Earth Sciences, University of Liverpool, PO Box 147, Liverpool L69 3BX, UK

Gregory J. Retallack, Department of Geological Sciences, University of Oregon, Eugene, Oregon 97403, USA

Peter J. Wagner, Department of Geology, Field Museum of Natural History, Roosevelt Road at Lake Shore Drive, Chicago, Illinois 60605–2496, USA

Introduction: Adequacy versus Incompleteness

> If sufficient fossils were available, it ought to be possible to reconstruct the actual course of a group's evolution, but Darwin warned about the 'imperfection of the fossil record' and believed that it would be difficult to obtain enough reliable information to make this kind of detailed reconstruction possible.
>
> (Peter J. Bowler, 1996, *Life's Splendid Drama*, University of Chicago Press, Chicago, p. 7)

Palaeontologists often give the impression that they don't believe in their own subject and the data that fossils provide. Indeed, palaeontology suffers from several misperceptions which, because they are often propounded by palaeontologists themselves, make us our own worst enemies. At least some palaeontologists give the impression that their subject involves little more than documenting fossils, describing their parts and assigning ages to the rocks that contain them. Then there is the legendary incompleteness of the fossil record which can be, and is, used as an excuse to reject any fossil evidence that runs counter to current perceptions. Many palaeontologists seem to have an inferiority complex and trot out 'imperfection' as an excuse whenever their data fall short of their ambitions. This bias dates back at least as far as *The Origin of Species*, in which Darwin argued that the record must be very incomplete (and, by inference, very inadequate) as it did not appear to provide appropriate evidence to test his theory. Occam's razor alone should lead us to accept the simplest hypothesis consistent with the facts – that the fossil record faithfully, if incompletely, preserves what actually happened. Yet practitioners of this art find it too convenient to abandon. Finally, palaeontology is said to offer less information than neontology (the study of living organisms). Indeed, to some palaeontologists, fossils preserve information so different from that derivable from living organisms that the two cannot be classified in the same way. Surely the real question is 'How much of this additional information is essential to investigate a particular idea?'

Adequacy and completeness are different concepts that should not be confused. The fossil record may be incomplete, but it is entirely adequate for many and most requirements of palaeontology, as well as answering wider questions in geology and biology. Incomplete knowledge is a reality of all areas of science – for example, see the essay by Ian Hughes discussing the limits of understanding in physics (*New Scientist*, 23 March 1996) – but that doesn't alter the significance of what we already know. The fossil record obviously doesn't preserve every organism or species that ever lived, perhaps not even a member of every major group; it only retains a sample that is biased in many ways, although we can often identify the nature of these influences.

The Adequacy of the Fossil Record is intended to be an up-to-date review that seeks to debunk these and other 'objections' that relate directly to the utility of palaeontological data. In such an area of study, broad theory must be backed up by the evidence of fossils. Thus, the first eight chapters are concerned largely with the broader issues of theory and interpretation. These are followed by four contributions that discuss particular fossil groups. These groups are diverse – foraminifers, bryozoans, bivalves and vertebrates – and have been specifically chosen to illustrate how the concepts of completeness and adequacy are influenced by intrinsic and extrinsic factors.

The original idea for this book was formulated during the period of a grant to the editors from the Leverhulme Trust, whose support is very gratefully acknowledged. We are grateful to Iain Stevenson of Wiley for his enthusiasm and encouragement during the early stages of the book's development. We would like to repeat the words of one of Iain's other editors, Ken McNamara, who acknowledged Iain's 'endless enthusiasm for all things concerning evolution'. Following Iain's departure to new pastures, we have been ably helped in our endeavours by Helen Bailey, Louise Portsmouth and Abi Hudlass, all of Wiley. Our contributing authors are thanked for their excellent contributions and their enthusiastic involvement in this project. Essential logistic support was provided by our departments in Kingston and Liverpool.

Stephen K. Donovan
and
Christopher R.C. Paul

1
Adequacy, Completeness and the Fossil Record

Christopher R.C. Paul

INTRODUCTION

Although it is universally agreed that the fossil record is incomplete, there is much misunderstanding as to the real significance of this indisputable fact. It is perhaps simplest to discuss the completeness of the fossil record in terms of biodiversity, although I shall point out other aspects later. In the context of biodiversity, I wish to consider three questions:

1. How complete is the fossil record in relation to total biodiversity, that is, what proportion of all species that have ever lived has been preserved in the fossil record?
2. How complete is our knowledge of this preserved proportion of past life?
3. How do both of the above relate to the adequacy of the fossil record as a history of past life?

In discussing these questions I shall endeavour to describe methods by which quantitative estimates of the answers to the first two questions may be derived. Before doing so, however, it is important to put the discussion in context and to clear up some past misunderstandings.

The Adequacy of the Fossil Record. Edited by S.K. Donovan and C.R.C. Paul.
© 1998 John Wiley & Sons Ltd.

DEFINITIONS OF COMPLETENESS

Are Complete Data Possible?

In reality, complete knowledge in any science, not just palaeontology, is unattainable. So long as a single fossil remains in the rock, it might add to our knowledge. Note, however, that the same argument is equally true of neontology (the study of Recent life). So long as a single living organism remains unexamined, it might add to our knowledge. It is immediately obvious that this 'problem' is more serious for neontology than for palaeontology, because reproduction is constantly adding to the numbers of living organisms at a rate with which scientists cannot possibly keep pace. Despite this, biologists seem unconcerned about the 'problem', whereas palaeontologists continually stress the incompleteness of the fossil record. While it is wise to be well aware of the limitations of one's data, this apparent difference of emphasis between neontologists and palaeontologists raises the most fundamental question about the adequacy of the fossil record. Yes, the fossil record is incomplete, but does this fact matter? For most practical purposes the answer must be 'no' and biologists are right in not worrying about the incompleteness of their data. What we do know about living or fossil organisms far outweighs what we do not know, and as new discoveries are made the balance is continually tipped in favour of knowledge and against ignorance. Despite this, understanding of the real value of the fossil record apparently advances slowly.

So What Do We Mean by Complete?

Completeness and adequacy can only be evaluated against some predetermined aim. Thus, even so apparently simple a task as defining completeness (or incompleteness) will not provoke universal agreement among scientists. Consider the simplest of palaeontological tasks: making a list of species present in a locality, or formation, or zone. (Ignore for the moment the problem that we can never know when the list is complete, because any unexamined fossil might add to the list.) To make such a list the information required is, in effect, just one identifiable fragment of each species present. However, if we were to extend the task to describing the fossils present thoroughly, we would need at least one 'complete' example of each species on which to base a 'complete' description. Should we wish to discuss the palaeoecology of this fossil biota, at the very least we would need a large enough sample to estimate how common each species was. I need go no further. The data required to compile a list of biota, to describe a biota, or to investigate the palaeoecology of a biota, are not the same. It follows that data which would allow us to 'complete' one task would not necessarily be adequate for another task. From this we can draw a second fundamental conclusion: completeness and adequacy are not the same thing. The fact that the fossil record is universally and correctly acknowledged to be incomplete does not mean it is an inadequate sample of past life. However, we still need to be clear about our aims before we can evaluate the adequacy of the fossil record for any particular purpose.

Comparisons with Other Sciences

Comparisons with the state of knowledge in other sciences are often instructive in evaluating the significance of any limitation of the fossil record. Neontology is an obvious choice, but any other science will do. I have made such comparisons before (Paul, 1982, 1985, 1992a). Here I will confine myself to just one point. Nicol (1977) estimated there were 1 239 129 living species of animals. Other estimates of total biodiversity in the 1960s and 1970s lay between one and two million species. More modern estimates are an order of magnitude higher, that is, 10–20 million species. I believe the more recent higher estimates result from biodiversity becoming a topical and political issue. Biologists have made serious estimates of what they do not yet know, rather than making estimates of published data and how many specific names are likely to be synonyms. Whatever the reasons for the higher estimates, if we accept them, then the known neontological record must be a tenth of what it was thought to be only 20 or 30 years ago. Despite this, biologists do not agonize about how incomplete or inadequate their knowledge is. Rather, they vigorously promote the need to document the living biota before too much more of it becomes extinct and lost to science for ever. Palaeontologists, on the other hand, seem to me to suffer from an inferiority complex. They always accept the limitations of their data whenever evidence from the fossil record conflicts with even theories, let alone evidence, from other sciences. This is doubly regrettable because it makes objective evaluation of data from the fossil record impossible. With the benefit of hindsight, I am amazed at how long we accepted that gradual morphological change was the norm when data to support this belief were so sparse and the discrepancy had been known since Darwin's time. Examples of gradual evolutionary trends in the fossil record can be counted on our fingers (*Gryphaea*, *Micraster*, *Zaphrentis*, the horse) and we simply ignored the countless examples that do not show the expected pattern. Lack of morphological change was equated with lack of data and the few examples of trends were equated with 'the truth'. There are sound theoretical arguments for believing that stasis (that is, no morphological change) dominates the fossil record (Paul, in press). One can even argue that, far from representing 'the truth', the few known examples of gradual trends are no more than random walks that just happen to be more or less linear.

In summary, complete knowledge of any science is impossible. Incompleteness of knowledge does not necessarily equate with inadequacy. Estimating both completeness and adequacy requires clear understanding of some predetermined aim: the same data may be more than adequate for one task and totally inadequate for another. Finally, comparisons with other sciences are highly instructive in evaluating the incompleteness of the fossil record. Scientists in most other disciplines accept readily that they deal with samples. Palaeontologists should do the same. Complete knowledge is unattainable; it is usually totally unnecessary, too.

Estimates of the completeness of the fossil record are of intrinsic interest in themselves. In the next sections I shall review some methods for obtaining such estimates and discuss the implications of the results. There are two aspects. First, how complete the fossil record itself might be, and second, how completely we know the resulting record.

COMPLETENESS OF THE FOSSIL RECORD

It is widely acknowledged that possession of a mineralized skeleton greatly enhances the chances of preservation in the fossil record. Palaeontologists accept that the fossil record is highly biased in favour of skeletized groups (see, for example, Conway Morris, 1986). Nicol (1977) exploited this fact as a means of estimating the numbers of living animals likely to be fossilized. Basically he assumed that all 'soft-bodied' animals would not be preservable, while all skeletized animals would. He estimated the total numbers of both and arrived at a proportion of preservable animal species. He calculated that, out of 1 239 129 living species, 99 800 were skeletized, hence approximately 8% of living animals were likely to be preservable. Notwithstanding changes in estimates of total biodiversity since Nicol's paper was published, it seems reasonable to assume that Nicol's estimate is not far wrong and something between 5 and 10% of living animals could easily become fossils. If we make a further assumption that the ratio of skeletized to soft-bodied animals has not changed significantly throughout the Phanerozoic, then we could take Nicol's estimate as representative of the entire Phanerozoic fossil record. Thus, the fossil record of animal species is likely to be between 5 and 10% complete in terms of numbers of species preserved.

However, soft-bodied organisms do get preserved under exceptional circumstances and *konservat Lagerstätten* can be used to test the assumption that the proportions of skeletized animals have not changed significantly. To be sure, Nicol's estimate was of global proportions, whereas *konservat Lagerstät-*

ten only conserve a small local sample of animals. Nevertheless, why should we assume that these samples are unrepresentative? In evaluating the Burgess Shale fauna, Conway Morris (1986) discussed precisely this point. That part of the Burgess Shale fauna which possesses mineralized skeletons is a typical Middle Cambrian, trilobite-dominated, shelly fauna. Thus, it is reasonable to assume that the Burgess Shale soft-bodied fauna is typical of what is not normally preserved in most Middle Cambrian assemblages. However, note that Conway Morris did accept that other Middle Cambrian *konservat Lagerstätten* with soft-tissue preservation seemed to represent a second community that differed from the community preserved in the Phyllopod Bed. Conway Morris (1986) calculated that 14% of the genera and perhaps 2% of the individuals preserved in the Phyllopod Bed were skeletized. Conway Morris kindly provided me with a more up-to-date table of data (personal communication, 22 November 1996). He would now revise the estimate of skeletized specimens to about 5% of total specimens. Also, as most genera are monospecific, the figure of 14% of skeletized genera is probably applicable at specific level. My own estimate of percentage skeletized taxa in the new list, which includes one genus with three species, is 13.3%. Thus, although based on a very small sample compared with Nicol's (1977) estimate (a total of 90 taxa), the proportion that would normally be preserved is quite close to that estimated by Nicol. It would not be outlandish to suggest that about 10% of all Phanerozoic species are likely to have been preserved in the fossil record, but see Chapters 8 and 9 for the effects of rare species that seem to dominate fossil and modern faunas.

A second possible method of estimating the completeness of the fossil record is to consider what proportion of species are endemic to *konservat Lagerstätten*. The assumptions here are that the *konservat Lagerstätten* preserve all or nearly all local species, whereas 'normal' contemporary deposits from similar environments only preserve the typical, skeletized animals. While these assumptions are open to evaluation in each specific example, the idea does offer an independent, alternative estimator to Nicol's approach. It is perhaps less applicable to the Burgess Shale fauna, since there are now several other examples of Middle Cambrian *konservat Lagerstätten* with soft-bodied preservation analogous to the original Phyllopod Bed, and therefore there is less likelihood that species of the Phyllopod Bed will be endemic to it. A possible solution here would be to consider all examples of Middle Cambrian *konservat Lagerstätten* together and compare endemism within them as a whole. This would overcome to some extent the difficulty of comparing Nicol's global estimate with *konservat Lagerstätten* local estimates of the ratio of skeletized to non-skeletized animals.

There is one final problem in estimating completeness. Nicol's argument related entirely to living organisms. I have explicitly extended it by a uniformitarian comparison, that is, I assumed that current proportions applied through time. Kier (1977) used the same argument to compare the fossil records of two skeletized groups, the regular and irregular echinoids. He noted that, whereas just over 50% of extant species were regulars, only about 20% of Tertiary fossil sea urchin species were. Kier concluded that regular echinoids had a poor fossil record compared with irregulars. There are various reasons for this difference, but it is interesting that, in collecting a very large sample of Cretaceous echinoids from Wilmington, Devon, England, Smith *et al.* (1988) found a nearly equal ratio of regular to irregular species. The fauna is dominated by a few species of irregular echinoids, *Discoides subuculus, Echinogalerus rostratus, Catopygus columbarius* and *Holaster nodulosus*, so that small samples do not yield a diverse fauna. However, when several thousand specimens were collected, the rarer and sometimes fragmentary regulars, none of which is truly abundant, were found to be present. In the 10 stratigraphic levels recognized, regulars never formed more than 25% of the specimens, but in seven of the ten levels they represented more than 50% of the species. At the levels where regular species formed less than 25% of the fauna, the total number of species was seven or less, about half the total number found in levels where regular species dominated the fauna. The inescapable conclusion is that, at least with the fauna at Wilmington, a greater search effort is needed to detect regular echinoids compared with irregulars, but, if the effort is made, the proportions of regular to irregular species are similar to those of the present-day fauna. Our knowledge of the fossil record of irregular versus regular echinoids is also biased by the way in which the record has usually been studied. Most palaeontologists have only reported on complete or fairly complete tests. This favours the record of irregulars because they often have an interlocking stereom, so their tests disintegrate less rapidly after death than those of regulars, and they tend to live in areas where sediment is accumulating, whereas regulars often live on rocky substrates that are being eroded (see Kier, 1977, for other reasons). Only those regulars which have large and obvious spines (mainly cidaroids) are commonly reported from fragmentary remains. However, if both fragments and complete specimens are studied, it is often found that regulars are common fossils (see, for example, Gordon and Donovan, 1992; Donovan and Paul, 1998). Part of our lack of knowledge of the fossil record results from our unwillingness to make full use of the data that are preserved.

COMPLETENESS OF KNOWLEDGE OF THE FOSSIL RECORD

For most practical purposes only well-skeletized taxa with a good fossil record are used. Hence, much past discussion of completeness and adequacy of the fossil record has centred on typical fossils, rather than on comparing the records of skeletized and soft-tissue fossils. Despite this, the almost universal opinion is that the fossil record of even well-documented groups is meagre. In this section I wish to discuss methods of estimating quantitatively the known fossil record and also to challenge this assumption that the record is poor. The topic has been discussed by a number of authors, of whom Durham (1967) and van Andel (1981) are fairly typical of conventional wisdom about the fossil and rock records, respectively, while Meehl (1983) has provided considerable insight, apparently from a non-palaeontological background. I shall start with Meehl's work, because he provides an additional important principle. Most quantitative methods require basic assumptions that are acknowledged to be unrealistic, for example, that chances of preservation and collection are unbiased or that modern proportions obtained throughout the Phanerozoic. The whole basis of Meehl's approach is that, if several different independent methods give more or less the same answer, even if individually based on implausible assumptions, the combination of similar answers enhances the plausibility of the result and reduces the consequences of the unreasonable assumptions. What Meehl set out to deliver was not so much an estimate of the completeness (or otherwise) of the fossil record, as a series of methods that could provide independent estimates. Regrettably, although Meehl provided four different tests, he did not actually apply them and compare real estimates of completeness for any fossil group. Nor, as far as I am aware, has anyone else applied Meehl's tests, although some have been 'discovered' independently and/or applied in slightly different ways. In addition, other estimates not considered by Meehl are available, so that his original principle of applying consistency tests can be used.

Meehl's Four Methods

The discovery asymptote (Figure 1.1)

This is based on the well-known phenomenon that when one collects a fauna or flora, the common species are soon encountered, but it takes progressively more effort to add to the number of species as the list grows. This can be expressed as a species–area curve (Cain, 1938), a species–specimens curve (Ager, 1963) or a growth of knowledge curve, that is, new taxa versus time (Paul, 1980). Meehl derived the following general equation:

$$y = N(1 - e^{-st})$$

where y is the number of species known at time t, N is the total number of

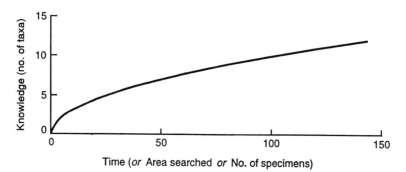

Figure 1.1   *A discovery curve. The curve tends to flatten out the longer the period of discovery*

species that have ever existed, s is the rate of discovery (in this case assuming a constant search effort) and e is the base of natural logarithms. y/N Meehl called the *completeness index*. He also provided a modification of the formula for the (more likely) case of increasing search effort, viz:

$$y = N(1 - e^{-(at + bt^2)})$$

The method of binomial parameters (Figure 1.2)

In effect, this method estimates the trajectory of a distribution curve back to the y axis where $N = 0$. Visually (Figure 1.2), it can be represented as the curve of frequency distribution for species known from 1, 2, 3...n finds. $N(0)$ is the total number of species not found, $\Sigma N(1)$, $N(2)$, $N(3)$...$N(n)$ is the total of all species that have been found. $\Sigma N(0)$, $N(1)$, $N(2)$, $N(3)$...$N(n) = N$, the total number of species that have ever existed. So:

$$(N - N(0))/N$$

is the completeness index.

Meehl also concluded that $\log N(0) = 2 \log N(1) - \log N(2) + \log (1/2)$ and he presented a more complex equation for a Poisson distribution.

The sandwich method

This is essentially a modified version of gap analysis (Paul, 1980). As Meehl himself pointed out, the concept is simple, but the derivation of a mathematical formula to estimate it quantitatively is much more difficult. I suspect this is because Meehl set himself the task of estimating N (the total number of species that have ever existed) in all four methods, but also because he did not appreciate the use of currently defined stratigraphic intervals as es-

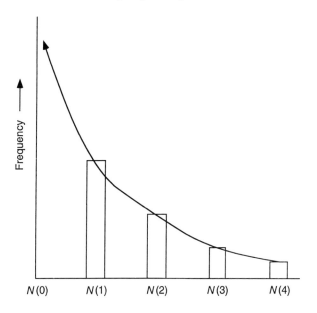

Figure 1.2   *A diagrammatic representation of Meehl's (1983) method of binomial parameters. If the trajectory of the distribution of N(1), N(2), N(3) values is known, one can estimate the intercept for N(0)*

timators of the size of gaps. Basically, Meehl assumed that the end members of ranges were defined and that gaps exist within them. Any absence between first and last known occurrences is treated as an 'incompleteness instance', irrespective of length. Decichrons are then defined to 'weight' large gaps. The approach is very complicated. A much more simple idea is just to sum the total number of gaps in conventional stratigraphic units (or as thickness of sediment) and express this as a proportion of the total stratigraphic ranges of all taxa including the gaps. Paul (1980, 1982) applied the simpler idea to the fossil record of cystoids and found 23% gap at family and series/stage level. However, the precise value derived depends on both the taxonomic and stratigraphic levels chosen. Percentage gap of cystoid genera at zone level would not be the same. At present, the simpler technique cannot be used to derive a completeness index comparable to Meehl's, but there are other ways to do this.

Method of extant forms

This is a very similar argument to the one Nicol (1977) used to estimate the number of living species likely to be fossilized, but extended throughout geological time. A completeness index for extant species (C^*) in a major taxon is defined as $C^* = Nf/Ne$, where Ne is the total number of extant species and Nf is the number of extant species also known as fossils. Then, for geological time, C^* is assumed to equal C, the completeness index for extinct species. If N^* is the total number of known extinct fossil species, $N^* = NC$ and we can solve for N, the total number of species that have ever existed.

Other Methods

Gap analysis

Historically, gap analysis was the first method to be used. Basically, this approach relies on the logic that in the fossil record, where the same taxon is known from below and above, but not actually within, a specific stratigraphic interval, this represents a situation where we know a fossil taxon existed, but it has not been preserved or has yet to be discovered. The method assumes that all gaps represent incompleteness, but this is only true of global estimates. Equally it can only represent a minimum estimate of incompleteness since gaps beyond known ranges cannot be detected by this method. A further assumption in the way in which I first applied the method (Paul, 1980, 1982) is that a given percentage 'gappiness' can be translated into an estimate of taxonomic incompleteness. Thus, I calculated that the fossil record of cystoid families at series/stage level included 23% gaps (Σfamily ranges = 107 stratigraphic intervals, 25 gaps) and translated this into an estimate that approximately 23% of families remained undiscovered. This last step involves a logical leap that may well not be justified. However, estimates derived by other independent methods (see below) were broadly similar and so Meehl's logic of consistency tests may mean that, empirically at least, percentage gap can be used as a crude estimator of taxonomic completeness index.

Gap analysis has been extended a little since its first description (Paul, 1980). At that time I presented stratigraphic ranges of taxa as discrete entities, and pointed out that gaps below and above first and last known occurrences might well exist, but could not be detected. However, if phylogenetic relationships are deduced so that stratigraphic ranges are presented as segments of phylogenetic trees, then, at least potentially, gaps between related families can be detected. For example, two families of Cambrian edrioasteroids are known and four in the Middle and Upper Ordovician (Paul, 1988), but none from the Tremadoc and Arenig. Phylogenetic analysis suggests that each of the Cambrian families gave rise to Ordovician descendants, so that at least two gaps beyond known ranges of families exist in both lineages. It may well be that some gaps between taxa may still go undetected, but this is a potential improvement.

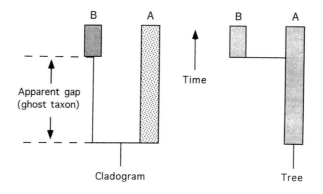

Figure 1.3   *Overestimation of gaps in the record using cladistic analysis. Bars represent known stratigraphic occurrence of two closely related taxa, A (long range) and B (short range). In the cladogram they are assumed to be sister groups with a common time of origin, suggesting a large gap exists in the range of B (that is, a 'ghost taxon' of Norrel, 1993). If the tree is the true situation, there is in fact no gap at all. Taxon A gave rise to taxon B late in its evolutionary history. Regular and irregular echinoids represent just such a pattern: regular echinoids (= A) first appeared in the Ordovician, whereas irregulars (= B) did not evolve until the Jurassic*

However, it does get into the contentious issue of how one derives phylogenetic relationships. Nowadays, cladistics would almost certainly be the preferred method, but cladistics alone would overemphasize the number of gaps because it assumes axiomatically that all pairs of sister taxa arose at the same time (Figure 1.3). This is manifestly not true at all taxonomic levels, but especially for supraspecific taxa as currently defined. Paraphyletic groups are unavoidable and greatly complicate estimation of gaps. If one accepts the allopatric speciation model, then even species can be paraphyletic. For a more thorough analysis of this problem see Chapter 7.

Monotypic taxa

Paul (1980, 1982) proposed that the frequency of monotypic taxa within a major taxon could be used as a crude estimator of incompleteness. The argument was that if the fossil record were very incomplete, most fossils would be distantly related to each other and likely to be assigned a relatively high taxonomic rank, such as a family. Excellent illustrations of this idea come from soft-bodied *konservat Lagerstätten*. Fossils such as *Anomalocaris* from the Burgess Shale or *Tulimonstrum* from the Mazon Creek fauna (Pennsylvanian) defy easy classification. Interpreting their morphology alone is a big enough challenge. The comparison between 'exceptional' and 'normal' faunas is again instructive. Applying this argument to the then known fossil record of cystoids, Paul (1980) found that eight of 24 families were monotypic and they were assumed to have 'incomplete' fossil records. At that time the proportion of 'incompletely known' families was 33% of the total. Since then, one genuinely new family has been recognized (Bockelie, 1984). In this sense, a genuinely new family is one that includes only species discovered since 1980, not one resulting from taxonomic refinements or reinterpretations of previously known species. This family, the Parasphaeronitidae, contains two species, so the proportion of 'incompletely known' families has declined slightly to 32%.

'Monotopic' taxa

This argument is analogous to that for monotypic taxa. Paul (1980, 1882) argued that if the record were very incomplete most fossil taxa would be known from only a single stratigraphic level. He applied the argument to the record of cystoid families and found that six of the 24 then known families were confined to a single series or stage. The proportion of 'monotopic' taxa was 25%, which again is close to the 23% gap in the record of cystoid families. One could well argue that a certain number of families might have been genuinely short lived and that their short known stratigraphic ranges reflect what actually happened, rather than the incompleteness of the record. However, taking the results at face value maximizes the apparent incompleteness of the fossil record of cystoid families. Hence again, although the assumptions behind the method imply that it cannot be applied rigorously, nevertheless they are not totally unwarranted and give results quite close to gap analysis. As both methods, gap analysis and analysis of monotopic taxa, involve the same taxonomic and stratigraphic levels of analysis, the coincidence of the results is quite striking.

Statistical modelling (bootstrapping)

This method involves reversing the normal approach. Here one defines a complete record and then samples it repeatedly at different levels to see either (1) at what point one gets a distribution similar to some known pattern, or (2) what confidence limits can be put on chances of having discovered, say, 50% of preserved fossils. The method requires a realistic data set. A large modern data set, such as the information on the occurrence of modern foraminifera around the North American continent referred to in Chapter 9, would be ideal. The method assumes random discovery, but again this need not be a totally unrealistic assumption and it does give insights into general problems of sampling the fossil record.

Stratigraphic ranges (survivorship analysis)

Foote and Raup (1996) have derived relationships between true taxonomic durations (the lifetime of a taxon), preservation probabilities and observed stratigraphic ranges. They used these relationships, among other things, to estimate taxonomic completeness (proportion of taxa preserved and known) based on nothing more than observed total stratigraphic ranges. Foote and Raup showed that, when preservation is random and the original distribution of true durations of taxa is exponential, their inferences about durations, preservation probability and completeness are exact. However, they did discuss relaxation of these criteria as well. They tested their methods against known data on species of trilobites, bivalves and mammals, and on genera of crinoids. Their estimates of preservation probability were similar to those derived from gap analysis, where data on gaps were available for comparison. So, as with Meehl's general approach, comparable results derived from different methods based on different assumptions suggest that Foote and Raup's assumptions are not all that unrealistic. Estimates of completeness varied from 60% to 90% for species and genera, which is far higher than most palaeontologists (including me) would expect, but closely similar to my own unexpectedly high estimates derived below by yet another method. Foote and Raup's method is comparable to survivorship analysis, which Paul (1980, 1982) suggested might be used for non-quantitative estimates of completeness. It has great appeal since the method is based on the most obvious and well-documented stratigraphic data, namely total stratigraphic ranges.

Methods That Do Not Give Reasonable Answers

It is perhaps appropriate to mention a couple of methods that have been used to estimate or at least indicate completeness, but which are likely to give unrealistic results for one reason or another.

Rate of description of taxa

Grant presented two graphs showing the rate of description of new brachiopod genera per decade and cumulative total brachiopod genera over the 200 years up to 1980 (Grant, 1980, figs 7 and 8, respectively). Both show exponential increase, which is in stark contrast to the shape of Meehl's 'discovery curve'. One possible interpretation is that we are still on the steepest part of the discovery curve, but that increasing research effort in more recent decades has made the curve concave up. The implication of this is that the fossil record of brachiopods is still extremely poorly known. However, Grant's raw data do not take any account of taxonomic refinement, although he was well aware of the activities of 'splitters'. The discovery curve can only be applied to the rate of discovery of 'genuinely new' taxa. With species there is usually no problem (but see the next section): a newly described species usually results from a newly discovered species, even if it is discovered in a museum collection where it has lain undetected for a century or more. Newly described higher taxa, genera and families, arise from at least two different phenomena. Genuinely new higher taxa, in the sense used here, result from the discovery of a new species which is so different from all known species that a new genus or family is created for it. However, many newly described genera and families arise from taxonomic refinement where old genera and families are simply divided up. No new species may be involved at all. Indeed, taxonomic revision may result in an apparent reduction in specific diversity due to the recognition of synonyms, while increasing generic or familial diversity. Paul (1980) made the distinction by assigning all known cystoid species to the then current classification. For each genus and family the first described species was identified and the genus or family deemed to date from that point, irrespective of the taxonomic assignment of the species at the time it was described. For example, Gyllenhaal (1772) described the first two species of cystoids known to science under the generic name *Echinus*. These two species are now assigned to the genera *Echinosphaerites* and *Sphaeronites*, and the families Echinosphaeritidae and Sphaeronitidae. These genera and families were effectively first discovered in 1772, because the first species that are now included within them were described then. Using this technique, discovery curves for genuinely new higher taxa can be derived (Figure 1.4). At least for cystoids, the curve for families shows distinct signs of flattening out despite the increased research effort since 1960. The curves for genera and species continue to climb. However, analyses such as this require extensive knowledge of the literature. Personally, I could not convert Grant's raw curve for brachiopod genera into a genuine discovery curve. However, the important point is that raw 'rate of description' curves are highly misleading in estimating the true rate of discovery.

Cladistic tests

The point has already been made that cladograms may overestimate stratigraphic gaps because they assume that all pairs of sister taxa arose simultaneously. However, ancestors really do exist and may well only spawn their descendants in their last death-throes, as seems to be true of several taxa that cross the Cenomanian/Turonian boundary (Jarvis *et al.*, 1988). Cladistic analysis may well identify the two most closely related taxa in a major group, but it will interpret them as sister taxa. If they are really ancestor and descendant, this will overestimate stratigraphic gaps and hence

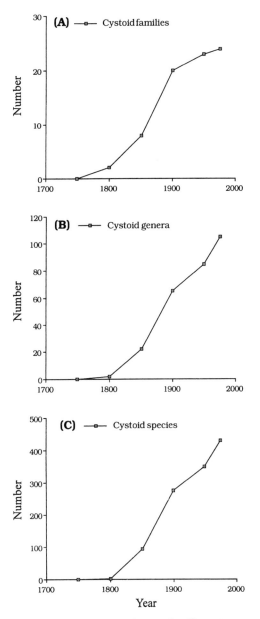

Figure 1.4 *Discovery curves for genuinely new families (A), genera (B) and species (C) of cystoids. Note that whereas the family curve has flattened out and its shape suggests that we have already discovered the majority of cystoid families, the curves for genera and species are still rising steeply*

incompleteness of the fossil record. Cladistic analysis is the first logical step in any phylogenetic reconstruction. However, converting cladograms to trees and comparing them with the known stratigraphic record of the branches is a valid test of cladistic interpretations (Paul, 1985, 1992a,b; Huelsenbeck, 1991, 1994; Wagner, 1995; Foote, 1996; Clyde and Fisher, 1997; Huelsenbeck and Rannala, 1997; and see Chapter 7). Use of cladistics to estimate incompleteness of the fossil record, or rates of molecular evolution, is fraught with danger because of the assumption that sister taxa arose simultaneously. Again, see Chapter 7 for a thorough discussion of the use of cladistics and biostratigraphy as estimators of completeness.

The Ultimate Test

Previous sections have outlined a variety of different methods for estimating our knowledge of the fossil record at various taxonomic and stratigraphic levels. They are at least partly of historical interest, and are included here mainly to illustrate potential tests and to suggest which might or might not be appropriate to a particular estimate of completeness. However, a simpler test exists that can be applied at all levels. Every time a fossil is collected and identified, wittingly or otherwise our knowledge of the fossil record is tested. Basically, the preserved fossil record consists of two parts: those taxa we already know about and have described, and those we do not know about. A newly discovered fossil either belongs to a new taxon or it does not. If we repeat this experiment many times, the proportion of new taxa will reflect the proportion of unknown taxa in the fossil record. To be sure, 'new' species will be described that subsequently prove to be synonyms, but this difficulty can be overcome by only considering genuinely new taxa, which I would define as taxa that meet the following criteria:

1. They are not synonyms.
2. They do not result from taxonomic refinement such as elevating sub-species to specific rank, etc.
3. They do not result from taxonomic renaming.

Most of these requirements can be met by considering monographic re-visions of fossil floras and faunas, rather than by the literal experiment of making a new collection. Monographs usually also have the advantage of being conducted by a single author and therefore of having consistent tax-onomic treatment. Using synonymy lists, one can check all three points and I would personally take the extreme view that any newly described taxon that has a synonymy is not genuinely new in the sense defined here. For example, Forbes (1848) described and illustrated an Ordovician cystoid under the name *Caryocystites litchi*. Re-examination of his material, and of much more collected since, showed that Forbes had included three distinct species within his concept of *C. litchi*, one of which, *Haplosphaeronis multifida* Paul, 1973, I described and named for the first time 125 years later. However, Forbes illustrated this 'new' species in his original work. Thus, although it was only named formally in 1973, it was known to science in 1848.

Clearly, our knowledge of well-studied parts of the fossil record in western Europe and North America will be better than our knowledge of poorly documented biotas from Antarctica, for example. However, the vagaries of plate tectonics and continental drift have ensured that the sample of the fossil record that is preserved in Europe and North America is a fairly random sample and includes, for example, marine and non-marine environments, shallow and deep marine environments, tropical temperate and polar environments, etc. Thus, although one of the faunas I have used to test this idea, Cretaceous echinoids from Wilmington, Devon, England (Smith *et al.*, 1988), contained not a single 'genuinely new' species (out of 37 described), reflecting our excellent knowledge of western European Cretaceous echinoid faunas, others provide a more realistic test of less well-known periods and major taxa. Among fossil echinoderms, two of the best examples are the Ordovician cystoid faunas of Morocco (Chauvel, 1966, 1977, 1978), a relatively unstudied area in a near-polar environment that contrasts strongly with the more temperate to tropical faunas of western Europe and North America, and the Ordovician echinoderm fauna from the Bromide Formation of Oklahoma, which was extensively collected in the 1960s and has since been thoroughly described (Sprinkle, 1982). I analysed both for 'genuinely new' families, genera and species, to provide data comparable to those provided by other methods discussed in the previous sections. I treated species identified tentatively or left under open nomenclature as a separate category. Thus, one can allow for the effects of these doubtful species in several ways. In Table 1.1 they are treated in three ways.

1. First they are treated as if they were 'old' species, which allows calculation of unambiguously 'new' taxa as a proportion of all taxa, and therefore provides a minimum estimate of the proportion of 'new' taxa .
2. Then, doubtful species are treated as if they were genuinely 'new' species, which provides a maximum estimate of the proportion of 'new' taxa.
3. Finally, I have taken the midpoint between these two extreme estimates as a more reasonable estimate of the true proportion of genuinely new taxa. The results (Table 1.1) are surprising.

Combining all the data shown in Table 1.1 and taking the median estimate for 'new' taxa as defined here, 1.8% of 249 families are genuinely new, 11.5% of 631 genera are genuinely new and 41.8% of 1158 species are genuinely new. Even at specific level, this suggests that we know more than half the fossil echinoderms and molluscs that have been preserved in the fossil record. The

Table 1.1 Numbers of genuinely new and old taxa, plus those of doubtful taxonomic identity from some monographic revisions of mollusc and echinoderm faunas

Families			Genera			Species			Source
New	Old	Uncertain	New	Old	Uncertain	New	Old	Uncertain	
0	36	0	4	110	0	73	94	13	1
0	2	0	0	9	0	4	10	5	1
0	76	0	26	191	3	169	139	97	2
0	*114*	*0*	*30*	*310*	*3*	*246*	*243*	*115*	*Total molluscs*
0	9	0	0	30	0	3	103	0	3
0	7	0	1	39	1	3	65	2	4
0	23	0	1	28	0	33	16	0	5
1	15	0	11	19	8	24	24	21	6
2	37	1	17	43	3	40	30	19	7
0	20	2	0	39	2	1	50	11	8
0	1	0	0	8	0	7	17	7	9
0	17	0	3	33	2	31	29	18	10
3	*129*	*3*	*33*	*239*	*16*	*142*	*334*	*78*	*Total echinoderms*
3	243	3	63	549	19	388	577	193	Grand total

Sources: (1) Woodring (1925; Jamaican Pliocene bivalves and scaphopods); (2) Woodring (1928; Jamaican Pliocene gastropods); (3) Kier (1984; Cuban spatangoid echinoids); (4) Macurda (1983; fissiculate blastoids); (5) Kier (1972; Saudi Arabian Mesozoic and Tertiary echinoids); (6) Chauvel (1966, 1977, 1978; Moroccan Ordovician echinoderms); (7) Sprinkle (1982; Oklahoman Ordovician echinoderms); (8) Smith *et al.* (1988; British Cretaceous echinoderms); (9) Smith and Paul (1982; all cyclocystoids); (10) Paul (1967a,b, 1973, 1974, 1984, 1997; all British cystoids). Apart from Smith *et al.* (1988), these examples are neither from geographical areas nor involve taxa that would be generally regarded as having an excellent fossil record.

two groups do not show all that much difference. Combining data for all molluscs gives 0% new families, 9.2% new genera and 50.2% new species, whereas values for the echinoderms are 3.3%, 14.2% and 32.7% of new families, genera and species, respectively. These estimates are comparable with Foote and Raup's (1996) data for species of trilobites, bivalves and mammals, and for genera of crinoids. However, the values contrast markedly with Tunnicliffe's data for new taxa in modern hydrothermal vent faunas (Tunnicliffe, 1992, table 2): 223 of 236 species (94.5%), 79 of 137 genera (57.7%) and 17 of 79 families (21.5%) were new. Vent faunas are a relatively new discovery and are highly endemic. If our knowledge of the fossil record were comparable with our knowledge of vent faunas, similar percentages of fossil taxa would be new. The fact that far smaller proportions of fossil taxa are new implies that we know the fossil record considerably better than we know vent faunas, and that we can anticipate the discovery of many more vent taxa as we investigate more vents.

Surprisingly, taken at face value, my data on fossils (Table 1.1) imply that at specific level the fossil record of molluscs is not as good as that of echinoderms. However, the one major mollusc monograph used was completed in 1928 and was for a fauna from a Caribbean island, not one of the best-explored parts of the fossil record. Counteracting that, however, was the fact that it was for a Pliocene mollusc fauna that contains a number of extant species which had already been described by neontologists. For an unbiased estimate of truly new *fossil* species, extant forms ought to be excluded.

In *The Natural History of Fossils* (Paul, 1980) I made the analogy with an iceberg. Does the fossil record represent the 10% tip of the iceberg of life, or do we have a fish's eye view of the iceberg with only the top 10% hidden from us? From this brief review of the completeness of the fossil record, and the completeness of our knowledge of it, I would conclude that the fossil record as a whole represents the human view of the iceberg. It would seem that roughly 10% of all species that have existed on Earth during the Phanerozoic are likely to have been preserved in the fossil record. However, for well-skeletized groups such as molluscs and echinoderms we already know about half of the total record at species level and only the top 10% or less is unknown at family level. Interestingly, this conclusion is in good agreement with Harper's quite independent estimate of completeness of the bivalve fossil record at family level (see Chapter 11).

ADEQUACY OF THE FOSSIL RECORD

As pointed out earlier, adequacy and completeness are not the same and both can only be estimated or quantified in relation to some predetermined aim. However, accepting the conclusions of the preceding section, a 10% sample of past life should be adequate for many biological and geological purposes. To be sure, we are never likely to make the ultimate biotic inventory, but, if we could, would it enhance our understanding of biology in the broadest sense? I suspect not. To my mind there is no doubt that the fossil record is quite adequate to assess evolutionary hypotheses, such as ancestor–descendant relationships (Paul, 1992b), and certainly it is adequate to reveal the history of life on Earth. Biostratigraphy is securely based. Elsewhere (Paul, 1982) I have shown that the order in which fossils are preserved in the fossil record is likely to reflect the actual order in which species evolved on Earth. Even with only a single known specimen of each species, between 95 and 99% of random comparisons between pairs of species are likely to give the correct answer to the question 'Which came first?' Add to this the possibility that we already know of 95% of bivalve families (see Chapter 11) or between 60% and 90% of trilobite, bivalve and mammal species, or of crinoid genera (Foote and Raup, 1996), and it becomes reasonable to argue that the fossil record of well-skeletized groups is as good for many purposes as the neontological record. Genuinely new species of living molluscs are still being described at a comparable rate to genuinely new fossil species. To my mind, the inescapable conclusion is that the fossil record is an invaluable repository of information relevant to most of the major concerns of biological science. The sooner we accept this fact and treat the fossil record accordingly, the better our understanding of the history of life on Earth will be.

CONCLUSIONS

1. Complete data are impossible to collect.
2. The significance of the incompleteness of the fossil record has been greatly overemphasized to the detriment of improved understanding of the history of life.
3. Estimates of completeness and adequacy require some predetermined objective. The same data may be complete and adequate for one task, but incomplete and inadequate for another.
4. Modern estimates of biodiversity suggest that the neontological database could be more incomplete than that for palaeontology. This is certainly true for modern hydrothermal vent faunas.
5. More than one estimate of the completeness of the fossil record suggests that about 10% of all species that have ever existed are likely to have been fossilized.
6. Independent estimates of the completeness of our knowledge of the fossil record of echinoderms, trilobites, mammals and molluscs suggest that > 60% of fossil species, > 80% of fossil genera and > 90% of fossil families have already been discovered.
7. The fossil record is perfectly adequate as a record of past life on Earth.

REFERENCES

Ager, D.V., 1963, *Principles of Paleoecology*, McGraw-Hill, New York: 371 pp.
Andel, T.H. van, 1981, Consider the incompleteness of the geological record, *Nature*, **294**: 397–398.
Bockelie, J.F., 1984, The Diploporita of the Oslo region, Norway, *Palaeontology*, **27**: 1–68.
Cain, S.A., 1938, The species–area curve, *American Midland Naturalist*, **19**: 573–581.
Chauvel, J., 1966, Échinodermes de l'Ordovicien du Maroc, *Cahiers de Paléontologie*: 120 pp.

Chauvel, J., 1977, Note complémentaire sur les cystoïdes rhombifères (Échinodermes) de l'Ordovicien marocain, *Notes et Mémoires du Service des Mines et du Carte Géologique du Maroc*, **38**: 115–139.

Chauvel, J., 1978, Compléments sur les échinodermes du Paléozoïque marocain (Diploporites, Eocrinoïdes, Edrioastéroïdes), *Notes et Mémoires du Service des Mines et du Carte Géologique du Maroc*, **39**: 27–78.

Clyde, W.C. and Fisher, D.C., 1997, Comparing the fit of stratigraphic and morphologic data in phylogenetic analysis, *Paleobiology*, **23**: 1–19.

Conway Morris, S., 1986, The community structure of the Middle Cambrian Phyllopod Bed (Burgess Shale), *Palaeontology*, **29**: 423–467.

Donovan, S.K. and Paul, C.R.C., 1998, Echinoderms from the Pliocene Bowden Shell Bed, southeast Jamaica, *Contributions to Tertiary and Quaternary Geology*, **35**: 19pp.

Durham, J.W., 1967, The incompleteness of our knowledge of the fossil record, *Journal of Paleontology*, **41**: 559–565.

Foote, M., 1996, On the probability of ancestors in the fossil record, *Paleobiology*, **22**: 141–151.

Foote, M. and Raup, D.M., 1996, Fossil preservation and the stratigraphic ranges of taxa, *Paleobiology*, **22**: 121–140.

Forbes, E., 1848, On the cystideae of the Silurian rocks of the British Isles, *Memoirs of the Geological Survey of the United Kingdom*, **2**: 483–538.

Gordon, C.M. and Donovan, S.K., 1992, Disarticulated echinoid ossicles in paleoecology and taphonomy: the last interglacial Falmouth Formation of Jamaica, *Palaios*, **7**: 157–166.

Grant, R.E., 1980, The human face of the brachiopod, *Journal of Paleontology*, **54**: 499–507.

Gyllenhaal, J.A., 1772, Beskrifning På de så kallade Crystall-åplen och kalkbollar, såsom petreficerade Djur af Echini genus, eller dess nårmaste slågtingar, *Konglige svenska Vetensk-Akademie Handlingar*, **33**: 239–261.

Huelsenbeck, J.P., 1991, When are fossils better than extant taxa in phylogenetic analysis? *Systematic Zoology*, **40**: 458–469.

Huelsenbeck, J.P., 1994, Comparing the stratigraphic record to estimates of phylogeny, *Paleobiology*, **20**: 470–483.

Huelsenbeck, J.P. and Rannala, B., 1997, Maximum likelihood estimation of phylogeny using stratigraphic data, *Paleobiology*, **23**: 174–180.

Jarvis, I., Carson, G.A., Cooper, M.K.E., Hart, M.B., Leary, P.N., Tocher, D.A., Horne, D. and Rosenfeld, A., 1988, Microfossil assemblages and the Cenomanian–Turonian (late Cretaceous) Oceanic Anoxic Event, *Cretaceous Research*, **9**: 3–103.

Kier, P.M., 1972, Tertiary and Mesozoic echinoids of Saudi Arabia, *Smithsonian Contributions to Paleobiology*, **10**: 242 pp.

Kier, P.M., 1977, The poor fossil record of the regular echinoid, *Paleobiology*, **3**: 168–174.

Kier, P.M., 1984, Fossil spatangoid echinoids of Cuba, *Smithsonian Contributions to Paleobiology*, **55**: 336 pp.

Macurda, D.B., Jr, 1983, Systematics of the fissiculate Blastoidea, *Museum of Paleontology, University of Michigan, Papers on Paleontology*, **22**: 290 pp.

Meehl, P.E., 1983, Consistency tests in estimating the completeness of the fossil record: a neo-Popperian approach to statistical paleontology, *Minnesota Studies in the Philosophy of Science*, **10**: 413–473.

Nicol, D., 1977, The number of living animal species likely to be fossilized, *Florida Scientist*, **40**: 135–139.

Norrel, M.A., 1993, Tree-based approaches to understanding history: comments on ranks, rates and the quality of the fossil record, *American Journal of Science*, **293–A**: 407–417.

Paul, C.R.C., 1967a, The British Silurian cystoids, *Bulletin of the British Museum, Natural History (Geology)*, **13**: 297–356.

Paul, C.R.C., 1967b, *Osculocystis*, a new British Silurian cystoid, *Geological Magazine*, **104**: 449–454.

Paul, C.R.C., 1973, British Ordovician cystoids: 1. Diploporita, *Paleontographical Society Monographs*, **127** (no. 536): 1–64.

Paul, C.R.C., 1974, *Regulaecystis devonica*, a new Devonian pleurocystitid cystoid from Devon, *Geological Magazine*, **111**: 349–352.

Paul, C.R.C., 1980, *The Natural History of Fossils*, Weidenfeld, London: 292 pp.

Paul, C.R.C., 1982, The adequacy of the fossil record. *In* K.A. Joysey and A.E. Friday (eds), *Problems of Phylogenetic Reconstruction, Systematics Association Special Publication 21*, Academic Press, London: 75–117.

Paul, C.R.C., 1984, British Ordovician cystoids: 2. Dichoporita, *Paleontographical Society Monographs*, **136** (no. 563): 65–152.

Paul, C.R.C., 1985, The adequacy of the fossil record reconsidered. *In* J.C.W. Cope and P.W. Skelton (eds), *Evolutionary Case Histories from the Fossil Record, Special Papers in Palaeontology*, **33**: 7–16.

Paul, C.R.C., 1988, Extinction and survival among the echinoderms. *In* G.P. Larwood (ed.), *Extinction and Survival in the Fossil Record, Systematics Association Special Publication 34*, Clarendon Press, Oxford: 155–170.

Paul, C.R.C., 1992a, How complete does the fossil record have to be? *Revista Española de Palaeontologia*, **7**: 127–133.

Paul, C.R.C., 1992b, The recognition of ancestors, *Historical Biology*, **6**: 239–250.

Paul, C.R.C., 1997, British Ordovician cystoids: 3. Fistuliporita, *Palaeontographical Society Monographs*, **151** (no. 604): 153–213.

Paul, C.R.C., in press, Selection and the fossil record. *In* J. Bintliff (ed.), *Structure and Contigency in the History of Life*.

Smith, A.B. and Paul, C.R.C., 1982, Revision of the class Cyclocystoidea (Echinodermata), *Philosophical Transactions of the Royal Society*, **B296**: 577–684.

Smith, A.B., Paul, C.R.C., Gale, A.S. and Donovan, S.K., 1988, Cenomanian and Lower Turonian echinoderms from Wilmington, SE Devon: taxonomy, biostratigraphy and palaeoecology, *Bulletin of the British Museum, Natural History (Geology)*, **42**: 246 pp.

Sprinkle, J. (ed.), 1982, Echinoderm faunas from the Bromide Formation (Middle Ordovician) of Oklahoma, *University of Kansas Paleontological Contributions*, **1**: 369 pp.

Tunnicliffe, V., 1992, The nature and origin of the modern hydrothermal vent fauna, *Palaios*, **7**: 338–350.

Wagner, P.J., 1995, Stratigraphic tests of cladistic hypotheses, *Paleobiology*, **21**: 153–178.

Woodring, W.P., 1925, Miocene mollusks from Bowden, Jamaica: I. Pelecypods and scaphopods, *Carnegie Institution of Washington Publications*, **366**: 222 pp.

Woodring, W.P., 1928, Miocene mollusks from Bowden, Jamaica: II. Gastropods and discussion of results, *Carnegie Institution of Washington Publications*, **385**: 564 pp.

2
Determining Stratigraphic Ranges

Charles R. Marshall

INTRODUCTION

The fossil record may be used to estimate two core temporal parameters: a taxon's time of origin and its time of extinction. Given these, a third parameter, its longevity, may be estimated. Values of these parameters may be used in a variety of ways, for example, for local or global correlation, for testing hypotheses of origination or extinction, for evaluating phylogenetic hypotheses and for calculating species turnover rates, among others. However, the fossil record is incomplete and it is generally expected that a literal reading of the fossil record will lead to inaccurate estimates of these core parameters. Here I review the quantitative methods available for correcting for the incompleteness of the fossil record. The efficacy of these methods depends in part on the characteristics of the fossil record of the taxa being analysed, as well as on the specific question being asked of the record. This review emphasizes those questions where an estimate of the true stratigraphic range is required, but this by no means exhausts the interesting questions that can be answered with stratigraphic data. For example, sometimes the relative sizes of true stratigraphic ranges are of interest, such as whether gastropods with planktotrophic larvae are longer lived than gastropods with non-planktotrophic larvae (Jablonski, 1986; Marshall, 1991). If there is an underlying theme to this review, it is that there must be an interplay between the use of quantitative methods, and geological and palaeontological intuition.

The Adequacy of the Fossil Record. Edited by S.K. Donovan and C.R.C. Paul.
© 1998 John Wiley & Sons Ltd.

SINGLE TAXA IN LOCAL SECTIONS

In local sections observed stratigraphic ranges are likely to underestimate true ranges. A way of estimating the stratigraphic horizons that correspond to the true end-points is desirable. Here I outline the quantitative methods for estimating the positions of these stratigraphic horizons.

Point Estimates of the Positions of True Stratigraphic End-points

Under the assumption of random fossilization, the average gap size in the observed range is an unbiased estimate of the distance (d) between the last occurrence and the true top of a taxon's range (Strauss and Sadler, 1989, equation 8). This result has been confirmed by computer simulation (Alvarez, 1983; Marshall, unpublished data). Thus, given a taxon with an observed range R_o known from H fossil horizons, and therefore ($H - 1$) gaps in its range, the unbiased estimate of the distance between the true and observed end-points is given by:

$$d = R_o/(H - 1) \tag{1}$$

The key difference between the use of this equation and the literal reading of the fossil record is that the amount of faith we put in the observed end-points of a taxon's range is tempered by the richness of the fossil record. With a rich fossil record (large H), d is small and we are confident that a literal reading of the record is a relatively good reading of the fossil record. However, with a poor fossil record (small H), d is large and a literal reading of the fossil record should be taken with a grain of salt, to the tune of the distance d.

Using the same logic used to derive equation 1, an unbiased estimate of the true range, R_t, is given by equation 11 of Strauss and Sadler (1989):

$$R_t = R_o(H + 1)/(H - 1) \tag{2}$$

Maximum likelihood methods are a standard way of estimating the most likely value of an unknown parameter. However, the maximum likelihood estimate of the true end-point of a stratigraphic range is the observed end-point; maximum likelihood estimation is equivalent to reading the fossil record literally and as such is biased (Strauss and Sadler, 1989). This perhaps counter-intuitive result is easily understood: all else being equal, the further one moves away from the observed end-point of a stratigraphic range, the less likely one is to recover fossils of that taxon. Thus, the single most likely place for the true end-point of a taxon's true range is the observed end-point.

Confidence Intervals on Positions of True Stratigraphic End-points

Equation 1 cannot be used to provide confidence intervals on the true end-point of a stratigraphic range because d is not distributed even approximately normally (Strauss and Sadler, 1989). Hence, an alternative approach is required. A way of calculating confidence intervals under the assumption of random fossilization has been provided by Strauss and Sadler (1989), following the pioneering work of Paul (1982), McKinney (1986a,b), and Springer and Lilje (1988). Strauss and Sadler (1989), using a simple and intuitively appealing approach, suggested that the length of the confidence interval (r_c) for the true end-point of a stratigraphic range be expressed as a fraction α of the observed stratigraphic range (R):

$$r_c = \alpha R \tag{3}$$

Under the assumption of randomly distributed fossil horizons, the value of the proportionality factor α depends on the number of fossil horizons within that range (H) and the confidence level (C) (Strauss and Sadler, 1989, equation 20; see also Marshall, 1990a):

$$\alpha = [(1 - C)^{-1/(H-1)} - 1] \tag{4}$$

The interpretation of the confidence interval is that the true end-point of the taxon lies within the distance r_c beyond the observed end-point with confidence C. Figure 2.1 shows the lengths of the 50% and 95% confidence intervals, as well as the unbiased estimate of the distance between the observed and true ends of a stratigraphic range for a variety of different richnesses of the fossil record. Strauss and Sadler (1989) also provided an equation for calculating confidence intervals on both end-points jointly (see Marshall, 1990a, for a less technical introduction), as well as a Bayesian formulation for estimating the posterior density distribution for the true end-point of a stratigraphic range.

Equations 3 and 4 have not been widely used, in part due to the lack of appropriately collected data: very few studies give the number or positions of all the fossils found within measured sections. However, considerable power can be gained by applying confidence intervals to the fossil reøcord (for example, see Marshall and Ward, 1996, and below); as palaeontologists we need to start collecting our data as uniformly as possible, as well as recording data with a much higher fidelity than is the usual custom.

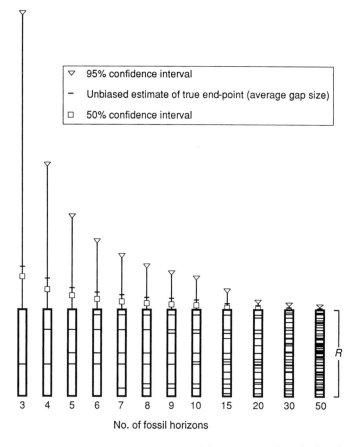

Figure 2.1 *Sizes of the 50% and 95% confidence intervals calculated under the assumption of randomly distributed fossil horizons (equations 3 and 4). The unbiased estimates of the true end-points of the stratigraphic ranges are also shown (equation 1). The bars within the stratigraphic range (R) represent fossiliferous horizons*

Randomness

The methods outlined above assume a random distribution of fossil horizons. However, for many sections this assumption is manifestly violated and it is crucial to test the assumption before applying the equations given above. None the less, I have been surprised how often observed distributions of fossil horizons are indistinguishable from random (see Strauss and Sadler, 1989; Marshall, 1990a, 1995a,b; Marshall and Ward, 1996; but see also Paul, 1982, and McKinney, 1986a, for examples of prevalent non-random distributions).

Plotnick *et al.* (1996) provided an interesting analysis of a small 'zoo' of distributions including the random distribution, as well as a way of distinguishing between them using lacunarity analysis. As different processes often result in different distributions of fossil horizons through time, the ability to distinguish between different distributions may enable the development of new models of fossilization and, concomitantly, more appropriate methods of calculating confidence intervals.

However, even if a distribution of fossil horizons is indistinguishable from random, this does not mean that the equations outlined above may be used with impunity; it should be borne in mind that, given limited data, non-random distributions may appear random. That is to say, with limited data, statistical tests have limited power (if β is the probability of committing a Type II error (failing to reject the null hypothesis when it is false), the power of a test is $(1 - \beta)$, the probability of rejecting the null hypothesis when it is false and some other alternative is true). Determining the power of a statistical test is not always straightforward and requires some specified alternative to the null hypothesis. In many cases the difference between two distributions will be greatest in the tails and thus the penalty for committing a Type II error will be greatest when estimating the size of confidence intervals for large values of C, such as 0.95 or 0.99 (Marshall, 1994).

Randomness and the Geological Setting

Clearly the geological setting, as well as the nature of the exposure and the way it was collected, will play a major role in determining whether fossils are randomly distributed. For example, the likelihood of randomly distributed fossils will be much lower for benthic taxa in an emptying cratonic sea, than for pelagic taxa in uniform deeper-water facies. In accord with this intuition, the null hypothesis of randomly distributed fossils may be rejected for almost all taxa in an analysis of Adegoke's (1969) data on the Neogene fossil record of benthic taxa in the Great Central Valley of California (Marshall, unpublished data). This result is not surprising given the complete recession of the sea from the valley during the Neogene. In contrast, the null hypothesis of randomly distributed fossils could not be rejected for most taxa, neither for ammonites nor for inoceramid clams, in the deeper-water facies from the Maastrichtian of the Bay of Biscay (Marshall and Ward, 1996). Here, the sections are dominated by some 300 rhythmites of approximately equal thickness, suggesting a relatively constant rate of deposition from one cycle to the next. Where appropriate, geological intuition should play a role in selecting the most appropriate sections and taxa to analyse. In all cases it is necessary to test each data set against the null hypothesis of randomly distributed fossil horizons.

Confidence Intervals with Non-random Fossilization

The ability to calculate confidence intervals without having to assume randomly distributed fossils would be desirable, given that: (1) in some geological settings non-random fossilization is the expectation; (2) in empirical studies non-random distributions are encountered; and (3) with the limited data available for many species, the power of statistical tests of the assumption of randomly distributed fossil horizon is limited. Two approaches have been developed.

Distribution-free confidence intervals

Different fossilization processes will lead to different distributions of gap sizes. If a stick (the true stratigraphic range) is randomly broken in H places, the length of the end piece will follow a Dirichlet distribution, the distribution reflected in Strauss and Sadler's equation for estimating the length of unobserved stratigraphic range (equations 3 and 4). Other processes will lead to different distributions of the length of the gap between the last occurrence and the true stratigraphic end-point. The distribution-free method for calculating confidence intervals on the end-points of stratigraphic ranges makes no assumptions about the nature of the underlying distribution of the distance between the observed and true end-points of a stratigraphic range, except that it is at least approximately continuous. However, the approach does assume that there is no correlation between the size of the gaps between adjacent fossiliferous horizons and stratigraphic position (Marshall, 1994). This restriction may be significant, because it means that the method is not applicable to exposures exhibiting secular trends in the probability of recovering fossils with stratigraphic position, such as those dominated by large-scale shallowing-up sequences, etc.

The distribution-free method exploits the fact that if enough data are collected it is possible to place upper and lower limits on any quantile of an unknown distribution. For example, noting that the median quantile corresponds to the 50% confidence interval ($C = 0.5$), there is only $1/2^6$ (1/64) chance that six consecutive random draws from a distribution will all be greater than, or all smaller than, the median, regardless of the distribution. Hence, at a significance level of 2/64, the size of the 50% confidence interval will be both larger than the smallest of the six observed gaps, and smaller than the largest of the six observed gaps. Using this logic, a look-up table for placing upper and lower bounds on the size of the confidence interval for differing confidence levels and richnesses of the fossil record may be developed (Marshall, 1994).

A two-fold price is paid for the generality of the method: (1) there are uncertainties associated with the sizes of the confidence intervals; and (2) unless one has an unusually rich fossil record, it is impossible to place an upper bound on confidence intervals for larger confidence levels (for example, 95%). The requirement for large amounts of data to assign an upper bound on the confidence intervals for large values of C is due to the difficulty in characterizing the tail of an unknown distribution with limited data.

Generalized confidence intervals

Marshall (1997) provided a generalized method for calculating confidence intervals with any known non-random distribution of fossil horizons, even those where there is a correlation between the gap size between adjacent fossil horizons and stratigraphic position. To apply the method, a graph of the probability of recovering fossils as a function of stratigraphic position is required (Figure 2.2). Strauss and Sadler's (1989) method for calculating confidence intervals under the assumption of randomly distributed fossils (equations 3 and 4) may be viewed as a special case of this approach: the case where the fossil recovery function is constant with stratigraphic position. With randomly distributed fossils the ratio of the length of the confidence interval (r) to the observed range (R), α, is given by equation 4. In Marshall's (1997) method, α gives the ratio of the area under the fossil recovery curve over the distance of the confidence interval to the area under the fossil recovery curve over the observed stratigraphic range (Figure 2.2). The beauty of the approach is that it takes into account not just the observed end-points of the stratigraphic range, but also the richness of the fossil record and where in the stratigraphic section those fossils occur with respect to where they are expected. To apply the method, the distribution of fossil horizons must first be tested against the fossil recovery curve.

The price paid for the generality of the approach is that one needs an explicit description of the probability of recovering fossils as a function of stratigraphic position, both within and beyond the observed stratigraphic range. In some cases simply using the exposure surface area as a proxy for the probability of recovering fossils may be appropriate. However, when major facies changes are present, more complicated considerations must be added, such as the relative probability of recovering fossils as a function of water depth, facies, etc. (Holland, 1995; Marshall, 1997). Constructing fossil recovery potentials as a function of stratigraphic position may rarely be easy and is a task that needs much work.

Confidence Intervals with Binned and Discrete Sampling

The methods discussed above assume implicitly that the exposure has been sampled completely and continuously, or nearly continuously. However, two commonly employed sampling regimes do not involve this type of

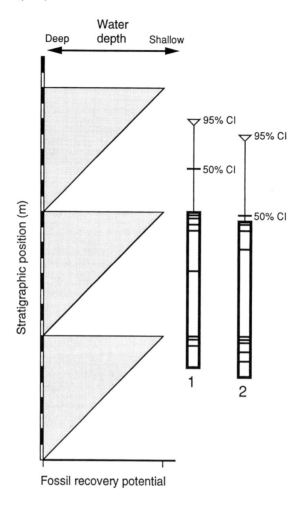

Figure 2.2 *Idealized fossil recovery potential for three shallowing-up parasequences for taxa with a preference for shallow-water facies. The lengths of the confidence intervals (CIs) were calculated using the generalized method of Marshall (1997). Both species 1 and 2 have stratigraphic ranges of the same length and are each known from 10 fossil horizons, but the sizes of the confidence intervals (especially the 50% interval) are of different lengths when calculated using the fossil recovery potential function shown on the left. (After Marshall, 1997)*

sampling: binned sampling and discrete sampling. Both may require a different statistical treatment from those discussed above. These treatments are currently being developed.

Binned sampling

In this type of sampling a section is divided into a series of contiguous bins, and the presence or absence of a taxon is recorded for each. Sometimes the number of specimens is recorded for each taxon in each bin. Usually the number of fossil horizons within each bin is not recorded (in some cases because they may not be discrete enough to be identified). Examples of this style of collecting include the descriptions by Erle Kauffman and his colleagues of the biotic changes in the mid-Cretaceous Western Interior Seaway (Kauffman, 1988; Elder, 1989; Harries, 1993).

In these situations a count of the number of bins containing fossils of a taxon will usually underestimate the number of fossil horizons (H in equations 1–4) that would have been observed had continuous sampling been employed. Thus, the estimate of the incompleteness of the fossil record using these equations will usually be too small. There are crude ways of correcting for the unrecorded fossil horizons in this type of data (Marshall, unpublished data). If the number of bins containing fossils is small compared with the total range, then using the observed number of bins containing fossils will be a good approximation for the number of bins that one would have expected to have observed with continuous sampling; that is, as the bins become small with respect to the total range, then in effect the sampling becomes continuous.

Discrete sampling

With discrete sampling, small samples are usually taken at regular intervals and the taxonomic composition of each sample is recorded. This is a standard way of sampling cores for microfossils. The relative abundances (either quantitative or qualitative) of each taxon in each sample may also be recorded. In this type of sampling most of the exposure, or core, is unsampled and the estimate of the richness of the fossil record is much smaller than it would have been with denser sampling. The equations presented in this chapter should only be applied to stratigraphic data derived from this type of sampling if it is felt that an increase in the sampling density would not affect the total number of samples yielding the taxon of interest. My suspicion is that this condition will not be met very often. Some of these issues are discussed further in Marshall (1994). An area of future work is the development of statistical techniques for handling discrete sampling regimes.

A further complication that may arise with this type of sampling is non-uniform collecting procedures. For example, if a worker feels that the presence or absence of a taxon is particularly important to establish in a given stratigraphic interval, perhaps because it is biostratigraphically important, he or she may search additional samples within that interval. Data collected in this way are compromised statistically, and comparison between data sets is made difficult by differing degrees of collecting intensity. Standard techniques for sampling cores that produce data suitable for statistical analysis also need to be established.

MULTIPLE TAXA IN LOCAL SECTIONS

Signor–Lipps and Jaanusson (Sppil–Rongis) Effects

Signor and Lipps (1982) recognized that, with incomplete preservation, sudden disappearances of taxa would appear as a smeared-out drop in diversity in the fossil record. The Signor–Lipps effect, as it has become known (Raup, 1986), has become a central issue in discussions of the rapidity of extinctions (see, for example, Jablonski, 1997; MacLeod *et al.*, 1997). Similarly, Jaanusson (1976) recognized that, with incomplete preservation, the first appearances of a set of taxa that all immigrated into a region at the same time would appear smeared out, giving the incorrect impression of sequential appearances. This has also been recognized by Springer (1990), as well as others such as Marshall (1995b) and Harries (personal communication), who have termed it the Sppil–Rongis effect.

The realization that incomplete preservation makes sudden extinctions appear gradual has changed the way many view the fossil record. For example, Ward et al. (1986) initially argued that the disappearance of ammonites in western Tethys suggested the end-Cretaceous impact did not play a major role in the ammonite extinctions. However, in a later discussion of the same sections, Ward *et al.* (1991) avoided drawing any firm conclusions from the observed gradual disappearance of ammonites, except to note the new data had revealed that at least 8 to 10 ammonites survived until the last metre of the Cretaceous, and possibly to the boundary. The Signor–Lipps effect has rendered the fossil record hard to interpret (though see Stanley and Yang, 1994, for some tests for distinguishing artificial from real extinction events, and Meldahl, 1990, for an important actualistic study), motivating Raup to argue that '[T]he challenge for future research is to develop a new calculus for treating biostratigraphic data' (Raup, 1989, p. 421). This calculus now exists, at least in rudimentary form (see also Chapter 5 for some heuristic methods).

Testing the Null Hypothesis of a Sudden Disappearance (or Origination)

Three methods have been used to distinguish between sudden and gradual extinctions.

Springer's method

Springer (1990) was the first to provide a quantitative method for distinguishing between sudden and gradual disappearances. The method is based on the realization that the confidence levels corresponding to the confidence intervals that extend from the observed end-points of the stratigraphic ranges of a set of taxa to a putative sudden disappearance horizon should be distributed uniformly on [0,1]. Any stratigraphic horizon where this condition holds is a candidate for a mass extinction horizon. Only taxa with last occurrences found below the putative extinction horizon are considered.

Marshall's method

Marshall (1995a) developed a similar method based on 50% confidence intervals. For a truncation event, the statistical expectation is that half the 50% confidence interval end-points should lie above the truncation horizon and half below. As with Springer's method, the null hypothesis of a truncation event is rejected if some of the putative victims have fossil records that extend beyond the truncation horizon.

The method is more easily applied than Springer's, though it has less statistical power because it does not take into account how far the end-points of the 50% confidence intervals lie from the proposed extinction horizon. The approach was developed with distribution-free confidence intervals in mind; Springer's (1990) method is not easily used with distribution-free confidence intervals due to the difficulties in accommodating the uncertainties in the lengths of the intervals, and because unrealistically rich fossil records are required to calculate upper bounds on distribution-free confidence intervals for all but intermediate confidence levels.

Solow's method

Solow (1996) pointed out that the methods of Springer and Marshall are conservative in that they only provide upper bounds on the significance level. He overcame this weakness by developing a maximum likelihood approach to testing the null hypothesis of concurrent end-points for which the significance level is known. The method is relatively easy to apply, and should supersede Marshall's and Springer's when the assumption of random fossilization appears to be a satisfactory approximation. The upper limit of a $(1 - p)$ confidence interval on the position of a truncation horizon, U_u, for n taxa can be found with the following equation:

$$-2 \ln \prod_{i=1}^{n} [R_i/(U_u - B_i)]^{(H_i-1)} = \chi^2_{2n}(p) \tag{5}$$

where R_i is the stratigraphic range, B_i the stratigraphic position of the base, and H_i the number of fossiliferous horizons for the ith taxon. $\chi^2_{2n}(p)$ is the upper p-quantile of the chi-squared distribution with $2n$ degrees of freedom. Equation 5 must be solved iteratively (different values of U_u are tried until the equation is satisfied), a task that is straightforward with a computer package such as *Theorist* or *Mathematica*. To test the null hypothesis that all n taxa have a common end-point, U_u is substituted with the stratigraphic position of the highest fossil in the data set and now the chi-squared value is based on $2(n - 1)$ degrees of freedom.

Comparison of the methods

All three approaches have been applied to Macellari's (1986) end-Cretaceous ammonite data for Seymour Island, Antarctica, and all three find Macellari's data consistent with concurrent end-points at the Cretaceous/Tertiary (K/T) boundary (but see below). However, Solow's method provides a narrower interval over where the truncation horizon might lie, by almost a factor of two (Solow, 1996).

Solow (1996) implied that the maximum likelihood approach will always give narrower confidence bands than the methods of Springer and Marshall. However, no significant difference is seen between the maximum likelihood analysis of the taxa known from the last 1.5 m of the Cretaceous in the Bay of Biscay and the analysis with 50% confidence intervals by Marshall and Ward (1996) (Figure 2.3). It appears that, compared with Solow's approach, Springer's and Marshall's methods will give wider confidence bands on the location of the putative extinction horizon when taxa are included that may well have become extinct prior to the truncation event, but otherwise the three methods will give similar results.

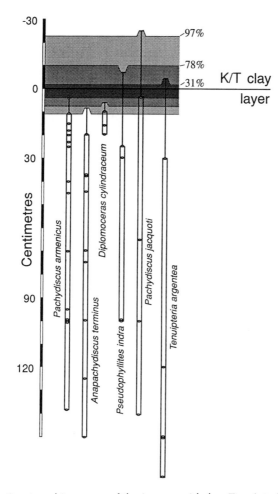

Figure 2.3 *Stratigraphic ranges of the inoceramid clam* Tenuipteria argentea *and ammonites known from two or more fossils in the last 1.5 m of the Cretaceous of the Bay of Biscay. Contours indicate predicted position of extinction horizon, assuming there was one, based on Solow's (1996) maximum likelihood method, except at the ends of the 50% confidence intervals, where the (usually) deflected position of the contour indicates the predicted position of the extinction horizon based on Marshall's 50% confidence-interval approach (Marshall and Ward, 1996). Both methods give very similar results and indicate that the fossil record of these species is consistent with their sudden extinction at the Cretaceous/Tertiary (K/T) boundary; the gradual decline in diversity over the last 30 cm appears to be due to the Signor–Lipps effect. The confidence that the truncation horizon lies within each shaded interval is given adjacent to the upper bound of each interval. Positions of the fossil horizons are indicated within ranges. The null hypothesis of randomly distributed fossil horizons could not be rejected for any species*

Failing to Reject the Null Hypothesis of a Truncation Event

Just because one fails to reject the null hypothesis of a sudden disappearance for a set of taxa does not mean they did disappear suddenly. Perhaps there was a gradual disappearance over a narrow stratigraphic interval that was undetected due to the 'graininess' of the raw data. Marshall (1995a) used a computer simulation of Macellari's (1986) study of the ammonite fossil record of the late Cretaceous on Seymour Island to show that there is a considerable range of gradual extinction scenarios that may also have produced a fossil record that, when analysed with 50% confidence intervals, would have been consistent with a mass extinction.

Generalizing from this analysis, it appears that, if a set of taxa had disappeared gradually over a distance roughly equivalent to, or less than, the average gap size within the observed stratigraphic ranges of those taxa, then statistical analysis of their fossil record would fail to reject the null hypothesis of a mass extinction for those taxa. This should not be surprising, because statistical methods cannot compensate for a lack of data; it is to be expected that the graininess of the data will be a limiting factor when trying to discriminate between hypotheses. A future task is to establish more formally the statistical power of the methods developed for discriminating between various extinction scenarios.

The danger of overinterpreting the failure to reject the null hypothesis of a mass extinction is emphasized by new data collected by Zinsmeister and colleagues on Seymour Island (Zinsmeister *et al.*, unpublished data). In Macellari's data there is no reason to reject the hypothesis that the ammonite *Pachydiscus ultimus* disappeared at the K/T boundary. The average gap size within its range is 15 m, which is also the size of the gap between the last fossil and the K/T boundary iridium anomaly. However, much more intensive collecting over the top 20 m of the Cretaceous has led to dozens of new fossils (compared with the original four), but the gap to the iridium anomaly has only decreased by a few metres (W. J. Zinsmeister, personal communication): more intensive collecting now suggests *Pachydiscus ultimus* disappeared from the sections prior to the bolide impact.

Distinguishing between Sudden and Other Disappearances

All the methods discussed above have assumed the simple case where two alternative disappearance scenarios are being tested: sudden versus gradual disappearance. However, in cases where a sudden disappearance is anticipated, it is not too unreasonable to presume that some taxa also disappeared prior to the truncation horizon.

With a truncation horizon in mind, it is relatively easy to make a statistical assessment of the taxa that most likely really disappeared prior to the truncation horizon (Marshall, 1990a; Marshall and Ward, 1996). A slight modification of equations 3 and 4 may be used to assess our confidence, C_e, that a taxon really disappeared before the truncation horizon. If G is the distance between the top of the observed range and the putative truncation horizon, R is the observed range and H the number of fossil horizons, C_e is given by:

$$C_e = (G/R + 1)^{-(H-1)} \tag{6}$$

For an ensemble of n taxa it is necessary to adjust the significance level in accordance with the number of taxa being analysed (a Bonferroni correction must be used: to achieve a significance value of p, the value of $(1 - C_e)$ for each taxon must be compared to $p' = p/n$, or $p' = 1 - (1 - p)^{1/n}$, rather than p; see Sokal and Rohlf, 1995).

Graphical Approach for Quickly Estimating True Extinction Patterns

Figure 2.4 illustrates a simple graphical approach that may be used to gauge the set of disappearance scenarios that might apply to given fossil records of an ensemble of taxa. The observed stratigraphic ranges, along with the 50%, 95% and 99% confidence-interval end-points, are plotted. The following rules of thumb are then applied:

1. Putative truncation horizons occur in stratigraphic intervals bracketed by approximately equal numbers of 50% confidence intervals, under the condition that roughly half of the end-points of those confidence intervals lie above the stratigraphically highest fossil of the ensemble of taxa. In Figure 2.4, the interval starting about 12 m below the K/T boundary to a short distance above the boundary is the only stratigraphic interval where a mass disappearance is admitted by the data. In this example, using confidence intervals to test the hypothesis of a sudden extinction is complicated by the fact that the rocks from about 8 to 1.5 m below the K/T boundary (indicated by the shading) are difficult to collect; there is a major inhomogeneity in the fossil record very close to the K/T boundary. Idiosyncrasies such as this must be dealt with in some way; in this case, the data were handled by conducting two separate analyses, one on either side of the shaded interval in Figure 2.4 (see Marshall and Ward, 1996). Note the pronounced Signor–Lipps effect for the 12 species of ammonite known from the last 1.5 m of the Cretaceous when the fossils from these strata are ignored (boxed taxa in Figure 2.4).

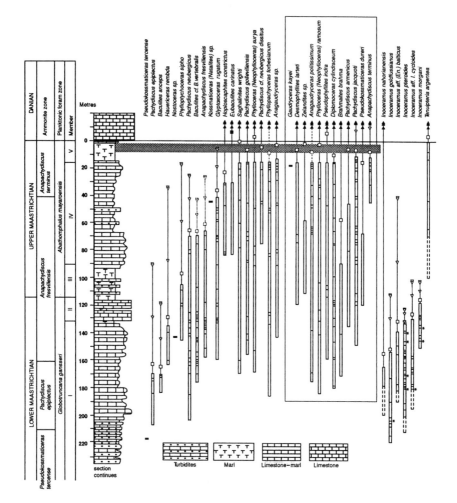

Figure 2.4 *Composite ranges of ammonites and inoceramid bivalves (seven species at the right) from several Bay of Biscay sections, projected on to the Zumaya measured section. Ammonite fossils from the last 1.5 m of the Cretaceous are not shown and the stratigraphic ranges of those species without these fossils are shown in the box. Note the pronounced Signor–Lipps effect for these taxa. The stippled region indicates strata from which it is particularly difficult to collect 8–1.5 m below the Cretaceous/Tertiary (K/T) boundary. Its closeness to the boundary complicates the statistical analysis of this fossil record (see Marshall and Ward, 1996). The dashed ranges for the inoceramids are range extensions based on other sections. These were not used in the calculation of the confidence intervals. Each fossiliferous bed is indicated by a horizontal bar. Two fossil horizons too close to resolve are indicated by asterisks. Confidence intervals were calculated under the assumption of randomly distributed fossils (equations 3 and 4). □, end of 50% confidence interval; ▽, end of 95% confidence interval; ▼, end of 99% confidence interval. ■, ▲ and ▲, 50%, 95% and 99% confidence intervals that extend more than 10 m above the K/T boundary, respectively. Dashed confidence intervals are approximations only – these taxa have non-random distributions of fossil horizons (see Marshall and Ward, 1996, for further details).*

2. Taxa that have last occurrences that lie stratigraphically higher than this interval identified by the 50% confidence intervals are excluded on the grounds that they have survived the truncation event. There are no taxa in Figure 2.4 that fall into this category.
3. Taxa with 95% or 99% confidence intervals that lie below a stratigraphic interval identified as potentially including a putative extinction horizon are also excluded on the grounds that they most likely disappeared before the time of the putative truncation event (though not all should be excluded, since by chance alone some victims of the truncation event, assuming it is real, are expected to have 95% or 99% confidence intervals that terminate prior to the truncation interval – see Bonferroni correction discussed above). In Figure 2.4, this includes the first half dozen or so of ammonite species at the left of the diagram, as well as five of the inoceramid clams. In effect, using the K/T boundary as the reference horizon, these taxa have values of C_e (equation 6) greater than 0.99.

Graphic Correlation and Confidence Intervals

Shaw's (1964) method of graphic correlation has become a primary tool for quantitative correlation (Mann and Lane, 1995). The simulated annealing approach of Kemple *et al.* (1995) is the most sophisticated application of the method, allowing multiple sections to be correlated simultaneously. A feature of their algorithm is the weighting of the observed end-points of the observed stratigraphic ranges by a measure of the richness of the fossil record. Specifically, they use the average gap size (equation 1) in weighting the observed end-point, though the median gap size (the 50% confidence interval) may be suitable as well (Kemple *et al.*, 1995).

Another potential use of confidence intervals in graphic correlation (or any other quantitative correlation method) is to eliminate range-through taxa. In any given section one might reasonably expect some taxa to have true stratigraphic ranges that extend beyond either (or both) ends of the sequence. Obviously these taxa should not be used in the correlation. However, they may be difficult to detect due to the Jaanusson and Signor–Lipps effects, especially if they are known from relatively few fossils. The inclusion of large numbers of range-through taxa may seriously mask the true signal of correlation reflected in those taxa whose real ranges end within the exposed sections. It should be possible to use the methods outlined above to identify range-through taxa prior to the process of correlation, thus improving the correlation. This approach has yet to be used.

A Question of Accuracy

Statistics does not concern itself with questions of accuracy, but rather with questions of precision. The accuracy of the raw data must be established prior to analysis; all the methods discussed above assume that fossils have been properly identified and that their stratigraphic positions were accurately recorded. The assumption of accuracy is important to keep in mind when conducting statistical analyses. For example, one of the important issues in assessing the patterns of extinction of microfossils is the question of reworked fossils (for example, see Olsson and Liu, 1993, as well as Huber, 1996, and MacLeod and Huber, 1996, for ways of distinguishing reworked from non-reworked fossils). This type of issue must be settled before the analysis.

GLOBAL DATA

Single Taxa

In global compilations, first and last appearances are typically used as proxies for times of origin and extinction. In standard global compilations (for example, *Treatise on Invertebrate Paleontology*; Sepkoski, 1992; Benton, 1993; Labandeira, 1994) global standard stratigraphic nomenclature is used to express times of first and last appearance. Estimated durations that include the stratigraphic intervals of the first and last appearances may overestimate (Paul, 1985) or underestimate true ranges, because the true ranges are unlikely to coincide with the ends of those intervals. Errors in assigning lithostratigraphic units that sport the first and last fossils of a taxon to the global standard stratigraphic time scale may result in reported stratigraphic ranges that over- or underestimate true ranges.

Assessing uncertainties in times of first and last appearance for global data is much more difficult than for local sections. Usually, quantitative data on the richness of the fossil record that makes up the range are not available and there are important sampling issues that need to be addressed. For example, key stratigraphic intervals may not be represented by rocks over wide geographical areas, or those rocks, if they exist, may not have been searched adequately (Marshall, 1989). The complexities involved in treating the incompleteness of the fossil record for global data are much greater than for local sections, and by and large have not been treated with any statistical rigour, although Valentine *et al.* (1991) provided a very thoughtful discussion of the relevant issues, and offered some quantitative constraints on the timing of the biological explosion associated with the Precambrian/Cambrian boundary. Crucial questions that need to be answered are: What level of precision does the question at hand require (maybe the incompleteness of the fossil record is not an issue) and what is the spatial distribution of the relevant rocks, and have they been sampled adequately (usually the sampling intensity is unknown)?

Multiple Taxa

While there seems to be little prospect of a simple method for estimating error on the times of origin and extinction of individual taxa based on global compilation, Foote and Raup (1996) have provided a simple method that may be used to estimate true extinction rate, and therefore the true median duration, of an ensemble of taxa. To apply the method, the stratigraphic scale is broken up into sequential bins. The only data the method requires are the observed ranges of each taxon, given as the number of sampling bins in the range. An estimate of the richness of the fossil record for each taxon is not required. This method is widely applicable and has the advantage that it needs no additional data beyond that usually recorded in standard compilations of observed stratigraphic ranges.

Solow and Smith (1997) have provided a similar approach that uses continuous data (rather than the discrete data used by Foote and Raup) and a maximum likelihood formulation. Solow and Smith's (1997) approach seems to give lower estimates of the median duration; the reason seems to be that by binning the observed stratigraphic ranges, Foote and Raup's (1996) method overestimates the observed durations.

INTERPRETING PATTERNS IN THE FOSSIL RECORD

The statistical methods outlined above may be used to test for different patterns in the fossil record, but additional information is required to interpret the meaning of the patterns. Below I outline some of the issues that must be considered when interpreting patterns in the fossil record.

Truncation Events

Identifying a putative truncation horizon in the fossil record is one thing, but interpreting its meaning is quite another. A truncation horizon may result from one of the following, sometimes interrelated, possibilities: (1) top of section/lack of collection effort; (2) facies change/taphonomic effect; (3) hiatus in deposition; (4) local/global extinction event. Clearly, additional information is required to distinguish between these possibilities.

Top of section/lack of collection effort

The simplest explanation for an observed putative truncation horizon is that it merely represents the limit of collecting effort. Without further information, all that can be concluded is that the relevant taxa were extant at least until the time of deposition of the last stratum collected.

Facies change/taphonomic effect

If a group of taxa is tied to a particular environment, and therefore to a single or limited set of facies, an abrupt change in the facies at a particular stratigraphic horizon will most likely coincide with the local disappearance of those taxa. In an elegant set of computer simulations, Holland (1995) has shown that artificial extinction events are introduced at abrupt facies changes caused by major changes in water depth, particularly at sequence boundaries. Without any further information, it is difficult to determine whether the taxa that disappeared at the truncation horizon became extinct, or simply migrated with the facies.

An abrupt change in the taphonomic regime may also introduce a truncation into the fossil record. I suspect that truncation horizons introduced by purely taphonomic effects are much less common than truncation events introduced by real biotic changes associated with facies changes.

Hiatus in deposition

It has long been recognized that a major hiatus will have the effect of artificially introducing the approximately (due to the Signor–Lipps effect) simultaneous disappearance of taxa preserved in the rocks that pre-date the hiatus (see, for example, Birkelund and Håkansson, 1982). Thus, even if the same environment is represented on either side of a major hiatus, many of the taxa that pre-date the hiatus will have become extinct during the time span of non-deposition/erosion and thus will not be expected in the rocks that post-date the hiatus. Obviously, hiatus effects will be most pronounced if the amount of unrecorded time is at least on the time scale of the duration of the taxa involved: small hiatuses will be relatively unimportant, large hiatuses most important.

Holland (1995) has explored this in some detail. In his computer simulations he selected simulation parameters so that the duration of the hiatuses in the rock record were of a similar duration to the longevities of the taxa involved, and he found artificially induced extinction events associated with sequence boundaries, especially where lowstand systems tracts were missing (note that in his simulations Holland deliberately selected relatively high preservation potentials, so that Signor–Lipps effects would not obscure the introduced mass-disappearance horizons).

Marshall (1997) provided a simple equation for estimating the probability that a taxon became extinct within a hiatus:

$$p = (1 - C_e)(1 - e^{\ln(0.5)t/T_m})　\tag{7}$$

where t is the duration of the hiatus, T_m the estimate of the median longevity of the group of taxa to which the taxon belongs, and C_e is our confidence that the taxon became extinct somewhere between the top of its range and the beginning of the hiatus (see equation 6).

Local/global extinction event

If all the previous alternatives can be ruled out, then this last possibility remains as the best explanation of an observed truncation event. Distinguishing between local and global extinction clearly requires data from as wide a geographical area as possible.

Multiple explanations for truncation events

The explanations listed above for a truncation horizon are not mutually exclusive, and are organized somewhat hierarchically. This may be illustrated by analysing the putative truncation event at, or near, the end of the Cretaceous in the Bay of Biscay (Figure 2.4). Several possible explanations are applicable: (1) collection bias – the post-Cretaceous strata are much more difficult to collect from than the Cretaceous strata; (2) there is certainly a major facies change at the K/T boundary; (3) global databases indicate that groups such as ammonites and inoceramid bivalves are not known from any other post-Cretaceous localities, so the hypothesis of a global extinction is tenable, though one cannot formally reject the hypothesis that at least a few taxa were locally extant in the Tertiary in the Bay of Biscay and were simply not preserved, or have eluded collection from Tertiary rocks.

I noted in the introduction that the significance of the incompleteness of the fossil record depends in part on the question being asked. Thus, for the Bay of Biscay K/T boundary sections, while it is difficult to determine what occurred after the K/T boundary due to the facies change and associated collecting difficulties at the boundary, the statistical analysis does indicate that the majority of ammonites survived at least until the boundary, rather than becoming extinct over a more protracted interval prior to the boundary as literal reading of the fossil record would indicate.

Summary

Four things must be known about a section in order to interpret a truncation event: (1) the collecting regime employed; (2) the sequence stratigraphic and facies architecture of the section; (3) the taphonomic controls on the taxa of interest; and (4) the global context of the section and its fauna.

Taxon Termination versus Lineage Termination

When interpreting disappearances in the fossil record, it is important to remember that the extinction of a taxon name does not necessarily mean that the evolutionary lineage to which that name belongs became extinct; that is, in cases of phyletic change where there has been sufficient morphological change to warrant a new species name, there is no lineage extinction even though a nominal species, the ancestral species, has ceased to exist.

This complication is probably relatively unimportant for species involved in mass extinctions. Their observed stratigraphic ranges will usually have extensive overlap (for example, see Figure 2.3) and thus ancestor–descendant pairs will be uncommon; the number of species names that became extinct will accurately reflect the number of lineage extinctions. However, for patterns of extinction where there is no stratigraphic overlap between species, the possibility of phyletic change, rather than lineage extinction, becomes important. For example, Marshall and Ward (1996) argued that approximately six species of ammonite probably became extinct well before the end of the Cretaceous in western European Tethys (see above and Figure 2.4). One of those species, *Anapachydiscus fresvillensis* disappears some 65 m below the K/T boundary, while another *Anapachydiscus* species, *A. terminus*, first appears 45 m below the K/T boundary (Figure 2.4). Perhaps *A. terminus* was the continuation of the *A. fresvillensis* lineage, and the disappearance of *A. fresvillensis* does not indicate an extinction. Similarly, *Pachydiscus epiplectus* does not stratigraphically overlap with four other later appearing *Pachydiscus* species. Without an adequate phylogeny of these taxa, substantiating the possibility that these taxon extinctions do not reflect lineage extinctions is difficult. However, note that the other species identified as becoming extinct well before the K/T boundary either stratigraphically overlap their putative descendant species or do not have candidates for phyletic continuation (under the assumption that phyletic changes were restricted to species in the same genus). Thus, their disappearances probably do represent lineage terminations.

In cases where a later-appearing species was derived from an earlier species, but both have stratigraphic overlap, the stratigraphic overlap means that the later-appearing species must have originated by a cladogenic event, and thus the disappearance of the earlier species must represent true lineage termination, even though the sister species survives.

Extinctions, Originations, or Simply Changes in Abundance?

Presumably there is a correlation between the abundance of a particular species and the likelihood that it will be recovered from the fossil record. Hence, if a species underwent a dramatic drop in abundance followed by its extinction some time later, the fossil record will more likely record the time of abundance drop (within the error bar of an appropriate confidence interval) rather than that time of extinction. This may or may not be of concern depending on the question being asked. For example, if one is trying to assess whether a major geological or ecological event was correlated with an observed change in the biota, it may be immaterial whether a species was decimated at that time, rather than becoming extinct (in the same way that the fact that there were survivors of the Hiroshima atomic bomb blast hardly diminishes our assessment of its devastating effect). Similarly, the discovery of a single Paleocene ammonite or dinosaur species would not seriously detract from the significance of the end-Cretaceous demise of the ammonites or dinosaurs.

However, if one is trying to ascertain whether a lineage is still extant some time after a major environmental change, then mistaking an abundance drop for extinction could be a serious problem. For example, the fossil record of coelacanths is quite rich and terminates in the Cretaceous. Regardless of how one estimates the richness of the fossil record (Cloutier, 1991; Cloutier and Forey, 1991), application of equation 6 would strongly suggest that living coelacanths should not be found if one were to make the mistake of failing to distinguish between extinction and a decrease in fossilization potential. The fact that there is only one living species of coelacanth, and that it is relatively rare, helps explain why the fossil record does not hint at the existence of living coelacanths. This is in contrast to the fossil record of platypuses, which does strongly indicate that these monotremes might be extant (Marshall, 1990a).

The same arguments hold for lineage originations. If we are interested in the time of origin (rather than just the time a species became ecologically important, for example), then the possibility of an extended period of cryptic existence of a species will make placing error bars on its time of origin most difficult (see below).

Times of Origin and Times of Divergence

One of the primary uses of the fossil record is estimating times of origin of major groups, such as the animal phyla, etc. At a glance, estimating the approximate time of origin of a group, such as birds, may seem straightforward. However, there are at least three major ways of defining the time of

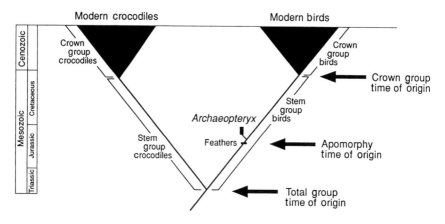

Figure 2.5   *Three different definitions for the time of origin lead to very different minimum estimates of the time of origin of birds. The estimates shown are based on the fossil record. Minimum estimates of the crown group time of origin based on molecular sequences fall deep in the Cretaceous or the Jurassic (Cooper and Penny, 1997)*

origin, each related to a different way of naming taxa (Jefferies, 1979; Patterson, 1994; Runnegar, 1996): (1) apomorphy-based definitions, that is, the time of first appearance of some chosen attribute; (2) stem-based or total group definitions, that is, the time of divergence between the group of interest and its extant sister clade; (3) node-based or crown group definitions, that is, the time of divergence of the most distantly related of the extant lineages. The estimate of the time of origin of a group may vary greatly with the definition used. For example, the minimum estimate of the time of origin of birds is the late Jurassic for the apomorphy-based definition, when feathers are used as the apomorphy; the basal Triassic for the total group definition, that is, the time of divergence from crocodiles; or the late Cretaceous (Chiappe, 1995) or early Paleocene (Feduccia, 1995), if the crown group definition, or last common ancestor of living birds, is used (Figure 2.5).

There are, of course, other definitions that might be used: for example, one might want to use the time of divergence of the group, defined by the presence of feathers, from its sister group, presumably an extinct group of theropod dinosaurs. Providing quantitative estimates of the uncertainties in times of origin estimated from the fossil record is difficult, and has not been adequately solved.

As the time of origin of an apomorphy within a lineage need bear no relation to the time divergence of that lineage from its sister lineage, I prefer to reserve the use of the 'time of origin' for apomorphy-based definitions, and the 'time of divergence' for speciation events that lead to separate lineages, regardless of whether those lineages are designated as total, crown, or some other group.

Apomorphy-based definitions (times of origin)

From a palaeontological standpoint, the easiest definition to handle is the apomorphy-based time of origin, though it is still fraught with difficulty. As discussed above, distinguishing the real time of origin from an increase in abundance, or some other factor that increased the chances of preservation and/or recovery well after the time of origin of the characteristic, remains a difficult and largely unsolved problem.

Even if one is satisfied with using the time at which a morphological attribute became readily 'fossilizable' as a proxy for its time of origin, care must be taken in estimating the richness of the fossil record when calculating the uncertainties on this proxy for the time of origin. For example, one must take into account any diversity changes in the clade characterized by the apomorphy. Presumably, the apomorphy had its origin in a single lineage, and presumably the number of lineages possessing that characteristic increased thereafter. So one might expect an increase in the number of fossils up-section simply as a function of the diversity increase. There are two different ways of handling the unwanted effects of diversity fluctuations when calculating confidence intervals on the end-points of stratigraphic ranges: (1) one could consider the fossil record of just the basal-most species of the group, or simply use the fossil record of one lineage from within the clade (Gingerich and Uhen, 1996), although this approach would of necessity require the identification of ancestral species; or (2) one could attempt to normalize for diversity effects, either by dividing the number of fossil horizons in each stratigraphic interval by the number of species present in the interval, or by making a model of species diversity change with time to derive a fossil recovery potential for use with generalized confidence intervals (see above). Regardless of the approach used, one must also be aware that estimating times of origin will usually require data collected from broad geographical areas. Correlation errors, gaps in the rock record, and uneven sampling will often introduce other errors in the analysis (Marshall, 1989, 1990a,b).

A preliminary simulation approach to estimating how much of the temporal duration of a clade is missing given a measure of the proportion of species represented in the fossil record has been developed by Tavaré and Martin (as reported in Martin, 1993). Using a relatively simple model for the diversification of a clade, and an estimate of the total number of primate species that have left a fossil record, they calculated that the observed time of origin of primates of a modern aspect is probably underestimated by about 40%! Using this correction factor, they estimated that the true time of origin is more likely some 80 Ma BP, rather than the observed time of origin of 55 Ma BP. This is a fairly dramatic increase in the estimated time of origin, although it is in good agreement with estimates based upon molecular clocks (Hedges *et al.*, 1996).

Crown or total group definitions (times of divergence)

Estimating divergence times is more difficult than estimating times of origin because a reliable phylogenetic tree must be available, not only of extant taxa, but also of extinct taxa as well (Marshall, 1990b). Once a phylogeny has been settled on, then most of the difficulties encountered with the apomorphy-based approach to estimating time of divergence must also be dealt with (Marshall, 1990b; Springer, 1995). Estimating times of divergence from the fossil record may often be much more difficult than estimating times of origin.

These difficulties have led some to conclude that the use of ghost taxa (range extensions on the bases of stratigraphic ranges to satisfy the condition that sister taxa share the same time of origin) provides the best estimate of the incompleteness of the fossil record (for example, Smith, 1994). However, the use of ghost taxa, even when properly applied, does not give any statistical sense of how effective such range extensions are. Many of these issues are taken up by Wagner in Chapter 7.

Molecular 'clocks' and times of divergence, versus the fossil record and times of origin

There has been a resurgence in the use of molecular 'clocks' to estimate the times of divergence between lineages, including the recent analyses of the animal phyla (Wray *et al.*, 1996), scleractinian corals (Romano and Palumbi, 1996), mammals (Hedges *et al.*, 1996) and birds (Cooper and Penny, 1997). These studies have attracted wide attention because they suggest much deeper times of divergence than inferred from the fossil record. Romano and Palumbi's scleractinian coral study illustrates the distinction between times of origin (typically measured with the fossil record) and divergence times (measured by molecular data).

Scleractinians first appear in the fossil record in the Middle Triassic. The origin of the group has always been somewhat problematic: the tabulate and rugosan corals disappear at the end of the Permian, and there are no known skeletonized corals in the Lower Triassic (see Oliver, 1996, and other papers in Stanley, 1996). Given the high preservation potential of corals, this gap in the fossil record has been puzzling and has led some to suggest that scleractinians arose from a group of unskeletonized cnidarians, rather from one of the Palaeozoic skeletonized corals. Romano and Palumbi's (1996) molecular data suggest that there were already two major clades of scleractinians by the late Carboniferous (that is, the crown group dates to at least this time), which immediately implies an extensive Palaeozoic history for the scleractinians in an unskeletonized form, and that the two extant clades independently skeletonized in the Mesozoic. If this scenario is correct, then the fossil record only gives a good date on the time of origin of the apomorphy of skeletonized forms, and is deeply uninformative on either the crown group or total group divergence times.

Molecular 'clocks' and the fossil record

At least one calibration point from the fossil record is required to estimate the divergence time between lineages from molecular divergence data. In discussions of rates of molecular change, the normal procedure is simply to acknowledge the incompleteness of the fossil record, and to note that because of this incompleteness rates of molecular change will be overestimated and, thus, divergence times calculated using molecular data will be underestimated.

Springer (1995) has undertaken the most comprehensive analysis of rates of molecular change in light of the incompleteness of the fossil record. Springer (1995) notes that a literal reading of the fossil record leads to the conclusion that the rate of single-copy DNA evolution, as measured by DNA–DNA hybridization techniques, varies by a factor of seven within marsupials, and by a factor of 17 for a combined marsupial and monotreme data set. However, after accounting for taxonomic uncertainties, the incompleteness of the fossil record (using the confidence interval approach outlined above), etc., he was able to demonstrate that the data are consistent with a molecular clock, and that the rate of DNA change is estimated at 0.4% divergence per million years, rather than an average rate of 1% divergence per million years calculated before the incompleteness of the fossil record is taken into account. By taking into account the incompleteness of the fossil record, a very different picture is revealed from that apparent in the raw data.

The reader is referred to Hillis *et al.* (1996), as well as Miyamoto and Fitch (1996), for a discussion of some of the important statistical issues relevant to using molecular data to estimate divergence times.

REFERENCES

Adegoke, O.S., 1969, Stratigraphy and palaeontology of the marine Neogene formations of the Coalinga Region, California, *University of California Publications in Geological Sciences*, **80**: 269 pp.

Alvarez, L.W., 1983, Experimental evidence that an asteroid impact led to the extinction of many species 65 million years ago, *Proceedings of the National Academy of Sciences*, **80**: 627–642.

Benton, M.J. (ed.), 1993, *The Fossil Record, 2*, Chapman and Hall, London: 845 pp.

Birkelund, T. and Håkansson, E., 1982, The terminal Cretaceous in Boreal shelf seas: a multicausal event. *In* L.T. Silver and P.H. Schultz (eds), *Geological Implications of*

Impacts of Large Asteroids and Comets on the Earth, Geological Society of America Special Paper, **190**: 373–384.

Chiappe, L.M., 1995, The first 85 million years of avian evolution, *Nature*, **378**: 349–355.

Cloutier, R., 1991, Patterns, trends, and rates of evolution within the Actinistia, *Environmental Biology of Fishes*, **32**: 23–58.

Cloutier, R. and Forey, P.L., 1991, Diversity of extinct and living actinistian fishes (Sarcopterygii), *Environmental Biology of Fishes*, **32**: 59–74.

Cooper, A. and Penny, D., 1997, Mass survival of birds across the Cretaceous–Tertiary boundary: molecular evidence, *Science*, **275**: 1109–1113.

Elder, W.P., 1989, Molluscan extinction patterns across the Cenomanian-Turonian stage boundary in the Western Interior of the United States, *Paleobiology*, **15**: 299–320.

Feduccia, A., 1995, Explosive evolution in Tertiary birds and mammals, *Science*, **267**: 637–638.

Foote, M. and Raup, D.M., 1996, Fossil preservation and the stratigraphic ranges of taxa, *Paleobiology*, **22**: 121–140.

Gingerich, P.D. and Uhen, M.D., 1996, Likelihood estimation of the time of origin of whales (Cetacea). *In* J.E. Repetski (ed.), *Sixth North American Paleontological Convention, Abstracts of Papers, Paleontological Society Special Publication*, **8**: 146.

Harries, P.J., 1993, Dynamics of survival following the Cenomanian–Turonian (Upper Cretaceous) mass extinction event, *Cretaceous Research*, **14**: 563–583.

Hedges, S.B., Parker, P.H., Sibley, C.G. and Kumar, S., 1996, Continental breakup and the ordinal diversification of birds and mammals, *Nature*, **381**: 226–229.

Hillis, D.M., Mable, B.K. and Moritz, C., 1996, Applications of molecular systematics: the state of the field and a look to the future. *In* D.M. Hillis, C. Moritz and B.K. Mable (eds), *Molecular Systematics* (2nd edn), Sinauer, Sunderland, Massachusetts: 515–543.

Holland, S.M., 1995, The stratigraphic distribution of fossils, *Paleobiology*, **21**: 92–109.

Huber, B.T., 1996, Evidence for planktonic foraminifers reworking versus survivorship across the Cretaceous–Tertiary boundary at high latitudes. *In* G. Ryder, D. Fastovsky and S. Gartner (eds), *The Cretaceous–Tertiary Event and Other Catastrophes in Earth History, Geological Society of America Special Paper*, **307**: 319–334.

Jaanusson, V., 1976, Faunal dynamics in the Middle Ordovician (Viruan) of Balto-Scandia. *In* M.G. Bassett (ed.), *The Ordovician System: Proceedings of a Palaeontological Association Symposium, Birmingham, September 1974*, University of Wales Press and National Museum of Wales, Cardiff: 301–326.

Jablonski, D., 1986, Larval ecology and macroevolution in marine invertebrates, *Bulletin of Marine Science*, **39**: 568–587.

Jablonski, D., 1997, Progress at the K–T boundary, *Nature*, **387**: 354–355.

Jefferies, R.P.S., 1979, The origin of chordates – a methodological essay. *In* M.R. House (ed.), *The Origin of the Major Invertebrate Groups, Systematics Association Special Volume 12*, Academic Press, London: 443–477.

Kauffman, E.G., 1988, Concepts and methods of high-resolution event stratigraphy, *Annual Review of Earth and Planetary Science*, **16**: 605–654.

Kemple, W.G., Sadler, P.M. and Strauss, D.J., 1995, Extending graphic correlation to many dimensions: stratigraphic correlation as constrained optimization. *In* K.O. Mann and H.R. Lane (eds), *Graphic Correlation, SEPM (Society for Sedimentary Geology) Special Publication*, **53**: 65–82.

Labandeira, C.C., 1994, A compendium of fossil insect families, *Milwaukee Public Museum Contributions in Biology and Geology*, **88**: 71 pp.

Macellari, C.E., 1986, Late Campanian-Maastrichtian ammonite fauna from Seymour Island (Antarctic Peninsula), *Paleontological Society Memoir*, **18**: 55 pp.

MacLeod, K.G. and Huber, B.T., 1996, Strontium isotope evidence for extensive reworking in sediments spanning the Cretaceous–Tertiary boundary at ODP site 738, *Geology*, **24**: 463–466.

MacLeod, N., Rawson, P.F., Forey, P.L., Banner, F.T., Boudagher-Fadel, M.K., Bown, P.R., Burnett, J.A., Chambers, P., Culver, S., Evans, S.E., Jeffery, C., Kaminski, M.A., Lord, A.R., Milner, A.C., Milner, A.R., Morris, N., Owen, E., Rosen, B.R., Smith, A.B., Taylor, P.D., Urquhart, E. and Young, J.R., 1997, The Cretaceous–Tertiary biotic transition, *Journal of the Geological Society, London*, **154**: 265–292.

Mann, K.O. and Lane, H.R. (eds), 1995, *Graphic Correlation, SEPM (Society for Sedimentary Geology) Special Publication*, **53**: 263 pp.

Marshall, C.R., 1989, *Sand Dollars, Fossils, DNA and Evolutionary Rates*, Unpublished Ph.D. thesis, University of Chicago, Chicago: 366 pp.

Marshall, C.R., 1990a, Confidence intervals on stratigraphic ranges, *Paleobiology*, **16**: 1–10.

Marshall, C.R., 1990b, The fossil record and estimating divergence times between lineages: maximum divergence times and the importance of reliable phylogenies, *Journal of Molecular Evolution*, **30**: 400–408.

Marshall, C.R., 1991, Estimation of taxonomic ranges from the fossil record. *In* N. Gilinsky and P.W. Signor (eds), *Analytical Paleobiology, Paleontological Society Short Courses in Paleontology*, 4, University of Tennessee, Knoxville: 19–38.

Marshall, C.R., 1994, Confidence intervals on stratigraphic ranges: partial relaxation of the assumption of a random distribution of fossil horizons, *Paleobiology*, **20**: 459–469.

Marshall, C.R., 1995a, Distinguishing between sudden and gradual extinctions in the fossil record: predicting the position of the Cretaceous–Tertiary iridium anomaly using the ammonite fossil record on Seymour Island, Antarctica, *Geology*, **23**: 731–734.

Marshall, C.R., 1995b, Stratigraphy, the true order of species originations and extinctions, and testing ancestor–descendant hypotheses among Caribbean Neogene bryozoans. *In* D.H. Erwin and R.L. Anstey, (eds), *New Approaches to Speciation in the Fossil Record*, Columbia University Press, New York: 208–235.

Marshall, C.R., 1997, Confidence intervals on stratigraphic ranges with non-random distributions of fossil horizons, *Paleobiology*, **23**: 165–173.

Marshall, C.R. and Ward, P.D., 1996, Sudden and gradual molluscan extinctions in the latest Cretaceous in western European Tethys, *Science*, **274**: 1360–1363.

Martin, R.D., 1993, Primate origins: plugging the gaps, *Nature*, **363**: 223–234.

McKinney, M.L., 1986a, Biostratigraphic gap analysis, *Geology*, **14**: 36–38.

McKinney, M.L., 1986b, How biostratigraphic gaps form, *Journal of Geology*, **94**: 875–884.

Meldahl, K.H., 1990, Sampling, species abundance, and the stratigraphic signature of mass extinction – a test using Holocene tidal flat molluscs, *Geology*, **18**: 890–893.

Miyamoto, M.M. and Fitch, W., 1996, Constraints on protein evolution and the age of the Eubacteria/Eukaryote split, *Systematic Biology*, **45**: 566–573.

Oliver, W.A., Jr, 1996, Origins and relationships of Paleozoic coral groups and the origin of the Scleractinia. *In* G.D. Stanley (ed.), *Paleobiology and Biology of Corals, Paleontological Society Papers*, **1**: 107–134.

Olsson, R.K. and Liu, C.J., 1993, Controversies on the placement of the Cretaceous–Paleogene boundary and the K/P mass extinction of planktonic foraminifera, *Palaios*, **8**: 127–139.

Patterson, C., 1994, Bony fishes. *In* D.R. Prothero and R.M. Schoch (eds), *Major Features of Vertebrate Evolution, Short Courses in Paleontology, 7*, University of Tennessee, Knoxville: 57–84.

Paul, C.R.C., 1982, The adequacy of the fossil record. *In* K.A. Joysey and A.E. Friday (eds), *Problems of Phylogenetic Reconstruction, Systematics Association Special Volume, 21*, Academic Press, London: 75–117.

Paul, C.R.C., 1985, The adequacy of the fossil record reconsidered. *In* J.C.W. Cope and P.W. Skelton (eds), *Evolutionary Case Histories from the Fossil Record. Special Papers in Paleontology*, **33**: 7–16.

Plotnick, R.E., Gardner, R.H., Hargrove, W.W., Prestegaard, K. and Perlmutter, M., 1996, Lacunarity analysis: a general technique for the analysis of spatial patterns, *Physical Review*, **E53**: 5461–5468.

Raup, D.M., 1986, Biological extinction in earth history, *Science*, **231**: 1528–1533.

Raup, D.M., 1989, The case for extraterrestrial causes of extinction, *Philosophical Transactions of the Royal Society, London*, **B325**: 421–435.

Romano, S.L. and Palumbi, S.R., 1996, Evolution of scleractinian corals inferred from molecular systematics, *Science*, **271**: 640–642.

Runnegar, B., 1996, Early evolution of the Mollusca: the fossil record. *In* J.D. Taylor (ed.), *Origin and Evolutionary Radiation of the Mollusca*, Oxford University Press, Oxford: 77–87.

Sepkoski, J.J., Jr, 1992, A compendium of fossil marine families (2nd edn), *Milwaukee Public Museum Contributions in Biology and Geology*, **83**: 156 pp.

Shaw, A.B., 1964, *Time in Stratigraphy*, McGraw-Hill, New York: 365 pp.

Signor, P.W. and Lipps, J.H., 1982, Sampling bias, gradual extinction patterns, and catastrophes in the fossil record. *In* L.T. Silver and P.H. Schultz (eds), *Geological Implications of Impacts of Large Asteroids and Comets on the Earth, Geological Society of America Special Paper*, **190**: 291–296.

Smith, A.B., 1994, *Systematics and the Fossil Record*, Blackwell, Oxford: 223 pp.

Sokal, R.R. and Rohlf, F.J., 1995, *Biometry* (3rd edn), W.H. Freeman, New York: 887 pp.

Solow, A.R., 1996, Tests and confidence intervals for a common upper endpoint in fossil taxa, *Paleobiology*, **22**: 406–410.

Solow, A.R. and Smith, W., 1997, On fossil preservation and the stratigraphic ranges of taxa, *Paleobiology*, **23**: 271–277.

Springer, M.S., 1990, The effect of random range truncations on patterns of evolution in the fossil record, *Paleobiology*, **16**: 512–520.

Springer, M.S., 1995, Molecular clocks and the incompleteness of the fossil record, *Journal of Molecular Evolution*, **41**: 531–538.

Springer, M.S. and Lilje, A., 1988, Biostratigraphy and gap analysis: the expected sequence of biostratigraphic events, *Journal of Geology*, **96**: 228–236.

Stanley, G.D. (ed.), 1996, *Paleobiology and Biology of Corals, Paleontological Society Papers*, **1**: 296 pp.

Stanley, S.M. and Yang, X., 1994, A double mass extinction at the end of the Paleozoic era, *Science*, **266**: 1340–1344.

Strauss, D. and Sadler, P.M., 1989, Classical confidence intervals and Bayesian probability estimates for ends of local taxon ranges, *Mathematical Geology*, **21**: 411–427.

Valentine, J.W., Awramik, S.M., Signor, P.S. and Sadler, P.M., 1991, The biological explosion at the Precambrian–Cambrian boundary, *Evolutionary Biology*, **25**: 279–356.

Ward, P.D., Wiedmann, J. and Mount, J.F., 1986, Maastrichtian molluscan biostratigraphy and extinction patterns in a Cretaceous/Tertiary boundary section exposed at Zumaya, Spain, *Geology*, **14**: 899–903.

Ward, P.D., Kennedy, W.J., MacLeod, K.G. and Mount, J.F., 1991, Ammonite and inoceramid bivalve extinction patterns in Cretaceous/Tertiary boundary sections of the Biscay region (southwestern France, northern Spain), *Geology*, **19**: 1181–1184.

Wray, G.A., Levinton, J.S. and Shapiro, L.H., 1996, Molecular evidence for deep Precambrian divergences among metazoan phyla, *Science*, **274**: 568–573.

3
Resolution of the Fossil Record: The Fidelity of Preservation

David M. Martill

INTRODUCTION

Fossils are the primary data set for investigating the history of life on Earth. Therefore, it follows that, to obtain the clearest picture of biological events gone by, the best-preserved fossils should be studied in the hope that they will yield the maximum amount of information. Unfortunately, the fossil record is an extremely biased data set in that not only is it very incomplete, but also it is highly skewed towards those organisms that have readily preserved components such as biomineralized tissues (bones, teeth and shells) or organic materials that have structural functions, such as wood and cuticle. Furthermore, the record is biased against those organisms that did not inhabit areas of sediment accumulation. Numerous authors have attempted to quantify this patchiness of the record (see other chapters in this volume) and in some studies it has proved that, for some types of data (for example, molluscan diversity based on shell morphology), the fossil record is really not at all bad (see Valentine, 1989). However, where a fossil-bearing formation might be a good recorder of hard parts of one group of organisms, it might be a poor data bank for another. By way of one example among many, the bone beds of the English Rhaetian (Upper Triassic, Westbury Formation) contain a diverse assemblage of fish and reptiles (phosphatic teeth and bones), but are almost devoid of molluscs and other invertebrates with calcareous shells. The bias here can be attributed to diagenetic removal of carbonates rather than to any original biological factor.

On the plus side, there are parts of the fossil record, usually restricted to

The Adequacy of the Fossil Record. Edited by S.K. Donovan and C.R.C. Paul.
© 1998 John Wiley & Sons Ltd.

small geographical areas and to relatively short stratigraphic intervals (= 'time spans'), in which preservation can be spectacular in terms of abundance of individuals, diversity of assemblage and/or quality (= fidelity) of preservation. Fossil accumulations in which preservational quality is extraordinarily high are known as *konservat Lagerstätten* (Allison, 1988a) and they have received a great deal of attention in the last 15 years. The fossils in such deposits are perhaps rare, but they may exhibit high degrees of skeletal element articulation, preservation of biomineralized tissues in their original mineralogy, preservation of non-mineralized tissues and retention of stomach contents, may demonstrate interorganismal relationships such as parasitism and may be preserved uncrushed. Not all *konservat Lagerstätten* exhibit all of these aspects of preservation, but a few do and are treasure troves for palaeontologists.

One particular mode of preservation that produces fossils with soft tissues of high fidelity is early diagenetic phosphatization (Martill, 1994). In this case, phosphatization aided by bacterial decomposition replaces and coats a variety of soft and hard tissue structures. Factors controlling this mode of preservation are still being studied and are currently poorly understood. (For reviews on mechanisms of phosphatization of fossil soft tissues, see Briggs and Wilby, 1996; Martill, 1994; Wilby, 1993; Lucas and Prévôt, 1991; and for studies on the timing of phosphatization see Briggs and Kear, 1993; Martill, 1989; Martill and Harper, 1990.) Soft tissues preserved by phosphatization can be examined by standard thin-section petrography or by scanning electron microscopy of acid-prepared samples. Extraction of phosphatized soft tissues can be achieved by the use of acetic and formic acids, but great care must be exercised as some phosphatizations are prone to collapse after removal of interstitial carbonates. An effective technique for extraction of delicately phosphatized arthropods has been devised by Müller (1985).

Phosphatization of soft tissues is by no means a rare phenomenon, but is easily overlooked, with many examples having been discovered largely by accident during acid extractions in search of vertebrate fossils. Examples of assemblages in which phosphatized soft tissues are reported occur from the Cambrian (Müller, 1985; Müller and Wallosek, 1985a), Ordovician (Andres, 1989), Silurian (van der Brughen *et al.*, 1996), Devonian (Dean, 1902; P.R. Wilby, personal communication), Carboniferous (Traquaire, 1884; Briggs *et al.*, 1983), Permian (Willems and Wuttke, 1987), Triassic (Weitschat, 1983, 1986), Jurassic (Owen, 1844; Allison, 1988b; Schultze, 1989), Cretaceous (Martill, 1988, 1990) and the Cenozoic (Wuttke, 1983). Phosphatization of soft tissues with high resolution of detail has also been found in archaeological sites (Piearce *et al.*, 1990) and has been achieved in laboratory experiments (Briggs and Kear, 1993), where the products compare remarkably well with fossil examples (Briggs *et al.*, 1993).

The geographical distribution of such occurrences is also widespread with examples known from most continents. The range of environments in which phosphatized soft tissues have been recorded includes fully marine and perhaps brackish conditions. No certain freshwater deposits have been documented with phosphatization, but at least one terrestrial example has been reported (Piearce *et al.*, 1990). Host lithologies for phosphatized fossil soft tissues are usually laminated mudrocks and carbonate concretions (Müller, 1985; Allison, 1988b; Martill, 1988), but limestone hosts are known (Wilby and Whyte, 1995).

A wide range of organisms has been reported with soft-tissue preservation by phosphate including prokaryote coccoid and filamentous bacteria (Martill and Wilby, 1994), bivalve molluscs (Harper and Todd, 1995; Wilby and Whyte, 1995), cephalopod molluscs (Weitschat, 1986; Allison, 1988b), annelids (Piearce *et al.*, 1990), trilobites (Müller and Wallosek, 1987), thylacocephalans (van der Brughen *et al.*, 1996), ostracods (Bate, 1972; Müller, 1979; Weitschat, 1983) and other crustaceans (Müller and Wallosek, 1985b), conodonts (Briggs *et al.*, 1983), elasmobranch fish (Dean, 1902; Brito, 1992), bony fish (Martill, 1988, 1990), reptiles (Willems and Wuttke, 1987) and archosaurs (Martill and Unwin, 1989; Kellner, 1996a,b). Although phosphatization has been recorded from numerous locations spanning the Phanerozoic, nowhere has it received more attention than in the Lower Cretaceous Santana Formation of north-east Brazil. Accordingly, I examine herein the nature of phosphatization in this deposit, emphasizing particularly the fidelity of the preservation.

THE SANTANA FORMATION

Locality and Stratigraphy

The Santana Formation is famous for both the abundance and quality of preservation of its fossils. Although the fossil-bearing parts of the formation have been known since the last century (Spix and Martius, 1823–1831), it is only in the last 20 years that it has achieved its status as one of the world's most spectacular Mesozoic fossil deposits (Maisey, 1991; Martill, 1993). The Santana Formation crops out on the flanks of the Chapada do Araripe, a plateau of some 600–800 m lying at the boundaries of the states of Ceará, Pernambuco and Piauhi in north-east Brazil. The plateau is 150 km east to west, and about 50 km north to south, and has been deeply incised, with scallop-shaped margins of high cliffs formed by sandstones of the overlying Exu Formation. The Santana Formation crops out discontinuously just beneath the foot of the cliff and also forms a few isolated outliers. There are probably more than 300 km of linear surface outcrop of the Santana Formation and it is spectacularly fossiliferous throughout.

Lithologically, the Santana Formation comprises a series of basal sandstones, siltstones and organic-rich mudstones, overlain by clays with carbonate concretions, the last containing abundant and very well-preserved vertebrate fossils (Figure 3.1). The sandstones and siltstones represent a series of fluvio-deltaic deposits with emergence surfaces, and rest disconformably on evaporites of the Ipubi Formation (formerly a member included within the Santana Formation). These gradually pass upwards into green and grey clays with abundant, thin, ostracod-rich limestones and concretion-bearing layers. This latter part of the sequence is referred to as the Romualdo Member. It is mined extensively for fossils by local people around the small towns of Santana do Cariri, Jardim, Porteiras and Abaiara (Martill, 1993).

The Santana Formation has proved difficult to date accurately as it is devoid of the more usual Mesozoic marine fossils. However, dates utilizing the fish, palynomorphs and, to a lesser degree, pterosaurs suggest an Albian to Cenomanian age (Berthou, 1990; Martill, 1990; D.M. Unwin, personal communication).

Biota

The most abundant fossils from the Santana Formation are ostracods of the genus *Pattersoncypris*, which are a major component of the rock in many of the carbonate concretions and thin limestones. However, in general the invertebrate fauna is restricted to a few species of molluscs and crustaceans, many known from only a few examples. This contrasts strongly with the vertebrate fauna from the same concretions which is diverse and includes the fish genera *Araripelepidotes*, *Araripichthys*, *Axelrodichthys*, *Brannerion*, *Cladocyclus*, *Enneles*, *Iamanja*, *Iansan*, *Leptolepis*, *Mawsonia*, *Notelops*, *Obaichthys*, *Ophiopsis*, *Oshunia*, *Paraelops*, *Procinetes*, *Rhacolepis*, *Santanaclupea*, *Tharrhias*, *Tribodus* and *Vinctifer*. Pterosaurs are also abundant (at least, relative to other pterosaur-bearing deposits) with the following genera reported: *Anhanguera*, *Araripedactylus*, *Araripesaurus*, *Brasileodactylus*, *Santanadactylus*, *Tapejara*, *Tropeognathus*, *Tupuxuara* and *Ornithocheirus*. Dinosaurs include a possible oviraptosaurid (Frey and Martill, 1995) and the enigmatic *Irritator* (Martill *et al.*, 1996), but dinosaur remains are far rarer than those of pterosaurs. The crocodilians *Itasuchus* and *Araripesuchus*, and the turtles *Araripemys* and an undescribed pelomedusid (Meylan, 1996; Gafney and Meylan, 1991), are also very rare. The pterosaurs are currently under study by numerous workers (Wellnhoffer, 1991; Kellner, 1996a; Unwin *et al.*, 1996) as they are exceptionally well preserved and offer the best opportunities for understanding the biomechanics of pterosaurian flight as well as their terrestrial locomotion. In addition, soft-tissue preservation in pterosaurs (Martill and Unwin, 1989; Kellner, 1996a), dinosaurs (Kellner, 1996b), fishes (Martill,

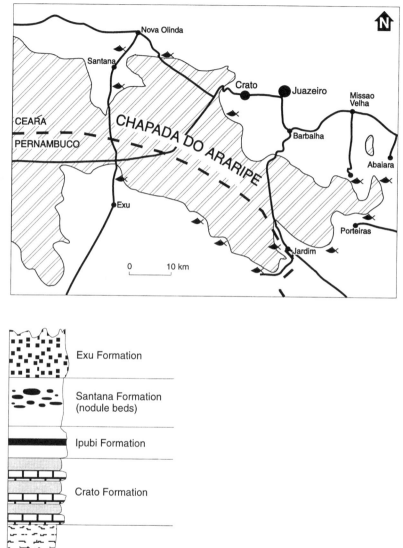

Figure 3.1 *Principal fossil fish localities in the Santana Formation at the eastern end of the Chapada do Araripe plateau in north-east Brazil. A simplified lithological log of the Araripe Group sedimentary rocks is shown. Fossils are found throughout the Santana Formation, but the highest quality of preservation is found within the concretions of the Romualdo Member of the Santana Formation*

1988, 1990) and microorganisms (Wilby, 1993; Martill and Wilby, 1994) has allowed detailed anatomical studies of tissues not usually encountered in the fossil record.

Fossil abundance in the Romualdo Member is high. Some bedding planes are covered in small fish and represent mass mortality horizons, with abundances reaching thousands of individuals per square metre (Martill and Brito, in preparation). In the case of the vertebrates, it is reasonable to assume that the fossil abundance might reflect the abundance of different taxa within the basin, not withstanding the effects of differing species longevity, as well as other taphonomic factors that may play a role, such as taxon-specific variations of post-mortem drifting.

However, the apparent scarcity of invertebrate remains is not a reflection of the low diversity of invertebrates *per se*. Rather, it is only those invertebrates with mineralized skeletons that are of low diversity. Gastropods are present, but they are very rare in the horizons yielding vertebrates and are usually small species. Bivalves are often abundant, but restricted to just one or two small species of corbulid. However, an examination of the stomach contents of the fishes reveals an abundance of small arthropods and benthic foraminiferans (Wilby and Martill, 1992). Diligent searching for invertebrates, especially for those forms that might be of biostratigraphic value, has singularly failed to reveal any 'normal' marine components. Although it is possible that some dissolution of aragonitic shells might have occurred, there is no calcitic marine invertebrate fauna and it must be assumed that the low diversity of shelled invertebrates is real. Thus, the taxonomic record for the Romualdo Member might be considered real for those organisms with readily preservable hard parts.

Preservation in Nodules

The Santana Formation fossils are usually encountered within carbonate concretions, but they also occur in the surrounding shales. Depending on which level the concretions are from, the enclosed fossils may be partly compacted or, when enclosed within early diagenetic concretions, they may be entirely undistorted with body cavities still maintaining void space (Martill, 1988, 1997). Skeletons may exhibit disarticulation due to post-mortem decay during drifting or internal collapse. Bones in concretions where the preservation of the skeleton is three-dimensional often still retain void space, but cavities may be lined or filled with diagenetic minerals. The bone shows little or no evidence of recrystallization or replacement.

Preservation of ostracods is in original calcite and usually as carapaces. Small gastropods are sometimes found in original white aragonite, but also occur as internal moulds of pyrite. These are usually found after acetic-acid preparation of concretions and so it is not unlikely that they also retained original shell material prior to digestion. Thus, for both bone and shell material, preservation is in original materials with little or no replacement.

Soft Tissues

Phosphatized soft tissues appear to be restricted to a few taxa, although the controls on this are not fully understood. For example, the abundant fish *Vinctifer comptoni* (L. Agassiz) only very rarely has preserved soft tissues, yet it often occurs in the same concretions as *Rhacolepis* specimens with considerable amounts of soft-tissue preservation. Soft tissues have been reported in an unnamed theropod dinosaur (Kellner, 1996b); in a pterosaur (Martill and Unwin, 1989), where dermis, epidermis, muscle fibres, 'structural' fibres and an enigmatic spongy tissue have been recognized; and in a number of arthropods, including parasitic copepods (Cressey and Patterson, 1973; Cressey and Boxshall, 1989), ostracods (Bate, 1972) and shrimps (Wilby and Martill, 1992). Phosphatized soft tissues have been found in the plant *Brachyphyllum* sp., but are as yet undescribed. Thus, the taxonomic spectrum in which soft tissues are found is extensive, but not complete for the assemblage. The range of tissues found preserved is also considerable, with muscle tissue, alimentary tract, gill lamellae, dermis, connective tissue and blood vessels being reported (Martill, 1988; Wilby and Martill, 1992).

Many hundreds of specimens of the common elopomorph fishes *Rhacolepis* and *Notelops*, and all specimens (that this author has seen) of the rare hybodont shark *Tribodus limae* Brito and Ferreira, have significant portions of their muscle tissue preserved. This material has been described extensively due to the remarkable levels of detail visible (Martill, 1988, 1990). Commonly the muscle material appears white or buff and reflects the original pattern of myotomes. It often appears to be fibrous and this is a direct replication of the original muscle fibres. It is preserved as a replacement by calcium phosphate (Martill, 1988; Wilby, 1993), usually in the form of aggregated microspheres between 0.3 and 2.0 μm diameter (Figure 3.2a) composed of crystallites between 10 and 100 nm long (Martill and Wilby, 1994); hollow microspheres of up to 3 μm diameter, and often with apertures of 0.25 μm diameter (Figure 3.2b); or as templated microcrystallites between 10 and 100 nm long (Figure 3.3a, b). The degree of detail visible is dependent partly on the morphology of the replacing mineral habit and also on the condition of the preserved tissue prior to preservation.

Martill and Harper (1990), and later Briggs and Kear (1993), showed that the onset of the mineralization of soft tissues was rapid (Martill, 1994), that it was protracted and that the best detail was preserved in those tissues that were preserved the quickest. It is generally believed that the fossilization of

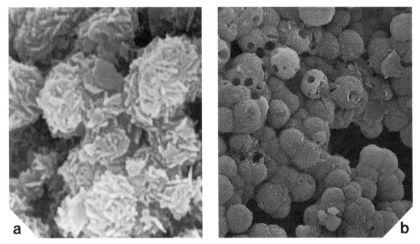

Figure 3.2   *Fabrics of calcium phosphate commonly found replicating soft tissues in fish from the Santana Formation of Brazil. Replication with these fabrics may result in detail at the cellular level when cells are large, as in the case of muscle fibres. (a) Subspherical aggregates of apatite crystallites. The spheres are in the size range 0.3–2.0 μm diameter, with individual crystallites generally between 10 and 100 nm in length. (b) Hollow spheres of phosphate with circular apertures. Martill and Wilby (1994) interpreted these as lithified bacteria. This fabric is commonly found replacing gill filament secondary lamellae, and afferent and efferent blood vessels to the gill filaments, in a manner similar to the autolithification described by Wuttke (1983) for the Eocene Messel Oil Shales of Germany. Fidelity is low, usually revealing only the gross morphology of organs. The crystallites are of the same size range as those found in the crystal aggregates. They may have nucleated within the extrapolymeric substance of the bacteria*

Figure 3.3 (opposite)   *Preservation at the cellular and organelle level of organization. High-fidelity preservation in striated muscle tissue of the elopomorph fish Notelops sp. from the Santana Formation of Brazil. (a) Phosphate has templated on specific sites in proteins in muscle fibrils. This has resulted in the striations both across the muscle fibre and along its length being visible. (b) Each of the muscle striae is seen to be composed of pairs of clusters of apatite crystallite aggregates. Interpretation of this banding is not easy. Possibly each pair of crystal aggregates reflects the A band of striated muscle, with the division between pairs equating to the M-line, but equally the pairs could represent an entire sarcomere, with the division between rows representing the Z-disc and the less-pronounced division between the pairs of crystal clusters representing the M-line. (c) A muscle fibre in which phosphatization has replaced cell organelles, in this case either cell nuclei or mitochondria. (d) High-power electron micrograph of a cell nucleus or mitochondrion showing the crystallites of apatite replicating the organelle and also irregularly overcoating it*

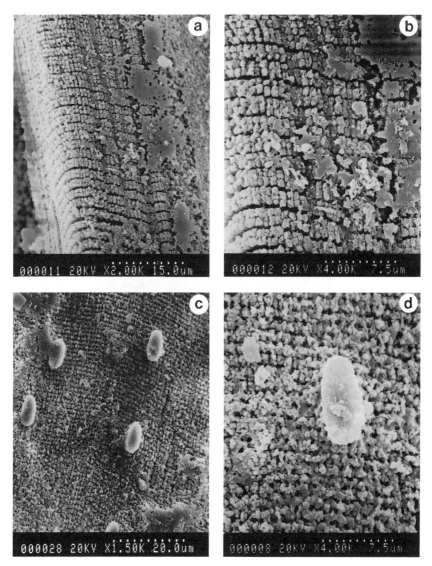

Figure 3.4   *A group of phosphatized, rod-shaped bacteria. These and other bacterial types are frequently found in association with phosphatized soft tissues and in some cases replicate them as a bacterial autolithification sensu Wuttke (1983)*

soft tissues was accompanied by active microbially mediated decay (Briggs *et al.*, 1993), and lithified bacteria (Figure 3.4) are often found in association with the preserved soft tissues (Martill and Wilby, 1994).

Fidelity

The most remarkable aspect of the phosphatized soft tissues is the level of observable detail, at the macro, micro and nanno (macromolecular) levels. Individual muscle fibres usually can be seen with the naked eye. Using the SEM, striations reflecting the A and Z bands of the sarcomeres of striated muscle are usually distinctive (Figure 3.3). The bands are usually best seen where crystallites of calcium phosphate have templated on to specific sites on pre-existing structural proteins such as actin and/or myosin. The bands are most distinctive where the crystallites are arranged with their crystallographic axes parallel with the structural proteins on to which they have templated. This may reflect not only the sarcomeres, but also the fibrils lying parallel within the muscle fibre (Figure 3.3a,b). The surface of some muscle fibres shows subcellular detail, including organelles such as mitochondria, T-tubules and cell nuclei, the last often preserved in discrete rows (Figure 3.3c,d). The myosepta separating the muscle myotomes are sheets of connect-

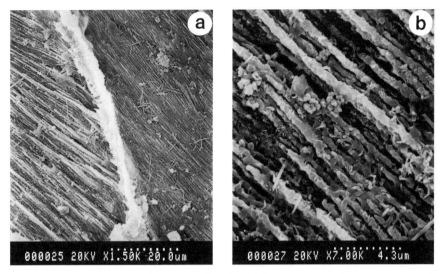

Figure 3.5 *Preservation with macromolecular fidelity. Scanning electron micrographs of myoseptum from elopomorph fish from the Santana Formation of north-east Brazil. (a) Two areas of myoseptum either side of a muscle fibre sarcolemmic membrane. The structure may represent a replaced membrane, but possibly represents a micro-void fill produced when the membrane had decayed, but prior to decay of the muscle filaments. The fibrous textures of the myoseptum itself are clearly seen. (b) High-magnification view showing the nature of the fibres. This probably represents phosphatized collagen fibrils from within the connective tissue. If this interpretation is correct, then preservation here has occurred at the macromolecular level. It is possible that original materials or their decay products are entrapped within the calcium phosphate replicas*

ive tissue composed largely of collagen filaments and are commonly preserved in the fish fossils (Figure 3.5).

Particularly intriguing is the preservation of secondary lamellae on the gill filaments of elopomorph fish. Observable detail is extremely variable, both between specimens and within a single specimen. The detail visible is dependent on the manner in which the calcium phosphate precipitates either on to the surface, or within the tissue, of the lamellae (Figure 3.6). In addition, the crystallite size and habit of the phosphate also affect the resolution of detail. Crystallite size is commonly around 50 nm, although both smaller and larger crystallites occur. These crystallites may nucleate on to specific organic templates in a manner remarkably similar to the ossification of collagen in bone formation. In such cases a high degree of fidelity is achieved in the fossilization of the soft tissue (Figure 3.6).

It appears that the preservation of soft tissues can occur in several ways. Templating of crystallites on to specific substrates, as mentioned above, can provide extremely high-fidelity resolution. However, phosphate also is

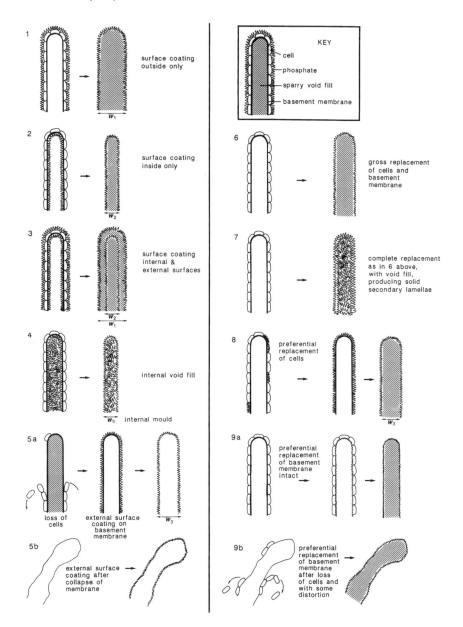

found preserving bacteria, which *en masse* can replicate an organ or tissue of a larger organism. This gives rise to a rather crude replication, which, although lacking ultrastructural detail, is nevertheless an important source of palaeobiological information. This mode of preservation has been discussed in detail by Wuttke (1983) and is found in other diagenetic minerals as well as phosphates. Phosphate preserving soft tissues also occurs as void fills and surface coatings (Figure 3.7). It is important that surface coatings are distinguished from replacement textures, as in very small structures dimensions can be exaggerated and topography may be enhanced or subdued. For example, the sensory setae on the appendages of arthropods can be replaced, coated or both. When simply replaced, they have very fine diameters that realistically reflect the original size. When preserved as a phosphate coating, they have exaggerated diameters, and may appear as hollow structures where the original setae have decayed away. When replacements and coatings co-occur, the impression can be of a thickened, solid seta.

DISCUSSION

Phosphatization of soft tissues during early diagenesis clearly offers an opportunity for tissues usually lost to decay processes to become incorpor-

---

Figure 3.6 (opposite)   *Preservation of secondary lamellae on gill filaments of* Notelops *sp. by phosphatization (1–5, phosphatization by surface coating; 6–9, phosphatization by tissue replacement). Differing forms of preservation may result in a variety of distinct textures and in differing levels of fidelity. These different modes of preservation may also result in dimensions that are exaggerated relative to the original. It is especially important to recognize this when determining the sizes of small structures (for example, cuticle thickness in small arthropods may be doubled if preservation includes surface coatings of phosphate). (1) Surface coating on external surface only. This will exaggerate the width (W$_1$) and may produce only a low-fidelity replication. (2) Surface coating of internal surface. This will produce a narrower width (W$_2$), but may show a very high-fidelity replication of the original surface as a mould. (3) Internal and external surface coatings combined. Such examples will allow the original thickness of the membrane to be determined (W$_1$ – W$_2$). (4) Internal void fill. Such infills may be more robust than the thin coating and withstand acetic-acid digestion better. High resolution of the internal surface is possible. (5) Cells of the epithelium of secondary lamellae fall away from the basement membrane during the very early stages of decay (Martill and Harper, 1990). (a) Phosphatization may post-date this event and act on the basement membrane, resulting in replication of the surface of the membrane. (b) Collapse of the secondary lamellae has usually occurred by the time that cells have fallen from the membrane, and such phosphatization may replicate the morphology of collapsed lamellae rather than erect ones. (6) Replacement by phosphate may include both epithelial cells and basement membrane. (7) Replacement may also occur in combination with a void infill. (8) In some examples, phosphate replacement has preferentially replaced the cells of the epithelium, providing high-fidelity preservation at the cellular level. (9) Some preferential replacement of basement membrane may occur both before collapse (a) and after collapse (b)*

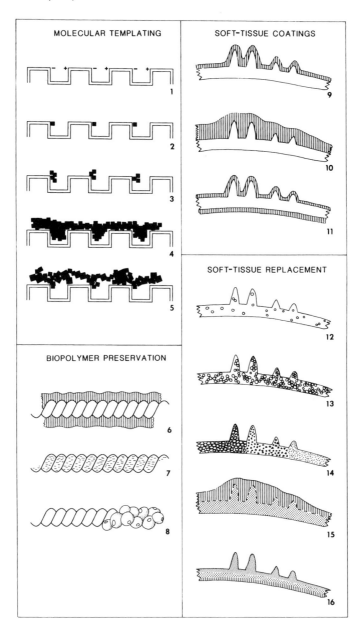

ated into the fossil record. This process appears not to be temporally restricted, for the Phanerozoic at least, and it is possible to obtain exceptionally well-preserved soft-tissue fossils from a time span of at least 540 Myr. Although the environmental distribution of this type of fossilization process appears to be limited, it occurs in both normal marine and restricted, perhaps brackish, systems, both of which are widespread, and are facies that are frequently rich in fossils. The fidelity of preservation, although somewhat variable due to the size and habit of the calcium phosphate crystallites, can be such that details of cell structure become available. Because phosphatization is often linked with the early precipitation of calcium carbonate (Briggs and Wilby, 1996), delicate phosphatized soft tissues are frequently overgrown by carbonate concretions and thus resist compaction. As a result many examples of phosphatized soft tissues are fully three-dimensional. It is unclear at this

---

Figure 3.7 (opposite)   *Possible modes of phosphatization at differing scales. Based on observation of phosphate microtextures found in specimens from the Santana Formation of north-east Brazil.*
Molecular templating   *This occurs where phosphate has an affinity for specific sites on a macromolecule. (1) Sites, indicated here simply as + or − charges. (2) The templating phosphate nucleates on to numerous specific sites. (3) Additional phosphate is attracted to the site and crystal growth commences. (4) Continued crystal growth in optical continuity occurs until the original template has been completely covered. In this case high-fidelity preservation may occur, with individual template molecule sites identifiable. (5) Crystal growth continued but with less-ordered structure. This will result in loss of fidelity.*
Biopolymer replication   *Here, large aggregations of macromolecules, such as collagen fibrils, may be overgrown or replaced. (6) Large fibril overgrown on external surface. This mode of preservation might offer opportunities for preservation of the original macromolecules by entrapment. (7) Replacement on a molecule-by-molecule basis such that the original material is lost, but replication is faithful with high fidelity. (8) Replacement occurs, but by poorly ordered crystal aggregates or autolithified bacterial replacements. Such preservation is common for collagen fibrils in the dermis of fishes from the Santana Formation.*
Soft-tissue coatings   *(9) Soft tissues such as arthropod cuticle may be coated on external surfaces as thin films just a few microns thick, resulting in high-fidelity replication. (10) Thicker coatings may blanket out fine surface detail giving only low-fidelity replication. (11) Coatings may occur on internal and external surfaces, leaving a high-fidelity mould.*
Soft-tissue replacement   *The soft tissues may be replaced in bulk by phosphatization. (12) Very low levels of phosphate precipitation result in only a few crystals, which probably are released during decay of the soft tissue, and thus no retention of soft tissue occurs. (13) Phosphatization can be partial, but with sufficient crystal sites in contact to preserve the soft tissue. Such fossils are extremely delicate, and are better not extracted from the matrix, but examined in thin section. (14) Replacement may be entire if sufficient phosphate is available, but may occur in several different growth habits, including hollow spheres, microsphere crystal aggregates and templated crystallites. (15) Both replacements and coating may occur together. This may result in significant loss of detail. (16) Composite replacement may occur where different crystal habits occur in the same specimen*

Figure 3.8 Pattersoncypris *sp., an ostracod from the nodules of the Santana Formation (Lower Cretaceous) of north-east Brazil, extracted by acetic-acid digestion. This specimen was just one of many individuals associated with a large fish in which muscle tissue was preserved. In this specimen the valves have been coated by calcium phosphate (the carbonate of the valves has dissolved during acid digestion) as well as the cuticle. This is an example of where mineralization has replaced the entire organism. Preservation of entire organisms is not as rare as previously believed, but is restricted to small fossils. The original specimen is about 1.5 mm long. Notice that 'sensory' hairs are preserved on the margins of the carapace. These probably represent phosphatic infills of pores within the originally calcitic valves. The velum can be seen as well as appendages and mouthparts. (Photograph and information on preservation courtesy of Mr Robin Smith, University of Leicester)*

stage whether phosphatization traps and incorporates original biomolecules within the phosphate, or if these become excluded during diagenesis. Attempts to extract and amplify DNA from phosphatized cell nuclei located in specimens of *Notelops* sp. from unweathered nodules from the Santana Formation have so far been unsuccessful.

Rarely is phosphatization in fossils complete, with only portions of soft tissues preserved in larger specimens. This is probably due to a shortage of available phosphate in bottom and pore waters, because in smaller organisms such as ostracods phosphatization may preserve the entire animal (Figure 3.8).

Although phosphatization does not produce the best-preserved fossils (this accolade must surely go to organisms trapped in resins; see Henwood, 1992), its distribution in time and space, and its occurrence in a wide variety of taxonomic groups, provides one of the most important taphonomic windows for palaeobiologists. Fossils incorporated in resins are known to have details of cell ultrastructures preserved, including mitochondria with cristae and ribosomes (Henwood, 1992; see Grimaldi, 1996, for stunning photographs). Claims have also been made that intact portions of DNA have been detected in insect inclusions in resins as old as Lower Cretaceous (Cano *et al.*, 1993). Resins, of course, are restricted to terrestrial environments and temporally restricted to that time when the first terrestrial resin-producing plants appeared in the Carboniferous. In terms of preservational fidelity, phosphatization can be considered as the marine equivalent of amber.

ACKNOWLEDGEMENTS

I thank Dr Philip Wilby (British Geological Survey, Keyworth) and Professor Derek Briggs (University of Bristol) for many valuable discussions over the years on matters concerning phosphatization of soft tissues. Thanks also to Mr Robin Smith (University of Leicester) for discussion and allowing me to use his data. Professor Stephen Donovan is thanked for his relentless encouragement.

REFERENCES

Allison, P.A., 1988a, Konservat-Lagerstätten: cause and classification, *Paleobiology*, **14**: 331–334.
Allison, P.A., 1988b, Soft bodied squids from the Jurassic Oxford Clay, *Lethaia*, **21**: 403–410.
Andres, D., 1989, Phosphatisierte Fossilien aus dem unteren Ordoviz von Sudschweden, *Berliner Geowissenschafte Abhandlungen*, **A106**: 9–19.

Bate, R.H., 1972, Phosphatized ostracods with appendages from the Lower Cretaceous of Brazil, *Palaeontology*, **15**: 379–393.

Berthou, P.Y., 1990, Le bassin d'Araripe et les petits bassins intracontinentaux voisins (NE du Brésil): formation et évolution dans le cadre de l'ouverture de l'Atlantique equatorial. Comparison avec les bassins ouest-africains situés dans le même contexte. *In* D. de A. Campos, M.S.S. Viana, P.M. Brito and G. Beurlen (eds), *Atas do Simpósio Sobre a Bacia do Araripe e Bacias Interiores do Nordeste, Crato, 14–16 de Junho de 1990*, DNPM, Crato: 113–134.

Briggs, D.E.G. and Kear, A.J., 1993, Fossilization of soft tissues in the laboratory, *Science*, **259**: 1439–1442.

Briggs, D.E.G. and Wilby, P.R., 1996, The role of the calcium carbonate–calcium phosphate switch in the mineralization of soft-bodied fossils, *Journal of the Geological Society, London*, **153**: 665–668.

Briggs, D.E.G., Clarkson, E.N.K. and Aldridge, R.J., 1983, The conodont animal, *Lethaia*, **20**: 107–115.

Briggs, D.E.G., Kear, A.J., Martill, D.M. and Wilby, P.R., 1993, Phosphatization of soft-tissue in experiments and fossils, *Journal of the Geological Society, London*, **150**: 1035–1038.

Brito, P.M., 1992, Nouvelles données sur l'anatomie et la position systématique de *Tribodus limae* Brito & Ferreira, 1989 (Chondrichthyes, Elasmobranchii) du Crétacé inferieur de la Chapada do Araripe (N-E Brésil), *Geobios, Mémoire Spéciaux*, **14**: 143–145.

Brughen, W., van der, Schram, F.R. and Martill, D.M., 1996, The fossil *Ainiktozoon* is an arthropod, *Nature*, **385**: 589–590.

Cano, R.J., Poinar, H.N., Pieniazek, N.J., Acra, A. and Poinar, G.O., 1993, Amplification and sequencing of DNA from a 120–135-million-year-old weevil, *Nature*, **363**: 536–538.

Cressey, R. and Boxshall, G., 1989, *Kabatarina pattersoni*, a fossil parasitic copepod (Dichelestidae) from a Lower Cretaceous fish, *Micropalaeontology*, **35**: 150–167.

Cressey, R. and Patterson, C., 1973, Fossil parasitic copepods from a Lower Cretaceous fish, *Science*, **180**: 1283–1285.

Dean, B., 1902, The preservation of muscle fibres in sharks of the Cleveland Shale, *American Geologist*, **30**: 273–278.

Frey, E. and Martill, D.M., 1995, A possible oviraptosaurid theropod from the Santana Formation (Lower Cretaceous, ?Albian) of Brazil, *Neues Jahrbuch für Geologie und Paläontologie, Monatshefte*, **1995**: 397–412.

Gafney, E.S. and Meylan, P.A., 1991, Primitive pelomedusid turtle. *In* J.G. Maisey (ed.), *Santana Fossils: An Illustrated Atlas*, TFH, Neptune, New Jersey: 335–339.

Grimaldi, D.A., 1996, Captured in amber, *Scientific American*, **274**: 70–77.

Harper, E. and Todd, J.A., 1995, Preservation of the adductor muscle of an Upper Jurassic oyster, *Paläontologische Zeitschrift*, **69**: 55–59.

Henwood, A., 1992, Exceptional preservation of dipteran flight muscle and the taphonomy of insects in amber, *Palaios*, **7**: 203–212.

Kellner, A.W.A., 1996a, Reinterpretation of a remarkably well preserved soft tissue from the early Cretaceous of Brazil, *Journal of Vertebrate Paleontology*, **16**: 718–722.

Kellner, A.W.A., 1996b, Fossilized theropod soft tissue, *Nature*, **379**: 32.

Lucas, J. and Prévôt, L.E., 1991, Phosphates and fossil preservation. *In* P.A. Allison and D.E.G. Briggs (eds), *Taphonomy: Releasing the Data Locked in the Fossil Record, Topics in Geobiology, 9*, Plenum Press, New York: 389–409.

Maisey, J.G., 1991, *Santana Fossils: An Illustrated Atlas*, TFH, Neptune, New Jersey: 459 pp.

Martill, D.M., 1988, Preservation of fish in the Cretaceous of Brazil, *Palaeontology*, **31**: 1–18.

Martill, D.M., 1989, The Medusa effect: instantaneous fossilization, *Geology Today*, **5**: 201–205.

Martill, D.M., 1990, Macromolecular resolution of fossilized muscle tissue from an elopomorph fish, *Nature*, **346**: 171–172.

Martill, D.M., 1993, *Fossils of the Santana and Crato Formations, Brazil, Palaeontological Association Field Guides to Fossils*, **5**: 192 pp.

Martill, D.M., 1994, La fossilisation instantanée, *La Recherche*, **25**: 996–1002.

Martill, D.M., 1997, Fish oblique to bedding in early diagenetic concretions: implications for substrate consistency, *Palaeontology*, **40**: 1011–1026.

Martill, D.M. and Brito, P.M., in preparation, Mass mortality of fishes in the Santana Formation (Lower Cretaceous) of north-east Brazil.

Martill, D.M. and Harper, E., 1990, An application of critical-point drying to the comparison of modern and fossilized soft tissues of fishes, *Palaeontology*, **33**: 423–429.

Martill, D.M. and Unwin, D.M., 1989, Exceptionally preserved pterosaur wing membrane from the Cretaceous of Brazil, *Nature*, **340**: 138–140.

Martill, D.M. and Wilby, P.R., 1994, Lithified prokaryotes associated with fossil soft tissues from the Santana Formation (Cretaceous) of Brazil, *Kaupia, Darmstädter Beiträge zur Naturgeschiste*, **4**: 71–77.

Martill, D.M., Cruickshank, A.R.I., Frey, E., Small, P.G. and Clarke, M., 1996, A new crested maniraptoran dinosaur from the Santana Formation (Lower Cretaceous) of Brazil, *Journal of the Geological Society, London*, **153**, 5–8.

Meylan, P.A., 1996, Skeletal morphology and relationships of early Cretaceous side-necked turtle, *Araripemys barretoi* (Testudines: Pelomedusoides: Araripemydidae), from the Santana Formation of Brazil, *Journal of Vertebrate Paleontology*, **16**: 20–33.

Müller, K.J., 1979, Phosphatocopine ostracodes with preserved appendages from the Upper Cambrian of Sweden, *Lethaia*, **12**: 1–27.

Müller, K.J., 1985, Exceptional preservation in calcareous nodules, *Philosophical Transactions of the Royal Society, London*, **B311**: 67–73.

Müller, K.J. and Wallosek, D., 1985a, A remarkable arthropod fauna from the Upper Cambrian 'Orsten' of Sweden, *Transactions of the Royal Society of Edinburgh, Earth Sciences*, **76**: 161–172.

Müller, K.J. and Wallosek, D., 1985b, Skaracarida, a new order of Crustacea from the Upper Cambrian of Västergötland, Sweden, *Fossils and Strata*, **17**: 1–65.

Müller, K.J. and Wallosek, D., 1987, Morphology, ontogeny and life habit of *Agnostus pisiformis* (Linnaeus, 1757) from the Upper Cambrian of Sweden, *Fossils and Strata*, **19**: 1–124.

Owen, R., 1844, A description of certain belemnites, preserved with a great portion of their soft-parts, in the Oxford Clay at Christian Malford, Wiltshire, *Philosophical Transactions of the Royal Society, London*, **1844**: 65–85.

Piearce, T.G., Oates, K. and Carruthers, W.J., 1990, A fossil earthworm embryo (Oligochaeta) from beneath a late Bronze Age midden at Potterne, Wiltshire, UK, *Journal of Zoology, London*, **220**: 537–542.

Schultze, H.-P., 1989, Three dimensional muscle preservation in Jurassic fishes of Chile, *Revista Geologica de Chile*, **16**: 183–215.

Spix, J.B. and Martius, C.F.P., 1823–1831, *Reise im Brasilien* (3 volumes and atlas), M. Lindauer, München.

Traquaire, R.H., 1884, Description of a fossil shark (*Ctenacanthus costellatus*) from the Lower Carboniferous rocks of Eskdale, Dumfriesshire, *Geological Magazine*, **21**: 3–8.

Unwin, D.M., Frey, E., Martill, D.M., Clarke, J.B. and Riess, J., 1996, On the nature of the pteroid in pterosaurs, *Proceedings of the Royal Society, London*, **B263**: 45–52.

Valentine, J.W., 1989, How good was the fossil record? Clues from the Californian Pleistocene, *Paleobiology*, **15**: 83–94.

Weitschat, W., 1983, Ostracoden (Myodocopida) mit Weichkörper-Erhaltung aus der Unter-Trias von Spitzbergen, *Paläontologische Zeitschrift*, **57**: 309–323.

Weitschat, W., 1986, Phosphatisierte Ammonoiden aus der Mittleren Trias von Central-Spitzbergen, *Mitteilungen Geologische Paläontologische Institut, Universität Hamburg*, **61**: 249–279.

Wellnhoffer, P., 1991, *The Illustrated Encyclopedia of Pterosaurs*, Salamander, London: 192 pp.

Wilby, P.R., 1993, *The Mechanisms and Timing of Mineralization of Fossil Phosphatised Soft Tissues*, Unpublished Ph.D. thesis, The Open University, Milton Keynes: 321+115 pp.

Wilby, P.R. and Martill, D.M., 1992, Fossil fish stomachs: a microenvironment for exceptional preservation, *Historical Biology*, **6**: 25–36.

Wilby, P.R. and Whyte, M.A., 1995, Phosphatized soft tissues in bivalves from the Portland Roach of Dorset (Upper Jurassic), *Geological Magazine*, **132**: 117–120.

Willems, H. and Wuttke, M., 1987, Lithogenese lakustriner Dolomit und mikrobiel induzierte Weichteil-Erhaltung bei Tetrapoden des Unter Rotliegenden (Perm, Saar-Nahe-Becken, SW-Deutschland), *Neues Jahrbuch für Geologie und Paläontologie, Abhandlungen*, **174**: 213–238.

Wuttke, M., 1983, 'Weichteil-Erhaltung' durch lithifizierte Mikroorganismen bei mittel-eozänen Vertebraten aus den Ölshiefern der 'Grube Messel' bei Darmstadt, *Senkenbergiana Lethaea*, **64**: 509–527.

NOTE ADDED IN PROOF

Xiao *et al.* (1998) recently announced the discovery of phosphatized embryos of algae and animals from the Neoproterozoic Doushantuo Formation of southern China (570±20 Ma). This is the first record of the Precambrian occurrence of phosphatization of labile tissue and extends this mode of fossilization back by at least 50 Myr. This is particularly significant, because most Precambrian organisms are considered to have lacked mineralized hard tissues and, in consequence, have a very poor fossil record.

REFERENCE

Xiao, S., Zhang, Y. and Knoll, A.H., 1998. Three-dimensional preservation of algae and animal embryos in a Neoproterozoic phosphorite, *Nature*, **391**: 552–558.

4

The Completeness of the Pleistocene Fossil Record: Implications for Stratigraphic Adequacy

Benjamin J. Greenstein, John M. Pandolfi and H. Allen Curran

INTRODUCTION

Nearly three decades after Lawrence (1968) published work emphasizing the importance of understanding the information lost via taphonomic processes, the palaeontological community has underscored a major change in its perception of the fossil record by publishing this volume. The emphasis on adequacy in the fossil record caps a renaissance in taphonomic research that began at a symposium entitled 'The Positive Aspects of Taphonomy', held at the 1984 south-east regional meeting of the Geological Society of America in Lexington, Kentucky. The symposium was a catalyst for renewed field and laboratory investigations of experimental and actualistic taphonomy (see Donovan, 1991, and Allison and Briggs, 1991, for synthetic reviews). A unifying theme to this research is that actuopalaeontological ('palaeontology of the recent', see Schäfer, 1972) investigations allow for an understanding of processes affecting the preservation of organisms. Applying results of ac-tuopalaeontological investigations to fossil material can therefore provide a great deal of palaeoenvironmental, palaeoecological and palaeobiological information applicable to palaeontological questions of varying scale.

Additionally, community ecologists and biologists are becoming increas-ingly aware that the fossil record is an exclusive and crucial database from

The Adequacy of the Fossil Record. Edited by S.K. Donovan and C.R.C. Paul.
© 1998 John Wiley & Sons Ltd.

which to interpret long-term community patterns (Ricklefs, 1987; Jackson, 1992; Jackson *et al.*, 1996). During the last decade, palaeoecological studies in terrestrial (for example, Davis, 1986; Delcourt and Delcourt, 1991; Davis *et al.*, 1994; Reed, 1994; Coope, 1995) and marine (for example, Jackson, 1992; Buzas and Culver, 1994; Allmon *et al.*, 1996; Jackson *et al.*, 1996; Pandolfi, 1996) systems have demonstrated that the fossil record possesses a wealth of information applicable to current concerns of both global change and environmental perturbations on a local scale. Thus, taphonomic studies take on an additional objective: to assess the accuracy with which the recent past history of modern ecosystems is preserved. Our ongoing studies of modern and Pleistocene reef taphonomy (for example, Greenstein and Moffat, 1996; Greenstein and Curran, 1997; Greenstein and Pandolfi, 1997; Pandolfi and Greenstein, 1997a,b) have such an assessment as their overarching objective, and the results are particularly relevant to a variety of dilemmas presently facing community ecologists and marine resource managers alike.

The Pleistocene fossil record of coral reefs over the last million years is a particularly valuable repository for biological data because of its generally spectacular preservation, and also, with few exceptions, the same coral taxa that inhabit modern shallow-water reef environments are present in Pleistocene deposits. Although this is in part due to the young geological age of the interval, a great deal of qualitative (for example, Mesollela, 1967; Mesollela *et al.*, 1970; Chappell, 1974; White *et al.*, 1984; White and Curran, 1987, 1995; White, 1989; K.G. Johnson *et al.*, 1995; Hunter and Jones, 1996) and quantitative (Greenstein and Moffat, 1996; Pandolfi, 1996) data suggest spectacular preservation is common for reef coral assemblages accumulating during at least the last 600 Ka in both the Indo-Pacific and Caribbean provinces. The Caribbean shallow-water coral (and mollusc) faunas have undergone little speciation or extinction since faunal turnover ended roughly a million years ago (Potts, 1984; Allmon *et al.*, 1993; Jackson *et al.*, 1993; Budd *et al.*, 1994, 1996; Jackson, 1994a,b), despite intensifying cycles in climate and sea level throughout the Pleistocene. Thus, Pleistocene fossil coral reef deposits can potentially be used as a database with which to address a variety of issues facing reef ecologists and marine resource managers, whose frustration over the temporally myopic view afforded by monitoring studies that rarely span a scientific career increasingly pervades the literature (for example, Done, 1992; Jackson, 1992; Hughes, 1994a; Bak and Nieuwland, 1995). Two such issues explored in this chapter are an assessment of the response of coral reef communities to environmental perturbations, and the reliability of the Quaternary fossil record of reefs for observing patterns of community assembly over human and geological time scales.

Important ecological influences on coral reefs may operate on a variety of temporal and spatial scales (Porter and Meier, 1992), including decadal time scales (Done, 1992; Bak and Nieuwland, 1995), and the need for long-term data sets has been recognized by a variety of workers (for example, Likens, 1987; D'Elia *et al.*, 1991; Jackson, 1992). Although the results of long-term (decadal scale) systematic monitoring studies of reef communities are becoming increasingly common (see, for example, case histories cited in Ginsburg, 1994), researchers generally acknowledge that patterns demonstrated to have occurred over 10, 20 and even 30 years (for example, Hughes, 1994a) may simply represent part of longer-term cycles that operate over geological time scales (Bak and Nieuwland, 1995), or at least over the multicentennial scale common to the generation time and longevity of corals, the main structural component of reefs.

Given the great disparity between human time scales on the one hand and the time scale over which corals survive and global change occurs on the other, marine scientists are increasingly looking to the Holocene and Pleistocene fossil record of coral reefs to assess the impact of environmental perturbations on the reef ecosystem (Jackson, 1992; Jackson *et al.*, 1996). Jackson (1992) suggested that the reef fossil record represents the exclusive database from which responses of coral communities to global change may be gauged. For example, in Barbados, preliminary qualitative data from the Pleistocene raised reef terraces suggested that similar coral communities and zonation patterns have prevailed for the past 600 Ka (Jackson, 1992). In an overview of mollusc, reef coral and planktic foraminiferal communities, Jackson (1994a) found little correlation between the magnitude of environmental change, and subsequent ecological and evolutionary response during the Pleistocene. In a detailed study that examined geographic and temporal changes in community structure, Pandolfi (1996) found limited species membership in Indo-Pacific reef coral communities from 125 to 30 Ka. These studies give a very different picture of coral reef community structure and stability from that derived from traditional, small-scale ecological studies in the Recent.

Additionally, the Pleistocene record of fossil reefs provides ecologists and palaeoecologists with the opportunity to utilize studies such as those listed above to assess patterns of community assembly occurring over human and geological time scales. A central debate in community ecology concerns species membership in ecological communities, and there exists a spectrum of definitions of community that range from essentially random aggregations of species inhabiting a specific space (for example, Newell *et al.*, 1959; Johnson, 1972) to the concept of the community as a superorganism (for example, Whittaker, 1975; Kauffman and Scott, 1976). Jackson *et al.* (1996) used the fossil record of reefs over the last 10 Ma to explore the link between temporal scale and various perspectives on community membership. The Pleistocene record of fossil reefs provides ecologists and palaeoecologists with the opportunity to document patterns in coral reef community structure, and to compare community studies occurring over human and geological time scales.

Given the great potential of the Pleistocene fossil record of coral reef communities for addressing the issues listed above, an understanding of the taphonomic bias likely to affect reef coral assemblages preserved in Pleistocene strata is essential. Over the last few years, we have been comparing live and dead coral assemblages in modern environments in the Florida Keys and Bahamas to Pleistocene reef facies exposed in the Florida Keys and Bahama Archipelago to determine the accuracy with which the fossil record represents the taxonomic composition of a once-living coral community. Our main objective has been to determine the degree to which Pleistocene material has suffered taphonomic bias, and thus to inform community ecologists and marine managers as to whether the long-term view afforded by Pleistocene material is pertinent to the problems they are trying to address. In this chapter, we report on the results of this work, and conclude that the Pleistocene record of fossil coral reefs is an extraordinary and unique database with which to address a variety of issues in basic and applied palaeobiological research. We present here a series of three case studies of comparisons between reef coral life, death and fossil assemblages. Each case study is followed by a discussion of the specific results obtained. We end this chapter with a more general discussion of the adequacy of the Pleistocene fossil record of coral reefs.

METHODS

Methods of data capture and analysis were essentially the same at all sites and locations, although the actual number of samples varied. The sampling protocol was developed by Pandolfi and Minchin (1995), and was first applied to reef coral life and death assemblages in the Indo-Pacific. Greenstein and Pandolfi (1997) and Pandolfi and Greenstein (1997b) subsequently applied the sampling strategy to shallow and deep reef coral life and death assemblages, respectively, in the Florida Keys. We refer the reader to those studies for additional details.

Field Methods

The linear point intercept (LPI) method was used (Lucas and Seber, 1977) and transects were constructed at each site. In order to estimate cover for the widest range of coral growth forms and colony sizes adequately, transects were 40 m long (Mundy, 1991), each separated by 20–50 m. At 20 cm intervals along each transect, the transect intercept was observed. In modern reef environments, the following data were recorded if the transect intercepted a coral: species, colony size, colony orientation, growth form, and whether the colony was alive or dead and whether whole or fragmented. The same data (with the obvious exception of whether the coral colony was alive or dead) were collected from transects laid across Pleistocene reef facies exposed in the Florida Keys and at Great Inagua and San Salvador Islands, Bahamas. All of the Pleistocene reef facies examined accumulated during the Sangamon interglacial stage. Radiometric dates indicate that the Bahamian Pleistocene reefs on San Salvador and Great Inagua flourished between 119 and 131 Ka BP (Curran *et al.*, 1989; Chen *et al.*, 1991), while the reefs preserved in Pleistocene strata exposed in the Florida Keys flourished approximately 120–140 Ka ago (Harrison and Coniglio, 1985).

In modern environments, the death assemblage is defined as *in situ* dead coral material encountered along each transect and the dead coral rubble accumulating adjacent to the reef framework. It is assumed that this assemblage represents a reasonable proxy for the material that ultimately becomes a fossil assemblage. Dead coral colonies encountered along the transect were identified to the specific level only if they could be recognized without breaking them open or peeling off any algae or other overgrowth. Rubble composed of dead coral was sampled at the 5, 15, 25 and 35 m marks of each transect. This methodology allowed for adequate sampling of the death assemblage as defined above. Rubble samples were placed in a 5 mm mesh bag constrained by a 10 litre bucket. Thus, coral species and growth form were recorded for each specimen > 5 mm in size that preserved colony structure sufficiently to allow for an identification. Taxonomic data obtained from the rubble samples were pooled with data obtained from dead corals encountered along each transect.

Data Analyses

We constructed species sampling curves to investigate whether our methodology adequately accommodated the diversity present in the coral assemblages studied. Comparison of taxonomic composition was calculated using the Bray–Curtis dissimilarity coefficient (Bray and Curtis, 1957), which has been shown to be one of the most robust coefficients for the analysis of taxonomic composition data (Faith *et al.*, 1987). Dissimilarity values range from 0 (for a pair of samples with identical taxonomic composition) to 1 (for a pair of samples with no taxa in common). Abundance data were transformed to their square roots prior to the analysis, to reduce the influence of occasional large abundance values for some taxa (Field *et al.*, 1982). In addition, the transformed abundance values for each taxon were standardized by the maximum attained by that taxon. This standardization equalizes the potential contributions of taxa to the overall dissimilarity in composition. Without standardization by taxon, the Bray–Curtis values are dominated by those taxa which attain high abundances (Faith *et al.*, 1987). The resulting Bray–Curtis matrix was subjected to an ordination technique that provided a visual summary of the pattern of dissimilarity values among the samples. The technique employed was global non-metric multidimensional scaling, or GNMDS (Kruskal, 1964), which has been shown to be one of the most effective methods available for the ordination of taxonomic composition data (Minchin, 1987). Each sample is represented as a point in a coordinate space with a given number of dimensions. The distances between each pair of points are, as far as possible, in rank order with the corresponding dissimilarities in taxonomic composition; points that are close together on the resulting scatter plot represent transects with similar coral constituents. The degree to which the distances on the scatter plot depart from a perfect rank-order fit is measured by a quantity known as 'stress'. The lower the stress value, the better the representation of the samples in the multidimensional space (stress values less than 0.2 generally result in interpretable results; Clarke and Warwick, 1994). The stress values we obtained decreased minimally after a two-dimensional analysis.

Species-richness patterns were compared between life, death and fossil assemblages. To compute species richness, the number of species in each sample was counted, and then corrected for sample size. Thus:

$$\text{species richness} = (S - 1)/(\log N)$$

where S = the number of species present in a sample and N = total number of specimens counted.

RESULTS

Case Study 1: Reef-Tract and Patch-Reef Environments of the Florida Keys and Pleistocene Analogues from Great Inagua Island, Bahamas

Modern life and death assemblages were systematically censussed at two reef-tract sites (Little Carysfort Reef, R1; Grecian Dry Rocks, R2) and two patch-reef sites (Horseshoe Reef, P1; Cannon Patch Reef, P2) (Figure 4.1). The reef-tract and patch-reef sites were selected because they represent differing wave energy regimes and show the classic coral zonation described for the Caribbean (see, for example, Goreau, 1959; Geister, 1977). Eight LPI transects were laid down in each site; thus the modern data set comprises 64 samples

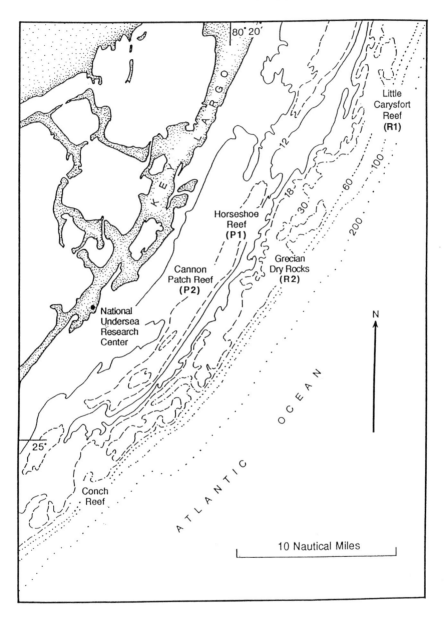

Figure 4.1 *The study area of modern reefs of the Florida reef tract. Little Carysfort Reef (R1) and Grecian Dry Rocks (R2) represent high-energy reef-tract sites; Horseshoe Reef (P1) and Cannon Patch Reef (P2) represent lower-energy patch-reef sites*

(8 transects × 2 environments (reef tract and patch reef) × 2 assemblages (life and death) × 2 sites) and 30 reef coral species.

A spectacularly preserved exposure of fossil corals occurs on Devil's Point, along the south-west coast of Great Inagua Island, Bahamas (Figure 4.2). Radiometric dates indicate a Sangamon age for the reef (Chen *et al.*, 1991). White and Curran (1995) suggested that the remarkable preservation is the result of rapid burial of a once-living barrier-reef/patch-reef system following the Sangamon interglacial interval. Similar taphonomic circumstances have been invoked for reefs of the same age exposed on San Salvador Island, Bahamas (Greenstein and Moffat, 1996). Many of the fossil coral colonies are in growth position, and subsequent weathering has, in places, produced a highly three-dimensional fossil reef surface that, except for its location approximately 2 m above present sea level, is extremely similar to a modern reef framework/coral rubble assemblage. Thus, the same LPI methodology employed on the modern reefs was feasible. We laid a total of 14 transects on reef facies exposed on Devil's Point and followed the environmental interpretation of White and Curran (1995). Ten transects were constructed in the shallow reef tract, and an additional four transects were placed over patch-reef facies. The data from the Pleistocene facies thus comprise 14 samples and the same coral taxa present in the modern environments.

The cumulative number of species encountered in each sample is plotted for life and death assemblages at each site of the Florida reef tract in Figure 4.3. Eight samples were sufficient to account for coral diversity in life and death assemblages in the patch reef (Figure 4.3A). This was generally the case for life and death assemblages in the reef tract, although the number of species counted from the death assemblage sampled at Little Carysfort Reef did not level off prior to the eighth transect (Figure 4.3B). Differences in diversity values shown by the species sampling curves are generally not significant (see below). Plots of cumulative coral species diversity versus number of transects level off for the reef-tract facies from Great Inagua (Figure 4.4A), indicating that sampling is adequate to estimate species richness and compare taxonomic composition between fossil, life and death assemblages. Plots of cumulative coral species diversity versus number of transects level off only between the final two transects for the patch-reef facies from Great Inagua, yielding some uncertainty as to the adequacy of the sampling regime (Figure 4.4B).

Ordination of the Bray–Curtis matrix reveals that, within environments, the taxonomic composition of modern reef coral life and death assemblages, and that of Pleistocene fossil assemblages, is very similar (Figure 4.5). Moreover, the distinction between reef-tract samples and patch-reef samples obtained in modern environments is matched by a distinction between samples obtained from Pleistocene reef-tract and patch-reef facies. Finally, while the taxonomic composition of the Pleistocene patch-reef assemblages is

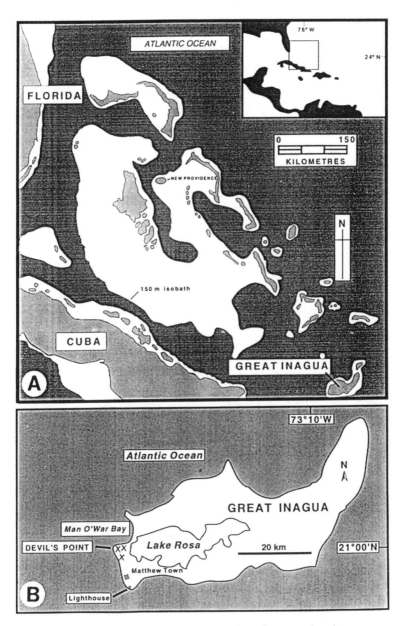

Figure 4.2 *(A) Location of Great Inagua Island in the Bahama Archipelago. (B) Transects (marked by Xs) constructed on the fossil coral reef at Devil's Point. (After White and Curran, 1995)*

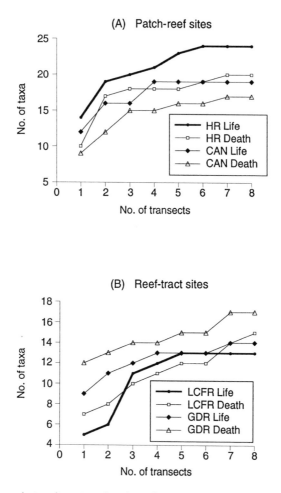

Figure 4.3 *Cumulative diversity of reef coral species versus number of sampling intervals (transects) for life and death assemblages from (A) patch-reef and (B) reef-tract sites of the Florida reef tract. Each line is a plot proceeding from the first to the eighth transect. HR, Horseshoe Reef; CAN, Cannon Patch Reef; CFR, Little Carysfort Reef; GDR, Grecian Dry Rocks*

similar to that of modern life and death assemblages within patch-reef environments, Pleistocene reef-tract assemblages are distinguished from the modern life and death assemblages of the Florida Keys.

Analysis of species richness revealed that no overall significant difference in diversity between life, death and fossil assemblages exists. However, the patch-reef life assemblage is more diverse than its dead counterpart, and the modern reef-tract death assemblage is more diverse than the reef-tract fossil assemblage preserved on Great Inagua (Figure 4.6).

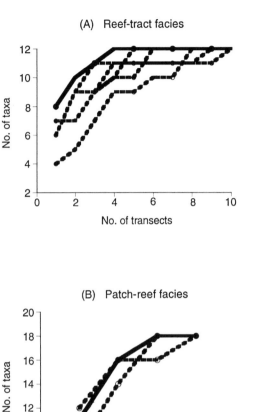

Figure 4.4 *Cumulative diversity of reef coral species versus number of sampling intervals (transects) for Pleistocene reef coral assemblages exposed on Devil's Point, Great Inagua Island. (A) Reef-tract facies; (B) patch-reef facies. For each plot, the solid line is a curve proceeding from the first to the last transect. The dashed lines are plots of five random sequences of transects*

Results of the ordination suggest that the community structure of modern patch reefs adjacent to Key Largo is accurately represented by analogous Pleistocene facies. Conversely, fossil reef-tract facies are less representative of their living counterparts. The fact that differences in the amount of ecological information preserved in the fossil assemblages exist between reef-tract and patch-reef environments indicates that taphonomic studies should be undertaken in a variety of modern reef environments, and the results applied

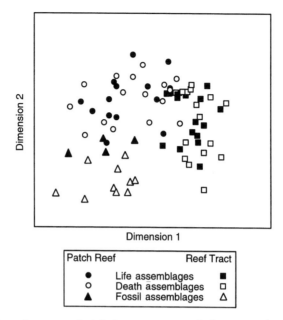

Figure 4.5   *Two-dimensional global non-metric multidimensional scaling (GNMDS) ordination of coral life and death assemblages from the Florida Keys and of fossil assemblages from Great Inagua Island, Bahamas. Points closest to one another represent samples (transects) that are more similar in taxonomic composition than points further away from one another. Samples obtained from the fossil assemblages are similar to samples obtained from the life and death assemblages occurring in modern patch-reef environments of the Florida Keys. Note also that the distinction between living reef-tract and patch-reef environments is tracked by both the death and fossil assemblages. The minimum stress value for the two-dimensional analysis was 0.20*

to palaeoecological studies of their suspected Pleistocene analogues. We also acknowledge that the geographical difference between the Florida Keys and Great Inagua, Bahamas may be partially responsible for our results. Geographical differences notwithstanding, our results clearly indicate that the community structure of the Pleistocene assemblages reflects the well-documented distinction between reef-tract and patch-reef coral communities (see, for example, Goreau, 1959; Geister, 1977) that exists on modern Caribbean reefs. The change in taxonomic composition present along the environmental gradient between living reef-tract and patch-reef coral assemblages also occurs in the corresponding death and fossil assemblages. This observation corroborates the suggestion of White and Curran (1995) that the north to south transition between *Acropora*-dominated to *Montastrea*- (and *Diploria*-) dominated facies observed along the south-west coast of Great Inagua represents a transition between once-living bank-barrier (that is, reef-tract) and lagoonal patch-reef systems.

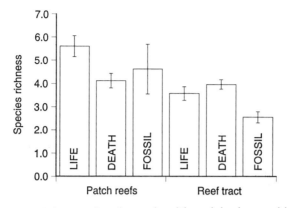

Figure 4.6 *Species-richness values for modern life and death assemblages in reef-tract and patch-reef environments of the Florida Keys, and for Pleistocene fossil assemblages from reef-tract and patch-reef facies exposed on Devil's Point, Great Inagua Island. Calculations to the first decimal place are appropriate given the formula we utilized (see text for further discussion). Only the fossil assemblage composing the reef-tract facies shows a significant decrease in diversity, owing primarily to the paucity of milleporids and of a variety of taxa that are relatively rare (for example,* Porites *furcata,* Dichocoenia stokesii *and* Mycetophyllia danaana; *see Figure 4.7). Error bars are standard errors of the mean*

The frequency distribution of the 25 most common coral species observed in the life and death assemblages in the Florida Keys and in the fossil assemblage on Great Inagua shows that coral colonies with branching growth forms predominate in the death assemblages, while massive colony growth forms predominate in the living and fossil assemblages (Figure 4.7A–C). This observation is more fully discussed later in the chapter. The diversity difference observed between life and death assemblages in the modern patch-reef is the result of the absence, in the death assemblage, of taxa that are rarely encountered alive (*Mycetophyllia lamarckiana, M. danaana* and *Dichocoenia stokesii*). Fossil reef-tract facies are less species-rich than their modern counterparts, primarily because of the paucity of milleporids, which are common components of the living reef tract, and secondarily because of the lack of a variety of taxa that are relatively rare on the living reef tract (for example, *Porites furcata, Dichocoenia stokesii* and *Mycetophyllia danaana* – compare Figure 4.7A–C). The lack of milleporids also accounts for the distinction between Pleistocene reef-tract facies and Holocene reef-tract life and death assemblages.

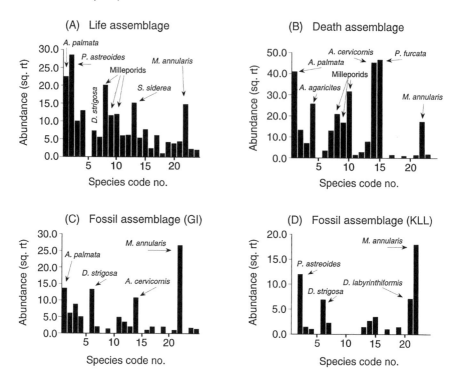

Figure 4.7 *Frequency distribution of the 25 most common coral taxa in life (A) and death (B) assemblages of the Florida Keys, and in fossil assemblages preserved on Great Inagua Island (C) and in the Key Largo Limestone (D). GI = Great Inagua; KLL = Key Largo Limestone. Abundance data are transformed to square roots. Note the relatively high abundance of branching coral growth forms in the death assemblage (for example, Acropora palmata, A. cervicornis and Porites furcata) and massive colony growth forms in the life and fossil assemblages (for example, Porites astreoides, Diploria strigosa, Siderastrea siderea and Montastrea annularis). Species are coded on the x-axis as follows:*

1. Acropora palmata	14. Acropora cervicornis
2. Porites astreoides	15. Porites furcata
3. Porites porites	16. Mycetophyllia lamarckiana
4. Agaricia agaricites	17. Montastrea cavernosa
5. Millepora sp.	18. Mycetophyllia danaana
6. Diploria strigosa	19. Colpophyllia natans
7. Favia fragum	20. Dichocoenia stokesii
8. Millepora squarrosa	21. Diploria labyrinthiformis
9. Millepora complanata	22. Montastrea annularis
10. Millepora alcicornis	23. Meandrina meandrites
11. Diploria clivosa	24. Solenastrea bournoni
12. Siderastrea radians	25. Solenastrea hyades
13. Siderastrea siderea	

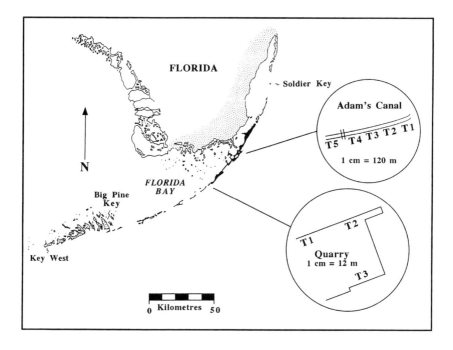

Figure 4.8 *Sample localities for the Key Largo Limestone exposed along Adam's Canal and in the Windley Key Quarry in the Florida Keys. Five transects (T1–T5) were constructed along Adam's Canal, and three transects (T1–T3) were constructed in the Windley Key Quarry. (After Stanley, 1966)*

Case Study 2: Reef-Tract and Patch-Reef Environments of the Florida Keys and Pleistocene Analogues from Key Largo, Florida

We examined fossil reef coral assemblages from outcrops of the Key Largo Limestone exposed along Adam's Canal (Key Largo) and on Windley Key for comparison with the corresponding life and death assemblages currently accumulating offshore (Figure 4.8). There has been some debate as to the reef environment represented by the Key Largo Limestone. Stanley (1966) recognized that the *Montastrea annularis*-dominated coral assemblage exposed on Windley Key and along the Key Largo Waterway (now known as Adam's Canal) could represent either the shallow patch-reefs or deep-water West Indian bank reefs described by Goreau (1959), and opted for the latter environment for the Key Largo Limestone. Hoffmeister and Multer (1968) agreed that the Pleistocene coral assemblage could reflect either environment, but preferred the interpretation of a shallow patch reef, pointing to stratigraphic constraints and, most importantly, the first description of the

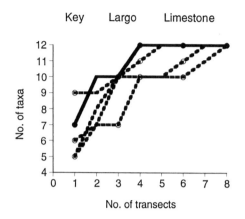

Figure 4.9   *Cumulative diversity of reef coral species versus number of sampling intervals (transects) for Pleistocene reef coral assemblages exposed along Adam's Canal and in the Windley Key Quarry in the Florida Keys. The solid line is a plot proceeding from the first transect sampled along Adam's Canal to the last transect sampled in Windley Key Quarry. The dashed lines are plots of five random sequences of transects*

reef-crest indicator species, *Acropora palmata*, at a similar stratigraphic level in cores obtained several miles east of the exposed *Montastrea*-dominated assemblages. Additional work by various researchers has supported the interpretation of shallow patch-reef facies for the Key Largo Limestone (for example, Dodd *et al.*, 1973; Hodges, 1980).

As with the exposures on Great Inagua, the outcrops allowed for LPI methodology. Five transects, each 40 m in length, were placed on vertical exposures of the Key Largo Limestone along the sides of Adam's Canal. Water in the quarry on Windley Key only allowed space for three transects. Data from transects from both localities were pooled, and compared with the data obtained from the reef-tract and patch-reef described above.

The cumulative number of species encountered on each transect has been plotted for the fossil assemblages preserved in the Key Largo Limestone exposed on Windley Key and along Adam's Canal (Figure 4.9). Six transects were sufficient to account for coral diversity in the Pleistocene facies, indicating that sampling is adequate to estimate species richness in the fossil assemblages and compare their taxonomic composition with that of the Holocene life and death assemblages occurring offshore.

Ordination of the Bray–Curtis matrix revealed different results from those obtained from Great Inagua. Comparison of the Key Largo Limestone with modern reef-tract and patch-reef environments (Figure 4.10) indicates that the fossil assemblages preserved on Key Largo and Windley Key are more similar to the life assemblages occurring offshore than they are to the death assemblages. Moreover, the fossil assemblages plot in close proximity to

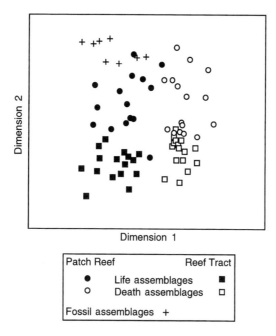

Figure 4.10   *Two-dimensional global non-metric multidimensional scaling (GNMDS) ordination of coral life and death assemblages from the Florida Keys and of fossil assemblages preserved in the Key Largo Limestone. Note that samples obtained from the fossil assemblages are more similar to samples obtained from the living patch reef than they are to samples from the death assemblage present in either the reef-tract or patch-reef environment. The minimum stress value for the two-dimensional analysis was 0.17*

patch-reef life assemblages, indicating that taxonomic composition and relative abundance of the living and fossil patch-reef are very comparable.

Analysis of species richness revealed no significant difference in diversity between life, death and fossil assemblages for the reef-tract localities and the Key Largo Limestone (Figure 4.11). Additionally, no significant difference in diversity between death assemblages sampled in patch-reef environments and the fossil assemblages was observed. Life assemblages in the patch-reef are more diverse than either their dead or fossil counterparts, owing primarily to the presence of corals that are relatively rare in the Florida reef-tract (for example, *Mycetophyllia lamarckiana, M. danaana, Dichocoenia stokesii* and *Solenastrea bournoni*; Figure 4.7D).

Results of the ordination (Figure 4.10) suggest that the taxonomic composition of reef corals preserved in the Key Largo Limestone more accurately represents the assemblage of corals currently living offshore (the patch-reefs, in particular) than the contemporary death assemblage. Although taphonomic analyses of the corals preserved in the Key Largo Limestone are

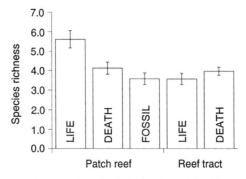

Figure 4.11   *Species-richness values for life, death and fossil assemblages in reef-tract and patch-reef environments of the Florida Keys. Except for the modern patch-reef life assemblage, no significant differences in diversity exist. Error bars are standard errors of the mean*

lacking, one would predict that such analyses would reveal spectacular preservation. This would be much like that anecdotally observed for the corals exposed on Great Inagua, and that demonstrated by Greenstein and Moffat (1996) for specimens of *Acropora palmata* and *A. cervicornis* preserved in a Pleistocene reef exposed on San Salvador, Bahamas (see below).

Case Study 3: Patch-Reef Environment in Fernandez Bay, and the Cockburn Town Fossil Reef, San Salvador, Bahamas

The Pleistocene reef exposed at Cockburn Town, San Salvador, Bahamas provides an excellent opportunity for a comparative taxonomic study because of its proximity to analogous modern reef environments (Figure 4.12). Here, a regressive sequence includes a coral rubblestone facies, composed predominantly of *Acropora cervicornis*, and a coralstone facies, which contains abundant *in situ Acropora palmata*. These facies have been interpreted to represent back-reef and reef-tract environments, respectively (White *et al.*, 1984). Telephone Pole Reef, located in Fernandez Bay, is a modern example of the coral rubblestone facies described by White *et al.* (1984), and is a mid-shelf patch reef that, in the past, was dominated by *A. cervicornis*. This branching coral has suffered a major decrease in abundance throughout the Caribbean region during the past two decades due to a combination of factors (see below). On Telephone Pole Reef, the once-abundant stands of *A. cervicornis* have been largely replaced by *Porites porites* (Curran *et al.*, 1994).

The limited size of Telephone Pole Reef allowed for construction of a total of four 40 m transects over modern life and death assemblages, while five transects were accommodated by the quarry that exposes the Cockburn Town fossil reef. The sampling curves indicate that we have not yet sampled

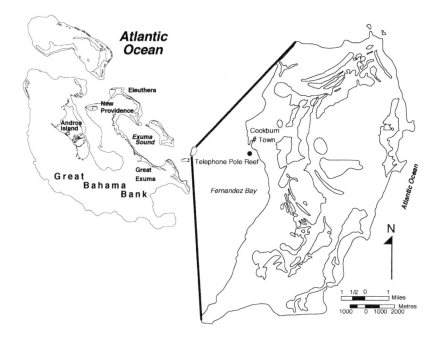

Figure 4.12 *Area of study, San Salvador Island, Bahamas. Telephone Pole Reef is located in Fernandez Bay, while the Cockburn Town fossil reef is located onshore at the north end of Fernandez Bay, in Cockburn Town*

in sufficient detail to account for the diversity of the life assemblage on Telephone Pole Reef (Figure 4.13A). In addition, sampling curves level off only between the final two transects for both the death and fossil assemblages (Figure 4.13B,C); we hesitate to claim adequate sampling based on these results.

Results of ordination reveal a different pattern from those obtained in the previous surveys. Samples from life, death and fossil assemblages are clearly distinct from one another, in marked contrast to the ordination plots generated for the previous two case studies (Figure 4.14). Moreover, the taxonomic composition of life assemblages is clearly less similar to that of the fossil assemblages than to that of the death assemblages (Figure 4.14). Analyses of species richness indicate that a significant decrease in species diversity occurs from life to death to fossil assemblages (Figure 4.15).

The discrimination between the life assemblage and the death and fossil assemblages by the ordination technique is interpreted to be the result of the recent change in coral community structure on Telephone Pole Reef that is part of a Caribbean-wide phenomenon. Beginning at least as early as the 1980s, *Acropora cervicornis* has suffered an extreme decrease in abundance as

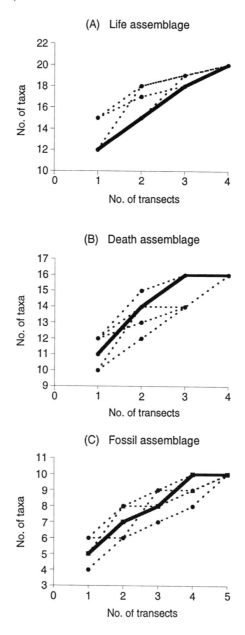

(A)   Life assemblage

(B)   Death assemblage

(C)   Fossil assemblage

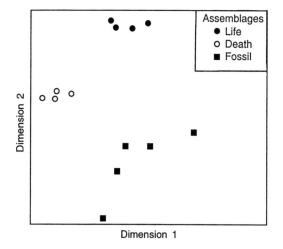

Figure 4.14 Two-dimensional global non-metric multidimensional scaling (GNMDS) ordination of coral life, death and fossil assemblages from San Salvador Island, Bahamas. Note the broad separation of each of the three assemblages. The minimum stress value for the two-dimensional analysis was 0.09

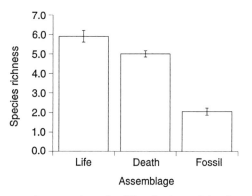

Figure 4.15 Species-richness values for life, death and fossil assemblages on San Salvador Island, Bahamas. Decreasing species-richness values are the result of the lack of milleporids and other taxa in the death and fossil assemblages that are relatively rare on the living reef. Error bars are standard errors of the mean

---

Figure 4.13 (opposite) Cumulative diversity of reef coral species versus number of sampling intervals (transects) from (A) life, (B) death and (C) fossil reef coral assemblages present on San Salvador Island, Bahamas. Four transects accommodated the diversity present in both death and fossil assemblages, but were insufficient to account for the diversity present on the living reef. For all sampling curves, the solid line is a plot proceeding from the first transect to the last transect sampled for each assemblage. The dashed lines are plots of five random sequences of transects

a result of a confluence of factors including hurricanes (Woodley *et al.*, 1981), spread of macroalgae consequent on sea urchin mass mortality (Lessios, 1988), coral diseases and coral bleaching (Brown and Ogden, 1993; Littler and Littler, 1996; Miller, 1996), and a variety of human-induced effects (Hughes, 1994b). On Telephone Pole Reef, *A. cervicornis* has been replaced by large colonies of *Porites porites*. The previous *A. cervicornis*-dominated community is now manifested in the death assemblage (Figure 4.16B), while *P. porites* is abundant only in the life assemblage (Figure 4.16A). Additionally, the paucity of milleporids in both the death and fossil assemblages relative to the life assemblage further segregates life assemblage samples from those obtained from the fossil and death assemblages (compare Figures 4.16A–C). Susceptibility of these hydrozoans to the variety of physical, biological and chemical processes that tend to destroy potential fossil material was also suggested by the results obtained for the first two case studies reported here (compare Figures 4.7A–D), and possibly implicates phylogenetic differences between hydrozoan skeletal microstructure and scleractinian skeletal microstructure as a source of difference in the preservation potential of these taxa. The decrease in species richness from life to death to fossil assemblage results from the absence of the three milleporids distinguished in the surveys (*Millepora complanata*, *M. squarrosa* and *M. alcicornis*) as well as of taxa that are rare in the life assemblage (for example, *Diploria clivosa*, *Montastrea cavernosa* and *Meandrina meandrites*; Figure 4.16). Finally, the greater similarity between life and fossil assemblages is primarily the result of the presence, in high abundance, of massive colonies of *Montastrea annularis*. The presence of

---

Figure 4.16 (opposite)   *Frequency distribution of common coral taxa in (A) life, (B) death and (C) fossil assemblages preserved on San Salvador Island, Bahamas. Abundance data are transformed to square roots. Note the relatively high abundance of* Porites porites, *milleporids and* Montastrea annularis *in the life assemblage. The death assemblage comprises* Acropora cervicornis, A. palmata *and lower abundances of* M. annularis. *Species are coded on the x-axis as follows:*

1. Acropora palmata	14. Acropora cervicornis
2. Porites astreoides	15. Porites furcata
3. Porites porites	16. Mycetophyllia lamarckiana
4. Agaricia agaricites	17. Montastrea cavernosa
5. Millepora *sp.*	18. Mycetophyllia danaana
6. Diploria strigosa	19. Colpophyllia natans
7. Favia fragum	20. Dichocoenia stokesii
8. Millepora squarrosa	21. Diploria labyrinthiformis
9. Millepora complanata	22. Montastrea annularis
10. Millepora alcicornis	
11. Diploria clivosa	
12. Siderastrea radians	
13. Siderastrea siderea	

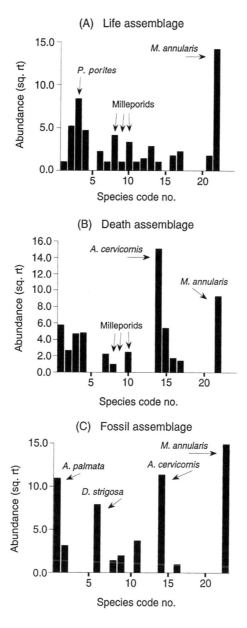

(A)   Life assemblage

(B)   Death assemblage

(C)   Fossil assemblage

recognizable massive colony growth forms in both life and fossil assemblages pervades this study, and is the result of a combination of taphonomic factors and sampling methodology. We elaborate on this statement in the following section.

DISCUSSION

An implicit assumption in the application of palaeoecological studies of Pleistocene reef communities to a variety of current ecological issues is some degree of congruence between fossil assemblages and their living predecessors. Studies of taphonomic processes in modern marine environments have dealt almost exclusively with the transition from life to death assemblages, primarily with molluscan skeletal remains (a great deal of this work is synthesized by Kidwell and Bosence, 1991), foraminifera (Martin and Wright, 1988) and, more recently, with reef corals (Pandolfi and Minchin, 1995; Greenstein and Moffat, 1996; Greenstein and Pandolfi, 1997; Pandolfi and Greenstein, 1997b). The death assemblage is viewed as an intermediate point during the transition from biosphere to lithosphere. The ecological information removed during the transition from the life assemblage to the death assemblage by a variety of physical and biological processes is collectively referred to as 'taphonomic bias'. Most authors assume that taphonomic bias evident in the death assemblage will, at best, simply be transferred to the fossil assemblage or, worse (from the standpoint of 'adequacy'), be augmented by subsequent processes during the later biostratinomic and diagenetic intervals. Examination of living, dead and analogous fossil assemblages in the former case might indicate a decrease in similarity in community structure and a change in diversity between the life and death assemblage, but no continuing differences when death and fossil assemblages are compared. The results of the latter case might reveal decreasing similarity of community structure and changes in diversity through each transition. The case studies we report here have allowed us to test these assumptions, and are thus unique in that they take actuopalaeontological investigations of life and death assemblages one step further, to include investigations of attendant Pleistocene facies composed of essentially the same coral taxa – the 'next step back' into the ancient. We have obtained surprising results: for the life and death assemblages accumulating in the Florida reef-tract and Fernandez Bay, San Salvador, taphonomic bias in the death assemblage has not translated to the fossil assemblage in a stepwise process. Rather, differences in the degree to which fossil assemblages reflect the community structure of life versus death assemblages exist between the Florida Keys and the Bahamas, and between reef-tract and patch-reef environments within the Florida Keys reef system. Thus, the death assemblage is not necessarily the predictor of what might become fossilized, and the assumption that a progression from life to death to fossil assemblage exists in all reef environments should be reconsidered.

Degradation and Coral Identification

The primary reason for the greater similarity of fossil assemblages preserved in the Key Largo Limestone (case study 2) to life assemblages instead of death assemblages in the Florida Keys is the prevalence of living and fossil massive coral colony growth forms (primarily *Montastrea annularis* and *Diploria strigosa*) (Figure 4.7A,B,D). This growth form is relatively rare in the death assemblages surveyed on the Florida reef-tract, but common in both the life and fossil assemblages. The differential representation of this growth form in life and fossil assemblages versus death assemblages is the result of three major factors.

The first factor is that large, massive colony growth forms, which cement directly on to the reef framework, suffer degradation by physical, biological and chemical processes disproportionately to branching and free-living colony growth forms. In a study of the effect of growth form and environment on the preservation potential of reef-building corals on the Great Barrier Reef (Pandolfi and Greenstein, 1997a), we demonstrated that massive colony growth forms in the death assemblage showed consistently higher levels of degradation than did branching and free-living growth forms. Because they possess more robust skeletons than their branching counterparts, massive coral colonies are able to survive for longer intervals of time in the taphonomically active zone (TAZ; see Davies *et al.*, 1989). They thus accumulate a variety of physical, chemical and biological agents of degradation that serve to obscure the details of the colony surface sufficiently to prevent discrimination between species. However, branching and free-living growth forms are more rapidly reduced to essentially unrecognizable grains of carbonate sand; when present in the death assemblage, they are found in less degraded condition because the skeleton does not survive long enough to accumulate extensive features of degradation. This fact influenced the results of our systematic comparisons between life and death assemblages accumulating in shallow (2 and 6 m) and deep (20 and 30 m) reefs of the Florida reef tract (Greenstein and Pandolfi, 1997, and Pandolfi and Greenstein, 1997b, respectively). We demonstrated a striking growth form bias in all depths and reef zones: branching coral growth forms are over-represented in the death assemblage, while massive colony growth forms are over-represented in the life assemblage. We refer the reader to these studies for a more detailed discussion of the nature and effect of this bias.

The second factor involves a combination of our sampling protocol and the way in which fossil corals present themselves in outcrop. First, the rubble samples obtained in modern environments were biased towards those taxa and growth forms not cemented directly on to the reef framework (that is, many branching coral species). Perhaps more importantly, our transect methodology included no way of identifying *in situ* massive coral colonies whose surficial degradation precluded accurate specific identification. When encountered, these surfaces were recorded as 'hard substrate'. Massive colonies preserved as fossils may readily be split open and identification can be accomplished by studying their internal microstructure. In fact, vertical outcrops present at all Pleistocene localities often presented massive colonies broken along the plane of the outcrop. If it had been logistically feasible to split open all the degraded massive colony growth forms that were encountered under water, we inevitably would have identified to the species level many more massive colonies in the death assemblages than we did using the present methodology, and the growth form bias would not have been as severe. However, even though our sample size of massive coral growth forms would have increased by breaking corals open, we cannot be sure whether increased identification would have resulted in greater representation in the death assemblages of the same massive coral species found in the life assemblages. In any event, the fossil assemblage, where coral microstructure is readily available for study, is more similar to the original life assemblage than the present death assemblage as we measured it.

The third factor involves the preservation history of the fossil reefs we examined. In addition to preserving microstructure that was readily available for study, the majority of fossil corals with massive growth forms could be identified on the basis of their surficial structures. This was particularly true on both Great Inagua and San Salvador Islands, where much of the original three-dimensional reef framework is exposed. The presence, in the fossil assemblages, of massive coral colonies that could be identified to the level of species suggests that rapid entombment of once-living reef communities has occurred. Several workers have outlined sedimentological (for example, White *et al.*, 1984; White and Curran, 1987, 1995; Curran *et al.*, 1989; White, 1989) and taphonomic (for example, Greenstein and Moffat, 1996) evidence for rapid burial of late Pleistocene bank-barrier and lagoonal reef systems of the Bahama Archipelago, perhaps as shallow subtidal sands encroached on them in response to the post-Sangamon regression. No such data are available for the Key Largo Limestone. For the Bahamian facies at least, we suggest that our results reflect the rapid entombment of once-living reef systems during the post-Sangamon regression. Thus, live and dead corals were buried concurrently, and a death assemblage as we define it (*in situ* dead coral as well as coral rubble accumulating adjacent to the reef framework) was essentially 'skipped' in the process. The observed similarity in taxonomic composition between the *Montastrea*-dominated facies preserved on Great Inagua and modern life and death assemblages in the Florida reef system (case study 1) also supports this interpretation.

Constancy and Change in Reef Community Structure

Zonation in distribution patterns of coral species has been well documented for Caribbean reefs in modern environments (Goreau, 1959) and their Pleistocene analogues (Mesollela, 1967; Geister, 1977; Jackson, 1992; Jackson et al., 1996). Our results suggest specifically that the distinction between reef-tract and patch-reef coral community structure is indeed preserved in fossil assemblages. However, does this mean that temporal patterns in reef community structure preserved within a particular reef facies over tens to hundreds of thousands of years (see Jackson, 1992, 1994a; Pandolfi, 1996) have been archived in Pleistocene strata? Our results from San Salvador, where a patch-reef currently undergoing a transition in community structure was compared with its Pleistocene analogue, might yield insight into the limits of the 'adequacy' of Pleistocene coral reef deposits.

As mentioned above, Telephone Pole Reef is currently undergoing a transition from an *Acropora cervicornis*-dominated coral assemblage to an assemblage with abundant *Porites porites*. The distinction between life, death and fossil assemblage samples by ordination (Figure 4.14) is primarily the result of the abundant stands of *P. porites* currently inhabiting the reef (Figure 4.16A), and, secondarily, the presence of hydrozoans. *Porites* is much less abundant in the death and fossil assemblages (Figure 4.16B,C). The death assemblage contains the once-abundant stands of *A. cervicornis* that are now extremely rare in the life assemblage and are also abundant in the fossil assemblage. Although insufficient sampling necessarily qualifies our results, they are strikingly different from those we obtained in the previous two case studies, and there are two alternative hypotheses that explain the apparent failure of the Pleistocene assemblage exposed on San Salvador to reflect the life assemblage accurately.

First, the demise of *A. cervicornis* in the Bahamas and Caribbean, and subsequent replacement by another coral species (on Telephone Pole Reef, *P. porites*), is without historical precedent. In Belize, the once-abundant stands of *A. cervicornis* have been replaced by *Agaricia agaricites* (Aronson, 1996; Aronson and Plotnick, in press). Careful examination of cores taken through the reef sedimentary record in Belize revealed no recognizable signals (abrupt changes in coral taxa or taphonomic evidence of an essentially monospecific death assemblage) of similar transitions, suggesting that the present drastic reduction of *A. cervicornis* has no precursor in the recent geological past (at least 3800 years; Aronson, 1996). In the Florida Keys, we purposely chose for our surveys modern reefs that conformed to earlier (pre-1980) descriptions (for example, Multer, 1977) of the majority of the reef-tract reefs (for example, abundant live *Acropora palmata* in the shallowest zones, grading to more diverse, deeper assemblages of living *Porites astreoides, Montastrea annularis* and *Diploria strigosa*). It is compelling that these 'healthy' reef communities were reflected by the fossil assemblages in the Florida Keys and Great Inagua, whereas the present *Porites*-dominated community on Telephone Pole Reef is not reflected by the fossil assemblage exposed on San Salvador. Moreover, it is sobering to consider the rapidity with which communities dominated by *A. cervicornis* have been altered. From this we can only conclude that the persistence of this coral association during Pleistocene climatic fluctuations (Mesollela, 1967; Jackson, 1992) is not an artefact of taphonomic bias.

An alternative hypothesis is that rapid changes in coral dominance within a community commonly occur, but the fossil record has not been studied in sufficient detail to observe these temporally short-term fluctuations in reef community structure. Short-term studies of living coral reefs have recorded fluctuations of dominant species at virtually all spatial scales, ranging from metre quadrats (for example, Hughes *et al.*, 1987; Bak and Nieuwland, 1995) through individual reefs (for example, Porter *et al.*, 1981; Woodley *et al.*, 1981) to entire provinces (for example, Lessios, 1988). Moreover, short-term fluctuations may be a prerequisite for long-term stability (Chesson and Huntly, 1989) and, thus, produce the type of long-term persistence of coral communities documented by Mesollela (1967), Jackson (1992) and Pandolfi (1996). With the tremendous age resolution now available for Quaternary reef deposits, it is possible to collect detailed information at much shorter time scales than previously thought. On the basis of our results here, we encourage short-term fossil studies as a complement to previous palaeoecological studies at larger time scales.

Adequacy of the Quaternary Fossil Record of Coral Reef Communities

Our results suggest that, for the Pleistocene strata we have examined, the stepwise transition from life to death to fossil assemblage cannot be assumed. Fossil assemblages may preserve modern life and death assemblages (case study 1), life assemblages (case study 2) or death assemblages (case study 3). Since the amount of taphonomic bias can be different in different reef environments, it becomes imperative to assess the amount occurring by conducting taphonomic analyses in each modern environment that is to be studied as a fossil assemblage. Based on the results of the three case studies we conclude that, with a synthesis of taphonomic studies in modern reef environments, the Quaternary record of fossil reefs becomes an extraordinary tool with which to address a variety of ecological issues, discussed below.

Response of reef communities to global change

Studies in Pleistocene reef community structure over geological time scales in both the Caribbean (for example, Mesollela, 1967; Mesollela *et al.*, 1970; Jackson, 1992; Jackson *et al.*, 1996) and Indo-Pacific provinces (Pandolfi, 1996) have shown remarkable persistence in reef coral community structure. These studies contrast with current observations of changes in reef communities over human time scales (see, for example, Porter *et al.*, 1981; Woodley *et al.*, 1981; Hughes *et al.*, 1987; Lessios, 1988; Hughes, 1994b; Bak and Nieuwland, 1995). Our results suggest that the pattern of persistence that is increasingly documented by palaeoecologists studying Quaternary marine ecosystems is not an artefact of an inadequate fossil record and that patterns on longer temporal scales may be very different from those observed on shorter temporal scales. We submit that, with companion taphonomic studies, the Pleistocene fossil record of coral reefs becomes a relevant reference for reef management policy decisions. Additional studies comparing the time over which individual Pleistocene (for example, Pandolfi *et al.*, 1994; Pandolfi, 1996) versus modern environmental perturbations occur are needed to document the resilience of the modern reef community system. Finally, our results from San Salvador suggest that fossil reefs preserved in Pleistocene strata may provide important baseline data with which to assess the taxonomic architecture of a 'pristine' reef assemblage. Thus, the fossil record becomes a useful resource to inform policy-makers seeking to target particular reef systems as natural preserves.

The nature of communities

A tremendous amount of ecological and palaeoecological research has been devoted to the spatial and temporal patterns of structure in communities. Two completely different viewpoints are: (1) communities are essentially random aggregations of species inhabiting a particular place at a particular time (for example, Newell *et al.*, 1959; Johnson, 1972); and (2) communities are highly integrated and may even behave as 'superorganisms' (for example, Whittaker, 1975; Kauffman and Scott, 1976). Some recent attempts to evaluate species membership in communities using the fossil record for marine ecosystems have favoured open communities with unlimited species membership (for example, Valentine and Jablonski, 1993; Buzas and Culver, 1994). In contrast, Pandolfi (1996) studied the taxonomic composition and diversity of Pleistocene reef coral communities during nine separate glacial cycles within two separate reef environments in raised terraces of the Huon

Peninsula, Papua New Guinea. Pandolfi (1996) showed that similar associations of reef coral species reassembled repeatedly through time, suggesting limited species membership. Jackson (1992) suggested that a similar pattern might exist throughout the past 500 000 years in Barbados. We believe our taphonomic work clearly shows that preservation bias is minimal for Quaternary fossil reefs. Moreover, in combination with attendant taphonomic analyses in appropriate modern environments, the Quaternary record of fossil reefs provides ecologists and palaeoecologists with a rich and reliable data resource to compare patterns of community assembly occurring over multiple time scales, both human and geological. However, we stress the importance of examining preservational attributes in palaeoecological studies, especially those which purport to examine the recent past history of modern ecosystems. This is especially true from our studies, since different reef environments responded differently in their preservational attributes.

Sequence stratigraphic controls on reef preservation

The life and death assemblages present adjacent to Key Largo have been accumulating during the Holocene transgression (Lidz and Shinn, 1991). Those preserved on Great Inagua and elsewhere in the Bahamas were buried during a late Pleistocene regression (White and Curran, 1995). Our analysis of Pleistocene fossil assemblages, combined with the previous work of Greenstein and Pandolfi (1997) and Pandolfi and Greenstein (1997b), suggests stratigraphic sequence (transgressive versus regressive) may be a first-order mechanism for determining preservational style of fossil reef systems. Specifically, regressive reefs might be more likely to be preserved by a rapid burial event as shallow subtidal sand and beach deposits encroach on them during a relative lowering of sea level. In contrast, transgressive reefs may be less prone to catastrophic burial and thus, if preserved at all, may exhibit poorer preservation than their regressive counterparts. We suggest that further taphonomic research on ancient reef communities be conducted in tandem with analyses of the sequence stratigraphic context in which the assemblages occur. This approach will allow palaeontologists to understand the implications of pristine versus highly degraded fossil coral material accumulated in the geological past, when the abundance and diversity of organisms that serve to degrade coral material may have been very different from those in Pleistocene or modern time.

ACKNOWLEDGEMENTS

A study of this magnitude could not have been completed without a great deal of help from many individuals. Fieldwork in the Florida Keys was entirely supported by two grants from the National Undersea Research Center (UNCW 9612, 9613) awarded to BJG and JMP. We acknowledge the facilities and staff of the centre on Key Largo, especially Dr Stephen Miller, for a variety of efforts on our behalf. Fieldwork on Great Inagua was facilitated by the efforts of Jimmy Nixon of the Bahamian National Trust. Logistical support on San Salvador Island was provided by the Bahamian Field Station. Additional financial support to BJG and HAC from Smith College is gratefully acknowledged. Several undergraduate students participated as field and laboratory assistants and were supported by the Schultz–Sherman Fairchild Foundation. The following students served as field and/or laboratory assistants: Tatiana Bertsch, Rebecca Falk, Elizabeth Gardiner, Lora Harris, Susan Timmons (Smith College), Kirsten Bannister (Whitman College) and Patrick Decowski (MIT). Janet Lauroesch lent her artistic expertise to many of the figures.

REFERENCES

Allison, P.A. and Briggs, D.E.G., 1991, *Taphonomy: Releasing the Data Locked in the Fossil Record*, Plenum Press, New York: 560 pp.

Allmon, W.D., Rosenberg, G., Portell, R.W. and Schindler, K.S., 1993, Diversity of Atlantic coastal plain mollusks since the Pliocene, *Science*, **260**: 1626–1629.

Allmon, W.D., Rosenberg, G., Portell, R.W. and Schindler, K.S., 1996, Diversity of Pliocene–Recent mollusks in the western Atlantic: extinction, origination and environmental change. *In* J.B.C. Jackson, A.F. Budd and A.G. Coates (eds), *Evolution and Environment in Tropical America*, University of Chicago Press, Chicago: 271–302.

Aronson, R.B., 1996, Community dynamics of Caribbean coral reefs: coordinated stasis or individualistic dynamics? *Geological Society of America Abstracts with Programs*, **28** (7): A177.

Aronson, R.B. and Plotnick, R.E., in press, Scale-independent interpretations of macroevolutionary dynamics. *In* M.L. McKinney (ed.), *Biodiversity Dynamics: Turnover of Populations, Taxa and Communities*, Columbia University Press, New York.

Bak, R.P.M. and Nieuwland, G., 1995, Long-term change in coral communities along depth gradients over leeward reefs in the Netherlands Antilles, *Bulletin of Marine Science*, **56**: 609–619.

Bray, J.R. and Curtis, J.T., 1957, An ordination of the upland forest communities of southern Wisconsin, *Ecological Monographs*, **27**: 325–349.

Brown, B.E. and Ogden, J.C., 1993, Coral bleaching, *Scientific American*, **268**: 64–70.

Budd, A.F., Stemann, T.A. and Johnson, K.G., 1994, Stratigraphic distribution of genera and species of Neogene to Recent Caribbean reef corals, *Journal of Paleontology*, **68**: 951–977.

Budd, A.F., Johnson, K.G. and Stemann, T.A., 1996, Plio-Pleistocene turnover and extinctions in the Caribbean reef-coral fauna. *In* J.B.C. Jackson, A.F. Budd and A.G. Coates (eds), *Evolution and Environment in Tropical America*, University of Chicago Press, Chicago: 168–204.

Buzas, M.A. and Culver, S.J., 1994, Species pool and dynamics of marine paleocommunities, *Science*, **264**: 1439–1441.

Chappell, J., 1974, Geology of coral terraces, Huon Peninsula, New Guinea: a study of Quaternary tectonic movements and sea-level changes, *Geological Society of America Bulletin*, **85**: 553–570.

Chen, J.H., Curran, H.A., White, B. and Wasserburg, G.J., 1991, Precise chronology of the interglacial period: ^{230}Th/^{234}U data from fossil coral reefs in the Bahamas, *Geological Society of America Bulletin*, **103**: 82–97.

Chesson, P. and Huntly, N., 1989, Short-term instabilities and long-term community dynamics, *Trends in Ecology and Evolution*, **4**: 293–298.

Clarke, K.R. and Warwick, R.M., 1994, *Change in Marine Communities: An Approach to Statistical Analysis and Interpretation*, Natural Environment Research Council, Swindon: 144 pp.

Coope, G.R., 1995, Insect faunas in Ice Age environments. *In* J.H. Lawton and R.M. May (eds), *Extinction Rates*, Oxford University Press, Oxford: 55–74.

Curran, H.A., White, B., Chen, J.H. and Wasserburg, G.J., 1989, Comparative morphologic analysis and geochronology for the development and decline of two Pleistocene coral reefs, San Salvador and Great Inagua Islands. *In* J. Mylroie (ed.), *Proceedings of the Fourth Symposium on the Geology of the Bahamas*, Bahamian Field Station, San Salvador: 107–117.

Curran, H.A., Smith, D.P., Meigs, L.C., Pufall, A.E. and Greer, M.L., 1994, The health and short-term change of two coral patch reefs, Fernandez Bay, San Salvador Island, Bahamas. *In* R.N. Ginsburg (ed.), *Proceedings of the Colloquium on Global Aspects of Coral Reefs: Health, Hazards, and History, 1993*, Rosenstiel School of Marine and Atmospheric Science, University of Miami, Florida: 147–153.

Davies, D.J., Powell, E.N. and Stanton, R.J., Jr, 1989, Relative rates of shell dissolution and net sediment accumulation – a commentary: can shell beds form by the gradual accumulation of biogenic debris on the sea floor? *Lethaia*, **22**: 207–212.

Davis, M.B., 1986, Climatic instability, time lags, and community disequilibrium. *In* J. Diamond and T.J. Case (eds), *Community Ecology*, Harper and Row, New York: 269–284.

Davis, M.B., Sugita, S., Calcota, R.R., Ferrari, J.B. and Frelich, L.E., 1994, Historical development of alternate communities in a hemlock–hardwood forest in northern Michigan, USA. *In* P.J. Edwards, R.M. May and N.R. Webb (eds), *Large-scale Ecology and Conservation Biology*, Blackwell Scientific, Oxford: 19–39.

Delcourt, H.R. and Delcourt, P.A., 1991, *Quaternary Ecology: A Paleoecological Perspective*, Chapman and Hall, London: 242 pp.

D'Elia, C.F., Buddemeier, R.W. and Smith, S.V., 1991, *Workshop on Coral Bleaching, Coral Reef Ecosystems, and Global Climate Change: Report of Proceedings*, Maryland Sea Grant Publication, University of Maryland, College Park, Maryland: 91 pp.

Dodd, J.R., Hattin, D.E. and Liebe, R.M., 1973, Possible living analog of the Pleistocene Key Largo reefs of Florida, *Geological Society of America Bulletin*, **84**: 3995–4000.

Done, T.J., 1992, Constancy and change in some Great Barrier Reef coral communities, *American Zoologist*, **32**: 655–662.

Donovan, S.K. (ed.), 1991, *The Processes of Fossilization*, Belhaven Press, London: 303 pp.

Faith, D.P., Minchin, P.R. and Belbin, L., 1987, Compositional dissimilarity as a robust measure of ecological distance, *Vegetatio*, **69**: 57–68.

Field, J.G., Clarke, K.R. and Warwick, R.M., 1982, A practical strategy for analysing multispecies distribution patterns, *Marine Ecological Progress Series*, **8**: 37–52.

Geister, J., 1977, The influence of wave exposure on the ecological zonation of

Caribbean coral reefs, *Proceedings of the Third International Coral Reef Symposium*, **1**: 23–29.

Ginsburg, R.N. (ed.), 1994, *Proceedings of the Colloquium on Global Aspects of Coral Reefs: Health, Hazards, and History, 1993*, Rosenstiel School of Marine and Atmospheric Science, University of Miami, Florida: 420 pp.

Goreau, T.F., 1959, The ecology of Jamaican coral reefs. I. Species composition and zonation, *Ecology*, **40**: 67–90.

Greenstein, B.J. and Curran, H.A., 1997, How much ecological information is preserved in fossil coral reefs and how reliable is it? *Proceedings of the Eighth International Coral Reef Symposium, Panama City, Panama*: **1**: 417–422.

Greenstein, B.J. and Moffat, H.A., 1996, Comparative taphonomy of Holocene and Pleistocene corals, San Salvador, Bahamas, *Palaios*, **11**: 57–63.

Greenstein, B.J. and Pandolfi, J.M., 1997, Preservation of community structure in modern reef coral life and death assemblages of the Florida Keys: implications for the Quaternary record of coral reefs, *Bulletin of Marine Science*, **19**: 431–452.

Harrison, R.S. and Coniglio, M., 1985, Origin of the Pleistocene Key Largo Limestone, Florida Keys, *Bulletin of Canadian Petroleum Geology*, **33**: 350–358.

Hodges, L.T., 1980, Coral size and relationships of the Key Largo Limestone, *Proceedings of the Third International Coral Reef Symposium*, **3**: 347–353.

Hoffmeister, J.E. and Multer, H.G., 1968, Geology and origin of the Florida Keys, *Geological Society of America Bulletin*, **79**: 1487–1502.

Hughes, T.P., 1994a, Coral reef degradation: a long-term study of human and natural impacts. *In* R.N. Ginsburg (ed.), *Proceedings of the Colloquium on Global Aspects of Coral Reefs: Health, Hazards and History, 1993, Case Histories*, Rosenstiel School of Marine and Atmospheric Science, University of Miami, Florida: C20–C25.

Hughes, T.P., 1994b, Catastrophes, phase shifts, and large-scale degradation of a Caribbean coral reef, *Science*, **265**: 1547–1551.

Hughes, T.P., Reed, D.C. and Boyle, M.J., 1987, Herbivory on coral reefs: community structure following mass mortalities of sea urchins, *Journal of Experimental Marine Biology and Ecology*, **113**: 39–59.

Hunter, I.G. and Jones, B., 1996, Coral associations of the Pleistocene Ironshore Formation, Grand Cayman, *Coral Reefs*, **15**: 249–267.

Jackson, J.B.C., 1992, Pleistocene perspectives on coral reef community structure, *American Zoologist*, **32**: 719–731.

Jackson, J.B.C., 1994a, Constancy and change of life in the sea, *Philosophical Transactions of the Royal Society of London*, **B344**: 55–60.

Jackson, J.B.C., 1994b, Community unity? *Science*, **264**: 1412–1413.

Jackson, J.B.C., Jung, P., Coates, A.G. and Collins, L.S., 1993, Diversity and extinction of tropical American mollusks and emergence of the Isthmus of Panama, *Science*, **260**: 1624–1626.

Jackson, J.B.C., Budd, A.F. and Pandolfi, J.M., 1996, The shifting balance of natural communities? *In* D.H. Erwin, D. Jablonski and J.H. Lipps (eds), *Evolutionary Paleobiology*, University of Chicago Press, Chicago: 89–122.

Johnson, K.G., Budd, A.F. and Stemann, T.A., 1995, Extinction selectivity and ecology of Neogene Caribbean reef corals, *Paleobiology*, **21**: 52–73.

Johnson, R.G., 1972, Conceptual models of benthic marine communities. *In* T.J.M. Schopf (ed.), *Models in Paleobiology*, Freeman, San Francisco: 148–159.

Kauffman, E.G. and Scott, R.W., 1976, Basic concepts of community ecology and paleoecology. *In* R.W. Scott and R.R. West (eds), *Structure and Classification of Paleocommunities*, Dowden, Hutchinson and Ross, Stroudsburg, Pennsylvania: 1–28.

Kidwell, S.M. and Bosence, D.W.J., 1991, Taphonomy and time-averaging of marine shelly faunas. *In* P.A. Allison and D.E.G. Briggs (eds), *Taphonomy: Releasing the Data Locked in the Fossil Record*, Plenum Press, New York: 115–209.

Kruskal, J.B., 1964, Multidimensional scaling by optimizing goodness of fit to a nonmetric hypothesis, *Psychometrika*, **29**: 1–27.

Lawrence, D.R., 1968, Taphonomy and information losses in fossil communities, *Geological Society of America Bulletin*, **79**: 1315–1330.

Lessios, H.A., 1988, Mass mortality of *Diadema antillarum* in the Caribbean: what have we learned? *Annual Reviews of Ecology and Systematics*, **19**: 371–393.

Lidz, B.H. and Shinn, E.A., 1991, Paleoshorelines, reefs, and a rising sea: South Florida, USA, *Journal of Coastal Research*, **7**: 203–229.

Likens, G.E., 1987, *Long-term Studies in Ecology: Approaches and Alternatives*, Springer-Verlag, New York: 214 pp.

Littler, M.M. and Littler, D.S., 1996, Black band disease in the South Pacific, *Coral Reefs*, **15**: 20.

Lucas, H.A. and Seber, G.A.F., 1977, Estimating coverage and particle density using the line intercept method, *Biometrika*, **64**: 618–622.

Martin, R.E. and Wright, R.C., 1988, Information loss in the transition from life to death assemblages of foraminifera in back reef environments, Key Largo, Florida, *Journal of Paleontology*, **62**: 399–410.

Mesollela, K.J., 1967, Zonation of uplifted Pleistocene coral reefs on Barbados, West Indies, *Science*, **156**: 638–640.

Mesollela, K.J., Sealy, H.A. and Matthews, R.K., 1970, Facies geometries within Pleistocene reefs of Barbados, West Indies, *American Association of Petroleum Geologists Bulletin*, **54**: 1899–1917.

Miller, I., 1996, Black band disease on the Great Barrier Reef, *Coral Reefs*, **15**: 58.

Minchin, P.R., 1987, An evaluation of relative robustness of techniques for ecological ordination, *Vegetatio*, **69**: 89–107.

Multer, H.G., 1977, *Field Guide to Some Carbonate Rock Environments: Florida Keys and Western Bahamas*, Kendall Hunt, Iowa: 415 pp.

Mundy, C., 1991, *A Critical Evaluation of the Line Intercept Transect Methodology for Surveying Sessile Coral Reef Benthos*, Unpublished Masters thesis, University of Queensland: 126 pp.

Newell, N.D., Imbrie, J., Purdy, E.G. and Thurber, D.L., 1959, Organisms communities and bottom facies, Great Bahama Bank, *Bulletin of the American Museum of Natural History*, **117**: 181–228.

Pandolfi, J.M., 1996, Limited membership in Pleistocene reef coral assemblages from the Huon Peninsula, Papua New Guinea: constancy during global change, *Paleobiology*, **22**: 152–176.

Pandolfi, J.M. and Greenstein, B.J., 1997a, Taphonomic alteration of reef corals: effects of reef environment and coral growth form. I. The Great Barrier Reef, *Palaios*, **12**: 27–42.

Pandolfi, J.M. and Greenstein, B.J., 1997b, Preservation of community structure in death assemblages of deep water Caribbean reef corals, *Limnology and Oceanography*, **42**: 1505–1516.

Pandolfi, J.M. and Minchin, P.R., 1995, A comparison of taxonomic composition and diversity between reef coral life and death assemblages in Madang Lagoon, Papua New Guinea, *Palaeogeography, Palaeoclimatology, Palaeoecology*, **119**: 321–341.

Pandolfi, J.M., Best, M.M.R. and Murray, S.P., 1994, Coseismic event of May 15, 1992, Huon Peninsula, Papua New Guinea: comparison with Quaternary tectonic history, *Geology*, **22**: 239–242.

Porter, J.W. and Meier, O.W., 1992, Loss and change in Floridian reef corals, *American Zoologist*, **32**: 625–640.

Porter, J.W., Woodley, J.D., Smith, G.J., Niegel, J.E., Battey, J.F. and Dallmeyer, D.G., 1981, Population trends among Jamaican reef corals, *Nature*, **294**: 249–250.

Potts, D.C., 1984, Generation times and the Quaternary evolution of reef-building corals, *Paleobiology*, **10**: 46–58.

Reed, K.E., 1994, Community organization through the Plio-Pleistocene, *Journal of Vertebrate Paleontology*, **14**: 43A.

Ricklefs, R.E., 1987, Community diversity: relative roles of local and regional processes, *Science*, **235**: 167–171.

Schäfer, W., 1972, *Ecology and Palaeoecology of Marine Environments*, Oliver and Boyd, Edinburgh: 569 pp.

Stanley, S.M., 1966, Paleoecology and diagenesis of Key Largo Limestone, Florida, *American Association of Petroleum Geologists Bulletin*, **50**: 1927–1947.

Valentine, J.W. and Jablonski, D., 1993, Fossil communities: compositional variation at many time scales. *In* R.E. Ricklefs and D. Schluter (eds), *Species Diversity in Ecological Communities*, University of Chicago Press, Chicago: 341–349.

White, B., 1989, Field guide to the Sue Point fossil coral reef, San Salvador Island, Bahamas. *In* J. Mylroie (ed.), *Proceedings of the Fourth Symposium on the Geology of the Bahamas*, Bahamian Field Station, San Salvador: 353–365.

White, B. and Curran, H.A., 1987, Coral reef to eolianite transition in the Pleistocene rocks of Great Inagua, Bahamas. *In* H.A. Curran (ed.), *Proceedings of the Third Symposium on the Geology of the Bahamas*, CCFL Field Station, Fort Lauderdale, Florida: 165–179.

White, B. and Curran, H.A., 1995, Entombment and preservation of Sangamonian coral reefs during glacioeustatic sea-level fall, Great Inagua Island, Bahamas. *In* H.A. Curran and B. White (eds), *Terrestrial and Shallow Marine Geology of the Bahamas and Bermuda*, Geological Society of America Special Paper, **300**: 51–61.

White, B., Kurkjy, K.A. and Curran, H.A., 1984, A shallowing-upward sequence in Pleistocene coral reef and associated facies, San Salvador, Bahamas. *In* J.W. Teeter (ed.), *Proceedings of the Second Symposium on the Geology of the Bahamas*, CCFL Field Station, Fort Lauderdale, Florida: 53–70.

Whittaker, R.H., 1975, *Communities and Ecosystems*, Macmillan, New York: 385 pp.

Woodley, J.D., Chornesky, E.A., Clifford, P.A., Jackson, J.B.C., Hoffman, L.S., Knowlton, N., Lang, J.C., Pearson, M.P., Porter, J.W., Rooney, M.L., Rylaarsdam, K.W., Tunnicliffe, B.J., Wahle, C.M., Wulff, J.L., Curtis, A.S.G., Dallmeyer, M.D., Jupp, B.P., Koehl, A.R., Neigel, J. and Sides, E.M., 1981, Hurricane Allen's impact on Jamaican coral reefs, *Science*, **214**: 749–755.

5

An Overview of the Completeness of the Fossil Record

Christopher R.C. Paul and Stephen K. Donovan

INTRODUCTION

Ideally we would like to present a systematic overview of the completeness of the fossil record throughout the Phanerozoic. However, at the current state of knowledge that is just not possible. This is almost entirely due to the fact that most compendia of fossil occurrences (*Treatise on Invertebrate Paleontology*; Harland *et al.*, 1967; Sepkoski, 1982; Benton, 1993) record total known stratigraphic ranges and omit any references to gaps within these ranges. Thus, gap analysis (Paul, 1982), though basically a simple procedure providing valid comparative data, involves a thorough search of the primary literature for all major fossil groups. It is true that one of us has presented some evidence for selected groups of Palaeozoic echinoderms (Paul, 1988), but this is a very limited data set. There is also no reason to suppose that the record of these particular echinoderms is typical of all Palaeozoic echinoderms, let alone all Palaeozoic or Phanerozoic fossils. Nevertheless, any comparative record represents a starting point and can be used as a baseline against which the records of other major taxa can be compared. Even such a limited database presents some intriguing information. The echinoderm fossil record was exceptionally poor across the Cambrian/Ordovician and Ordovician/Silurian boundaries. Smith (1988) has already argued that the poor fossil record of echinoderms across the Cambrian/Ordovician boundary has influenced perceptions of the 'Cambrian fauna'. Here we would suggest further that the real magnitude and significance of the late Ordovician mass extinction are difficult to evaluate fully in the absence of a

The Adequacy of the Fossil Record. Edited by S.K. Donovan and C.R.C. Paul.

good early Silurian (= Llandovery) fossil record (Donovan, 1994). Just how many echinoderm taxa really became extinct in the Hirnantian and how many survived into the early Silurian, but have not been found because the record of this interval is so poor? There is little doubt that the quality of the record affects our perception of patterns of evolution and extinction. Objective analysis of the quality of the record throughout the Phanerozoic should be a major priority for any serious analysis of the history of life on Earth.

An alternative approach would be to develop a generalized predictive model that can be tested at various stratigraphic levels where the quality of the record is known or can easily be investigated. Here we argue that the most vulnerable parts of the marine fossil record are deposits laid down on the margins of shallow basins at times of high eustatic sea level. This is because every time sea level falls such marginal deposits are liable to destruction by erosion. Since the vast majority of invertebrate fossils are preserved in deposits of shallow-shelf seas, systematic loss of these deposits by subsequent erosion will seriously affect the quality of the record. It is this effect, we suspect, that contributes at least partly to the poor record of echinoderms in the Lower Silurian. Well-studied, echinoderm-bearing deposits of the Upper Ordovician in North America, Europe and North Africa are generally of shallow-water shelly facies. However, most Lower Silurian deposits of these areas are deep-water graptolitic shales. In rare instances where shallow-water, shelly, Lower Silurian deposits are preserved (such as the Cincinnati Arch, USA and the Pentland Hills, Scotland), echinoderm faunas intermediate between those of the Upper Ordovician and the Middle Silurian reef facies have been found (see, for example, Ausich, 1987; Brower, 1975; Donovan, 1993), extending the ranges of some taxa and filling in gaps in those of others. Thus, we can raise a more general question: is the current perception of radiations and extinctions at least partly an artefact of the quality of the fossil record caused by fluctuations in sea level?

Finally, while there is no doubt that the way in which ranges (or even gaps) are recorded tends to exaggerate biotic turnover at stratigraphic boundaries (Paul, 1985), another serious question affecting the quality of the record is: just how sudden and selective were mass extinctions? Opinions differ as to how the quality of the record might affect perceptions of the 'suddenness' of significant faunal change. On the one hand it is often argued that cryptic gaps in the sedimentary record (hiatuses) may enhance the apparent rate of extinction by truncating ranges and making many more taxa seem to terminate at the same level. On the other hand the 'Signor–Lipps effect' (Signor and Lipps, 1982) has been argued to cause the 'smearing back' of last-occurrence data (LADs) from the real extinction level, thus making truly sudden events appear to have been more gradual.

Thus, this chapter has three principal aims:

1. To present some limited data on the comparative completeness of the echinoderm fossil record in the Palaeozoic to exemplify a simple method of evaluating changes in the quality of the record through time and to present a pattern with which patterns for other taxa may be compared. We do this unashamedly in the hope that it will stimulate workers on other fossil groups to undertake similar analyses.

2. To present a generalized model of the relationship between the quality of the fossil record of shallow-water organisms and changes in eustatic sea level. We also use this model to test the supposition that marine mass extinctions are an artefact of the quality of the record.

3. To present detailed data on the occurrences of fossils across marine-mass extinction levels and adjacent to known stratigraphic breaks to investigate the effects of stratigraphic completeness and the 'Signor–Lipps effect'. We use these data, together with artificial truncations of sections, to demonstrate a method of analysing such records of occurrence to reveal the underlying true patterns of range terminations. However, see Chapter 2 for alternative approaches.

GAP ANALYSIS AND THE QUALITY OF THE PALAEOZOIC RECORD OF ECHINODERMS

Paul (1980, 1982) presented some data on occurrences of gaps in the records of 18 families of cystoids at series/stage level from the Tremadoc to the Frasnian in an attempt to estimate the completeness of their fossil record. Two unanticipated, but manifestly obvious, developments (one taxonomic, the other stratigraphic) arose from this initial analysis. First, it was possible to compare the quality of the records of individual families. Some, such as the Cheirocrinidae, had extensive and continuous ranges (Tremadoc–Ashgill, no gaps). Others had significant gaps: the worst families had as much gap as record, while two important families (the Sphaeronitidae and Pleurocystitidae) were known from the Ordovician and Devonian but are still entirely unknown from the Silurian. In both families, one genus occurs in both the Upper Ordovician and the Devonian. Furthermore, both families have several autapomorphies and there is no possibility whatsoever that the Devonian forms are homeomorphs rather than true descendants of the Ordovician forms. In a later analysis (Paul, 1988), the longest single stratigraphic gap recorded in the literature was shown to be an artefact of misclassification. The Ordovician edrioasteroid genera *Cyathotheca* and *Cyathocystis* share no common character with the Devonian genera *Timeischytes* and *Hadrochthus* that they do not share with all other edrioasteroids, other than small size. All four had been classified in the family Cyathocystidae by Bell (1980), largely because their small size resulted in simplified morphology.

However, Bockelie and Paul (1983) were able to show that, whereas the Ordovician genera were closely related, the Devonian genera belonged to an entirely separate lineage and probably represented neotenous agelacrinitids. Thus, the longest gap in any family investigated by Paul (1988) proved to be artificial.

The second development, more relevant to this chapter, was that the records of different stratigraphic intervals could be compared. Even with the tiny subset of data available in 1980, it was immediately obvious that the fossil record of cystoids in the Caradoc was excellent: 15 families were known from the Caradoc and all 15 were represented by actual fossils. No Lazarus taxa of cystoids occurred in the Caradoc. Equally obvious was the appalling quality of the record of cystoids in the Llandovery: eight families could be inferred to have existed in the Lower Silurian because they were recorded from the Upper Ordovician (Caradoc or Ashgill) and from the Middle Silurian (Wenlock) or younger strata, but only one family was actually represented by fossils of Llandovery age. Seven of eight cystoid families are Lazarus taxa in the Lower Silurian. The one family that is known, the Callocystitidae, is represented by three genera. *Brockocystis* includes five possible named species (Bassler and Moodey, 1943; Broadhead and Etten-sohn, 1981), although it is doubtful that all five are valid. All five were described from very limited material, although the type species, *B. tecumseth*, is known from abundant isolated plates. *Osculocystis* is known from a unique type species (Paul, 1967b) of which less than 20 specimens are now known and they all come from the topmost Llandovery. The third genus, *Anartiocystis* (Ausich and Schumacher, 1984), is known from three specimens of the type species and possibly some disarticulated cup plates. The fossil record of cystoids in the Lower Silurian is meagre in the extreme. It is true that cystoids were never as abundant and diverse after the end of the Ordovician, but the Wenlock Limestone of Dudley (England) alone has a fauna of seven species, some known from hundreds of articulated specimens (Paul, 1967a), while Schuchert (1904) recorded 17 species from the Upper Silurian–Lower Devonian Keyser Formation of Keyser, West Virginia, again with some species represented by hundreds of specimens. Since then, Donovan (1993, p. 3, figs 2A and 3A) has extended the range of another cystoid family, the Cheiro-crinidae, into the Lower Silurian with the description of a single thecal plate of *Homocystites*? sp. However, all other subsequent research has confirmed the appalling record of Lower Silurian echinoderms.

Figure 5.1A shows the raw numbers of families and numbers of Lazarus families for the data set of Paul (1988). In this data set there are no Lazarus taxa in the Cambrian (but numbers of families are very small), the Caradoc (the main peak) and the Lower Carboniferous (Tournaisian and Visean). On the other hand, Lazarus taxa account for the entire record in the Sakmarian. Figure 5.1B shows the percentage of Lazarus taxa in this data set, together with some of the key peaks of extinctions (1–8) recognized by Sepkoski (1986). Apart from the Upper Arenig and Guadalupian (events 1 and 8 in Figure 5.1B), all the Palaeozoic mass extinctions coincide with very low percentages of Lazarus taxa and therefore with a good fossil record. Furthermore, the periods of good record are immediately followed by peaks of Lazarus taxa, suggesting that a succeeding poor fossil record contributes to the severity of the mass extinctions. The last mass extinction differs from this pattern simply because none of the families in this particular data set survived into the Guadalupian. The difference is an artefact. Nevertheless, care should be taken in making inferences from Figure 5.1B. The data set is small, it only involves selected families of echinoderms, and need not be typical of Palaeozoic echinoderms as a whole, let alone the entire Palaeozoic fossil record. In addition, the later peaks of Lazarus taxa are based on relatively small numbers of families. However, the pattern is intriguing and it would be very interesting to see if gap analysis of other major Palaeozoic fossil groups confirmed or modified it.

GENERAL MODEL

The second approach is to try to develop a general model for preservation of fossils in the record over Phanerozoic time. Can we predict when the record will be good and when bad, and then test these predictions against actual data from the fossil record? Here we suggest a general model, the assumptions of which are as follows:

1. That most fossils are preserved in shallow shelf-sea deposits, reflecting the modern biogeographical distribution of organisms.
2. That such shelf-sea deposits will be extensive at times of high sea level, but immediately vulnerable to erosion and destruction during any subsequent period of low sea level. Conversely, although shelf sea deposits will be less extensive during periods of low sea level, they are more likely to be buried permanently by offshore deposits when sea level rises again (Figure 5.2).

Thus, the predictions of the model are that the record will be good when low sea level is followed by highstand, but it will be poor when high sea level is followed by lowstand. Several independent lines of evidence support this model.

Cenomanian Pulse Faunas

Jeans (1968) coined the term 'pulse faunas' for stratigraphically widely separated occurrences of long-ranging calcitic macrofossils in the Ceno-

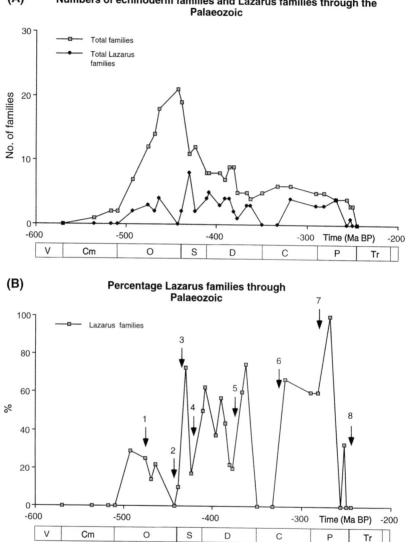

(A) **Numbers of echinoderm families and Lazarus families through the Palaeozoic**

(B) **Percentage Lazarus families through Palaeozoic**

manian of north-west Europe. He argued that they appeared at the bases of several fining-upwards sequences in the Cenomanian chalk, that is, they tend to accompany sequence boundaries. The elements of the pulse faunas are bivalves, brachiopods and belemnites. The fossils are calcitic, unlike gastropod and cephalopod faunas, so their sporadic occurrence is unlikely to be due to special requirements of preservation or to post-depositional dissolution. Solid calcitic fossils, such as the rostra of belemnites in particular, are unlikely to have been overlooked at intermediate stratigraphic levels and most unlikely to have been destroyed by diagenesis after deposition. The inescapable conclusion is that the sporadic occurrence of these fossils reflects their original occurrence. Various lines of evidence, sedimentological, geochemical and faunal (Paul *et al.*, 1994), suggest that the pulse faunas inhabited shallow water and were brought into the Cenomanian chalk basins by periodic sea-level falls. The pulse faunas represent intervals when shallow-water fossils were preserved nearer the basin centres and have subsequently been preserved by the covering of deeper water, offshore deposits. This interpretation of pulse faunas proved of considerable predictive value. The best-documented case occurs in the Plenus Marls where Jefferies (1962, 1963) recorded the lithostratigraphy and faunas in great detail. The Plenus Marls are a thin series of more siliciclastic-rich marls and marly chalk beds sandwiched between more pure chalks in the Anglo-Paris Basin. This stratigraphic level is now known to be very close to the Cenomanian/Turonian boundary, to involve a significant mass extinction event (Sepkoski, 1986) and to be accompanied by a major positive $\delta^{13}C$ excursion (see, for example, Scholle and Arthur, 1980; Schlanger *et al.*, 1987). Similar faunal changes occur in other basins, and there is no doubt that the Cenomanian/Turonian Boundary Event (CTBE) is a world-wide phenomenon. Detailed bed-by-bed collecting led Gale to recognize elements of the CTBE pulse fauna at the Lower/Middle Cenomanian boundary. He hypothesized that this occurrence might also be accompanied by a $\delta^{13}C$ excursion. A detailed programme of investigation confirmed a double-peaked positive $\delta^{13}C$ excursion, although faunal changes were found to be temporary (Paul *et al.*, 1994). Furthermore,

---

Figure 5.1 (opposite) *(A) Numbers of echinoderm families and of Lazarus families and (B) percentage Lazarus families in the data set of Paul (1988) through the Palaeozoic. Arrows indicate positions of peaks of extinction identified by Sepkoski (1986), as follows: (1) Upper Arenig (Lower Ordovician); (2) Middle Caradoc and (3) Upper Ashgill (both Upper Ordovician); (4) Wenlock (Middle Silurian); (5) Givetian (Middle Devonian); (6) Upper Visean (Lower Carboniferous); (7) Stephanian (Upper Carboniferous); (8) Guadalupian (Upper Permian). Note that many mass extinctions precede intervals with a high percentage of Lazarus taxa and therefore a poor fossil record. The last mass extinction does not appear to precede an interval with a poor record only because families in this data set did not extend into the Guadalupian. V, Vendian; Cm, Cambrian; O, Ordovician; S, Silurian; D, Devonian; C, Carboniferous; P, Permian; Tr, Triassic*

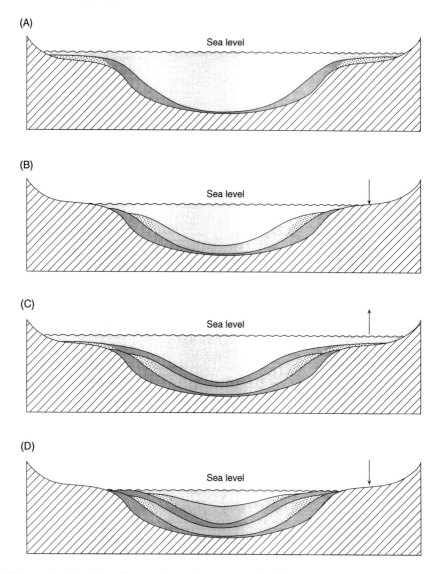

Figure 5.2   *Model to illustrate how during periods of low sea level (B and D) deposits formed on shallow shelves during high sea level (stippled in A and C) tend to be eroded away and destroyed. However, shallow-shelf deposits formed during periods of lowstand (stippled in B and D) tend to be preserved by overlying highstand deposits. Thus, the general model predicts that periods of high sea level will have a poor fossil record of shallow-water faunas, whereas periods of low sea level will have a better record of shallow-water faunas. Arrows indicate change in sea level from previous diagram*

an extended programme of faunal and geochemical investigation led Mitchell and Paul (1994) and Mitchell *et al.* (1996) to conclude that in the mid-Cretaceous not only does $\delta^{13}C$ track background eustatic sea level, but also significant $\delta^{13}C$ excursions accompany rapid changes in sea level. Thus, the pulse fauna model, originally developed for the Cenomanian, seems to have broader application and strongly suggests that both absolute sea level and fluctuations in sea level affect the quality of the fossil record in general.

The Ordovician/Silurian Boundary

We have already mentioned that significant facies changes across the Ordovician/Silurian boundary make it very difficult to compare faunas above and below the boundary. In many European successions, Upper Ordovician, shallow-water, shelly faunas are rapidly followed by Lower Silurian, deep-water, graptolitic facies (see, for example, Brenchley, 1988). Not surprisingly, many of the common Upper Ordovician fossils are unknown from the succeeding Lower Silurian. A number of Upper Ordovician taxa reappear in the Middle Silurian and so the faunal change across the Ordovician/Silurian boundary appears to be exaggerated by the facies change. This, in turn, has consequences for the severity, or even the reality, of the late Ordovician (Hirnantian) mass extinction. If the Lower Silurian fossil record is indeed very poor, how can we be sure all taxa that are last known from the Hirnantian really became extinct then and not in the succeeding Llandovery? The shallow-water shelf deposits that could have contained the last survivors of these taxa were presumably deposited high on the margins of Lower Silurian basins and have since been lost by subsequent erosion. Thus, the situation across the Ordovician/Silurian boundary also supports the idea that sea-level change affects the quality of the fossil record.

Stratigraphic Modelling

Holland (1995) modelled the stratigraphic distribution of fossils in a series of increasingly more realistic steps. Initially, Holland assumed that if a taxon existed it would have been preserved. He then ran a series of computer simulations in which taxa could originate at any time from 20 steps before a stratigraphic succession 80 horizons thick, and in which extinction could occur at any time from the base of the succession to 20 steps beyond the top. This model established stratigraphic ranges for 50 taxa that were exactly equivalent to their true periods of existence. Holland then added increasingly realistic modifications to the model by simulating the effects of incomplete sampling, facies control (largely related to changes in sea level) and possible taphonomic gradients.

Sampling effects were simulated by assuming different levels of sampling (for example, 10%, 50% and 90%), and randomly selecting stratigraphic levels where fossils would and would not be found. This produced more realistic, patchy stratigraphic records for all taxa, in which the known range is often considerably shorter than the total period of existence. Facies control was modelled first by assuming all species had a preferred depth, a depth tolerance (or depth range) and a level of peak abundance. These were defined differently for each taxon, with the further assumption of a Gaussian (normal) distribution. Once the depth distributions of the fossils had been established, variations in water depth were simulated. The model simulated two equal cycles of sea-level change in which water depth decreased gradually and linearly over 40 stratigraphic levels, but increased instantaneously. Then a sequence stratigraphic model was superimposed on the sea-level cycles. This last simulation then compared the numbers of first (FAD) and last (LAD) occurrences per stratigraphic level with those expected from random sampling. Statistically significant concentrations of FADs and LADs occur at particular levels in the depositional sequences. Spikes of FADs occur immediately above the sequence boundaries and the flooding surfaces of the transgressive systems tracts. Spikes of LADs occur immediately below the sequence boundaries and the flooding surfaces. Repeated simulations showed that tops and bases of preserved stratigraphic ranges tend to be concentrated at sequence boundaries. The final modification to the model was to add a taphonomic gradient to the chances of collection to simulate preferential preservation either onshore or offshore. Holland did this by multiplying the chance of collection by a taphonomic factor that ranged either from 1.5 at 0 m depth to 0.5 at 65 m depth (preferential onshore preservation) or from 0.5 at 0 m depth to 1.5 at 65 m depth (preferential offshore preservation).

To test the model, Holland compared it against actual fossil occurrences in the Upper Ordovician of south-east Indiana using data from Cumings and Galloway (1913). While there were some peaks of FADs and LADs at unexpected levels in Cumings and Galloway's data, several peaks did occur at sequence boundaries as predicted by Holland's model. Some details of the model Holland used are obviously open to different interpretation and/or modification; nevertheless, the general conclusion that sea-level fluctuations affect preserved stratigraphic ranges of fossils seems reasonably well established.

The very general model proposed here, which relates the quality of the fossil record to changes in eustatic sea level and to fluctuations in sea level, has considerable predictive potential. At its simplest, high sea level should equate with a poor-quality fossil record and low sea level with a good-quality record. Thus, any sea-level curve can be used to predict periods of good and bad record. Truncation of the Haq *et al.* (1988) relative coastal onlap curve at 0.75 (for example) suggests a poor record for shallow marine faunas in the Kimmeridgian, the entire Upper Cretaceous, the Paleocene and the Lower Eocene. Truncation of the long-term sea-level curve at +150 m also suggests a poor record in the Lower Cretaceous and mid-Cretaceous to late Eocene (Figure 5.3). Examination of the short-term curve provides further details of periods when lowstand is followed by highstand (and vice versa). The model proposed here suggests the former will result in a poor record and the latter in a good record, at least for shallow shelf faunas. On this more restricted set of circumstances, relatively poor records should be frequent events, but one wonders how global they will have been given the relatively short duration of many sea-level falls.

Although this model has obvious limitations, it does raise the possibility of testing whether or not mass extinctions are artefacts of, or seriously influenced by, the quality of the fossil record as suggested by gap analysis of Palaeozoic echinoderm families. Sepkoski (1986, fig. 3) recognized as many as 29 periods of above-average rate of extinction of marine fossil families through the Phanerozoic. If these were to correlate well with major transgressions (that is, lowstand followed by highstand), one would seriously suspect that the quality of the record contributed to the apparent concentration of range terminations. Comparisons are not all that easy since the Haq *et al.* (1988) curve is related to a different geological time scale from the conventional geological column. Thus, comparisons can only be made using radiometric age estimates with all their uncertainties. However, assuming that radiometric ages do not introduce a systematic bias, the two data sets can be compared for the Mesozoic and Tertiary. In this period Sepkoski (1986) identified 10 major extinctions, and these show no consistent relationship with the long-term sea-level curve (Figure 5.3). Relative stability of sea level does not seem to correlate with mass extinctions either. Some periods of significant, rapid changes in sea level do occur immediately prior to a mass extinction event such as the Cenomanian/Turonian boundary mass extinction, but there is no consistent relationship. Altogether, we can detect no regular relationship between any feature of the Haq *et al.* (1988) sea-level curve and mass extinctions as recognized by Sepkoski (1986). This suggests that the mass extinction events are real, even if susceptible to modification by the quality of the fossil record.

TESTING FOR THE SIGNOR–LIPPS EFFECT

Signor and Lipps (1982) suggested that a combination of incomplete preservation and incomplete sampling might cause recorded last occurrences to be

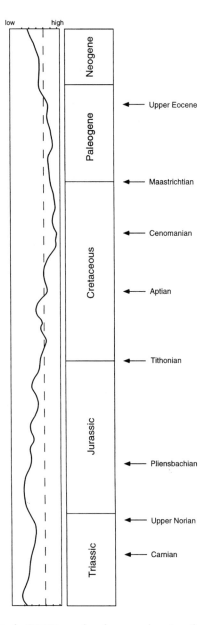

Figure 5.3   *The Haq* et al. *(1988) sea-level curve showing the stratigraphic levels at which a poor fossil record might be expected under the simple model. The dashed line represents sea level at 150 m above present-day sea level. Assuming this as a cut-off level, a poor record might be expected in the Lower and Upper Cretaceous and Lower to Middle Paleogene. Positions of peaks of extinctions recognized by Sepkoski (1986) are indicated by arrows. Note there is no obvious correlation between sea level and occurrence of Mesozoic and Tertiary mass extinctions*

'smeared back' from a genuinely sudden mass extinction event, producing an apparent pattern of gradual extinctions. This has since become known as the 'Signor–Lipps' effect. In Chapter 2, Marshall has presented a thorough analysis of how one might test for the Signor–Lipps effect and distinguish between a truly sudden event with many coeval extinctions and a more gradual event with sequential extinctions. It is clearly also important to distinguish between a stepwise pattern of mass extinctions, with several sudden events, and a genuinely gradual reduction in diversity. The difference between sudden and gradual patterns is clearly relevant to possible causes of mass extinctions. For example, an impact model for the Cretaceous/ Tertiary boundary mass extinction predicts a very sudden event and requires extinctions concentrated in a very short time interval compared with normal geological processes. However, a model related to the Deccan traps, which erupted over a period of about 500 000 years, merely requires a concentration of extinctions adjacent to the Cretaceous/Tertiary boundary. Most of the methods discussed by Marshall involve the assumption of random preservation and collection. Marshall specifically points out that these methods are not generally applicable to microfossil data because the data points (samples) are rarely randomly distributed through the sections. Any gaps are controlled by the sampling programme at least as much as by the actual occurrences of fossil taxa. Here we want to present two examples that illustrate (1) how to test for the Signor–Lipps effect in microfossil data and (2) how to compensate for the Signor–Lipps effect in randomly collected macrofossil data. In both cases analysis is only possible if raw data on actual occurrences of fossils in sections are published. Recording total ranges renders such analyses impossible.

Microfossils from the Cenomanian/Turonian Boundary Event (CTBE)

Vaziri (1997) has presented data on microfossil (foraminifera and ostracods) occurrences across the CTBE in three sections in England. We use his data for benthic foraminifera at Eastbourne – the largest data set across the thickest succession. Diversity decreased significantly, with 52 taxa known from before the onset of the event, but only 17 afterwards, including four species that appear or reappear in the top two samples (Figure 5.4). Diversity appears to decrease in two major steps, through bed 1 of the Plenus Marls, and through beds 4–9 (Figure 5.4). Legitimate questions are: 'How real is this pattern of last occurrences?' and 'Is the pattern an artefact of the incompleteness of the fossil record?' The reduction in diversity could have been gradual and continuous, as the pattern of last occurrences through the thinner succession at Westbury, Wiltshire, suggests (Vaziri, 1997). Alternatively, it could have

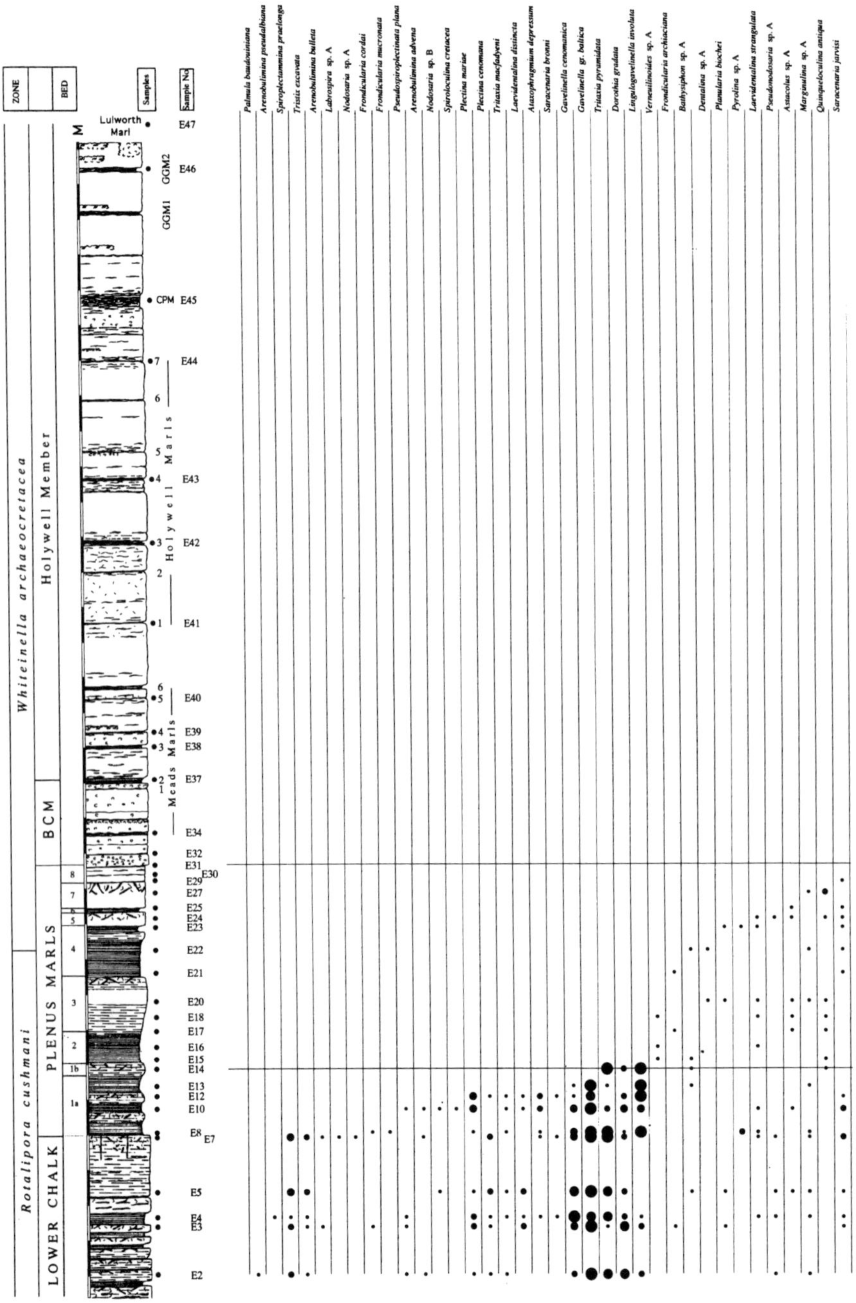
ZONE
BED
Samples
Sample No.
Whiteinella archaeocretacea
Rotalipora cushmani
Holywell Member
BCM
PLENUS MARLS
LOWER CHALK
Lulworth Marl
GGM2
GGM1
CPM
Holywell Marls
Meads Marls
E47
E46
E45
E44
E43
E42
E41
E40
E39
E38
E37
E34
E32
E31
E30
E29
E27
E25
E24
E23
E22
E21
E20
E18
E17
E16
E15
E14
E13
E12
E10
E8
E7
E5
E4
E3
E2
Palmula bouduiniana
Arenobulimina pseudalbiana
Spiroplectammina praelonga
Tristix excavata
Arenobulimina bulleta
Labrospira sp. A
Nodosaria sp. A
Frondicularia cordai
Frondicularia mucronata
Pseudospiroplectinata plana
Arenobulimina advena
Nodosaria sp. B
Spiroloculina cretacea
Plectina mariae
Plectina cenomana
Tritaxia macfadyeni
Laevidentalina distincta
Ataxophragmium depressum
Saracenaria bronni
Gavelinella cenomanica
Gavelinella gr. baltica
Tritaxia pyramidata
Dorothia gradata
Lingulogavelinella involuta
Verneuilinoides sp. A
Frondicularia archiaciana
Bathysiphon sp. A
Dentalina sp. A
Planularia biochei
Pyrolina sp. A
Laevidentalina strangulata
Pseudonodosaria sp. A
Astacolus sp. A
Marginulina sp. A
Quinqueloculina antiqua
Saracenaria jarvisi

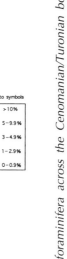

Figure 5.4 *Occurrences of benthic foraminifera across the Cenomanian/Turonian boundary at Eastbourne, southern England, arranged in order of last occurrence. Note the marked decrease in diversity through the Plenus Marls, which takes place in two steps. The pattern of last occurrences, whereby rare and patchy species apparently go first whereas common and consistent species go last, is exactly what one might expect if the Signor–Lipps effect were 'smearing back' last occurrences from two rapid extinction events. This suggests that the Cenomanian/Turonian boundary mass extinction took place in two sudden steps, indicated by the horizontal lines. The first event was at the top of bed 1 of the Plenus Marls and the second at the top of bed 8. (Data from Vaziri, 1997)*

been sudden, with one instantaneous extinction event, the steps resulting from incomplete preservation and/or sampling. Finally, the pattern of extinction could genuinely have been stepped, with more than one significant and sudden event. In particular, can we detect the Signor–Lipps effect and, if so, by how much has it altered the pattern of last occurrences?

In Figure 5.4, benthic foraminiferal taxa are arranged in order of last occurrences and the size of the dots reflects their relative abundance within a sample, larger dots indicating more abundant taxa. Two features are immediately apparent.

First, several of the more abundant species are consistently recorded, occurring in every sample or almost every sample throughout their entire stratigraphic range (for example, *Gavelinella reussi, Marssonella trochus* var. *turris* and *Lingulogavelinella globosa*). Presumably their ranges are well established. Other taxa occur only in one or a few samples and generally as minor constituents of the fauna. One assumes their ranges are less well established.

Secondly, as one approaches the putative extinction events, it is noticeable that the rarer species with more patchy occurrences go first, whereas the commoner and more consistently occurring species go last. This is exactly the pattern of last occurrences that one would expect if the Signor–Lipps effect were 'smearing back' ends of ranges of taxa from a genuinely sudden extinction event. The chances of detecting a rare, patchy taxon in the last sample below a real extinction event would be small, whereas it would be highly likely that a common, consistent species would be found in the last sample before its extinction. Thus, we argue here that the last occurrences of *Tritaxia pyramidata, Dorothia gradata, Lingulogavelinella involuta* and *Gavelinella baltica* define the first extinction event and that the other 10 species with last recorded occurrences in bed 1 of the Plenus Marls probably became extinct at the same time. Similarly, we argue that the last occurrences of *Eggerellina mariae, Laevidentalina gracilis* and *E. brevis* (together with the drastic reduction in abundances of *Lenticulina* spp., *Textularia chapmani* and *Textularia* sp. A) define the second extinction event, and that the other 24 species with last recorded occurrences in beds 4–9 also probably became extinct at this time. Such a pattern of last occurrences can be exploited to test for the Signor–Lipps effect. In a genuinely sudden extinction there should be a good correlation between order of last occurrences and consistency of occurrence, and between order of last occurrence and relative abundance. In a more gradual extinction event there is no reason to suppose that the order of extinction should correlate with abundance or consistent occurrence. Hence, any correlation should be weak. We presume that some statistical technique could be used to test the significance of any relationship between order of last occurrence and abundance, or between order of last occurrence and consistency of occurrence.

In our second example, Lee (1988) presented data on occurrences of macro-

Last occurrences per 10 m

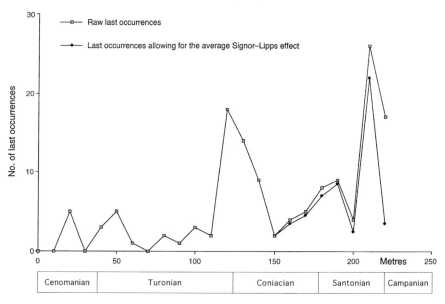

Figure 5.5 *Numbers of last occurrences of macrofossils per 10 m interval in the chalk of the North Downs, southern England. The lower curve from 150 m upwards is a modification of the raw data to allow for the average Signor–Lipps effect due to truncation of the section by erosion at 220 m. Note that this dramatically reduces the number of last occurrences in the top 10 m (210–220 m), but has only a slight effect on the peak of last occurrences between 200 and 210 m. We conclude that this peak of last occurrences is real and that there was a significant faunal turnover at the high Santonian. (Data from Lee, 1988)*

fossils through the chalk succession (late Cenomanian to basal Campanian) of the North Downs, England. Here, sampling interval was purely random, but first and last occurrences of taxa have been grouped into 10 m intervals, analogous to the standard stratigraphic intervals used to detect Sepkoski's (1986) mass extinction events, and absolute numbers of first (FAD) and last (LAD) occurrences plotted on a standard scale (Figure 5.5). The largest peak of LADs occurs in the penultimate 10 m interval. Does this represent a real peak in faunal turnover or is it an artefact of the erosional truncation at the top of the succession? Incidentally, the top of this succession is the local Cretaceous/Tertiary boundary with virtually all the Campanian, all the Maastrichtian and some of the Danian missing. Thus, it is possible to investigate the effects of stratigraphic gaps on patterns of last occurrence as well. To do so, the section was artificially truncated at 10 m intervals from 210 m down to 170 m to investigate the effects of erosive truncation. After each truncation the pattern of last occurrences was recalculated using the highest

Table 5.1   Mean number of additional terminations of ranges due to five arbitrary truncations of the chalk section in the North Downs, England (data from Lee, 1988)

Level below truncation (m)	Mean additional terminations
0–10	14.4
10–20	3.8
20–30	1.4
30–40	0.4
40–50	0.4
50–60	0.8
60–70	0.4
70–80	0.4
80–90	0.2

occurrences below the level of truncation (Table 5.1). Then, by adding the new last occurrences created by each truncation to those already recorded for each 10 m interval, we can estimate the average increase in last occurrences per 10 m interval caused by truncation. This shows, not surprisingly, that the maximum effect of truncation is always in the ultimate 10 m interval below the truncation level, and that the effect decreases away from the truncation level. However, in one case the effect was translated back through nine 10 m intervals. Using the table of mean additional last occurrences caused by truncation (Table 5.1), we can estimate what proportion of the original last occurrences recorded from between 200 and 210 m was due to truncation of the section at 220 m. This shows that roughly four of the 26 last occurrences recorded between 200 and 210 m were probably caused by the truncation, and that the extinction event is almost certainly real. In this example, not only can we detect the Signor–Lipps effect, but we can quantify it for any given level below the erosion surface at the top of the local section.

SUMMARY

We present an analysis of percentage Lazarus taxa as a measure of completeness of the fossil record for a selected set of Palaeozoic echinoderm families. This suggests that the Lower Silurian has the worst record, and the Caradoc and Lower Carboniferous have the best records. We invite others to test these predictions by undertaking similar gap analysis for other major taxa.

We also present a generalized model that the fossil record of 'shelly' shelf faunas will be worst when sea-level highs follow periods of lowstand, because shelf sediments are likely to be reworked whenever sea level falls. Assuming this simple model is relevent, the Haq *et al.* (1988) sea-level curve for the Mesozoic and Tertiary suggest that poor records occur in the Kimmeridgian, the Upper Cretaceous, the Paleocene and the Lower Eocene. Again, these predictions can be tested against the known fossil record by gap analysis. This model further suggests that peaks of marine extinctions are not artificial and do not result from poor records at the levels of the mass extinction events.

Finally, we present methods to recognize and estimate quantitatively the Signor–Lipps effect both for microfossils, where gaps are largely determined by sampling, and for macrofossils, occurrences of which can be randomly distributed. These suggest that the Cenomanian/Turonian boundary event was a two-stage process with sudden extinction events at the tops of bed 1 and bed 8 of the Plenus Marls. Rhythmic bedding in the Plenus Marls suggests that these events were separated by approximately 80 000 years. Repeated truncation of the Upper Cretaceous section in the North Downs, England, produces an average of 14.4, 3.8 and 1.4 additional last occurrences due to the truncation in the ultimate, penultimate and antepenultimate 10 m intervals, respectively. The Signor–Lipps effect may extend back 90 m or more, but is not responsible for the peak of last occurrences between 200 and 210 m in the chalk of the North Downs.

ACKNOWLEDGEMENTS

We are extremely grateful to C.P. Lee and R. Vaziri for permission to use their unpublished data on fossil occurrences in the Upper Cretaceous of England.

REFERENCES

Ausich, W.I., 1987, Brassfield Compsocrinina (Lower Silurian crinoids) from Ohio, *Journal of Paleontology*, **61**: 552–562.
Ausich, W.I. and Schumacher, G.A., 1984, New Lower Silurian rhombiferan cystoid (Echinodermata, Callocystitidae) from southwestern Ohio, *Journal of Paleontology*, **58**: 9–15.
Bassler, R.S. and Moodey, M.W., 1943, Bibliographic and faunal index of Palaeozoic pelmatozoan echinoderms, *Geological Society of America Special Paper*, **45**: 733 pp.
Bell, B.M., 1980, Edrioasteroidea and Edrioblastoidea. *In* T.W. Broadhead and J.A. Waters (eds), *Echinoderms: Notes for a Short Course, University of Tennessee, Studies in Geology*, **3**: 158–174.
Benton, M.J. (ed.), 1993, *The Fossil Record, 2*, Chapman and Hall, London: 845 pp.
Bockelie, J.F. and Paul, C.R.C., 1983, *Cyathotheca suecica* and its bearing on the evolution of the Edrioasteroidea, *Lethaia*, **16**: 257–264.
Brenchley, P.J., 1988, Environmental changes close to the Ordovician–Silurian bound-

ary. *In* L.R.M. Cocks and R.B. Rickards (eds), *A Global Analysis of the Ordovician-Silurian Boundary, Bulletin of the British Museum, Natural History, (Geology),* **43**: 377–385.

Broadhead, T.W. and Ettensohn, F.R., 1981, A new *Brockocystis* (Echinodermata: Rhombifera) from the Lower Silurian (Llandovery) of Kentucky, *Journal of Paleontology,* **55**: 773–779.

Brower, J.C., 1975, Silurian crinoids from the Pentland Hills, Scotland, *Palaeontology,* **18**: 631–656.

Cumings, E.R. and Galloway, J.J., 1913, The stratigraphy and paleontology of the Tanners Creek section of the Cincinnati Series in Indiana, *Indiana Department of Geology and Natural Resources Annual Report,* **37**: 353–479.

Donovan, S.K., 1993, A Rhuddanian (Silurian, Lower Llandovery) pelmatozoan fauna from south-west Wales, *Geological Journal,* **28**: 1–19.

Donovan, S.K., 1994, The late Ordovician extinction of the crinoids in Britain, *National Geographic Research and Exploration,* **10**: 72–79.

Haq, B.U., Hardenbol, J. and Vail, P.R., 1988, Mesozoic and Cenozoic chronostratigraphy and cycles of sea-level change. *In* C.K. Wilgus *et al.* (eds), *Sea Level Changes – An Integrated Approach, SEPM Special Publication,* **42**: 71–108.

Harland, W.B., Holland, C.H., House, M.R., Hughes, N.F., Reynolds, A.B., Rudwick, M.J.S., Satterthwaite, G.E., Tarlo, L.B.H. and Willey, E.C. (eds), 1967, *The Fossil Record: A Symposium with Documentation,* Geological Society, London: 827 pp.

Holland, S.M., 1995, The stratigraphic distribution of fossils, *Paleobiology,* **21**: 92–109.

Jeans, C.V., 1968, The origin of the montmorillonite of the European chalk with special reference to the Lower Chalk of England, *Clay Minerals,* **7**: 311–329.

Jefferies, R.P.S., 1962, The palaeoecology of the *Actinocamax plenus* Subzone (lowest Turonian) in the Anglo-Paris Basin, *Palaeontology,* **4**: 609–647.

Jefferies, R.P.S., 1963, The stratigraphy of the *Actinocamax plenus* Subzone (Turonian) in the Anglo-Paris Basin, *Proceedings of the Geologists' Association,* **74**: 1–33.

Lee, C.P., 1988, *Macrobiostratigraphy of the Chalk of the North Downs, Southern England,* Unpublished Ph.D. thesis, University of Liverpool: 286 pp.

Mitchell, S.F. and Paul, C.R.C., 1994, Carbon isotopes and sequence stratigraphy. *In* S.D. Johnson (ed.), *High Resolution Sequence Stratigraphy: Innovations and Implications, Abstract Volume,* University of Liverpool: 20–23.

Mitchell, S.F., Paul, C.R.C. and Gale, A.S., 1996, Carbon isotopes and sequence stratigraphy. *In* J.A. Howell and J.F. Aitken (eds), *High Resolution Sequence Stratigraphy: Innovations and Applications, Geological Society Special Publication,* **104**: 11–24.

Paul, C.R.C., 1967a, The British Silurian cystoids, *Bulletin of the British Museum, Natural History (Geology),* **13**: 297–356.

Paul, C.R.C., 1967b, *Osculocystis,* a new British Silurian cystoid, *Geological Magazine,* **104**: 449–454.

Paul, C.R.C., 1980, *The Natural History of Fossils,* Weidenfeld, London: 292 pp.

Paul, C.R.C., 1982, The adequacy of the fossil record. *In* K.A. Joysey and A.E. Friday (eds), *Problems of Phylogenetic Reconstruction, Systematic Association Special Volume 21,* Academic Press, London: 75–117.

Paul, C.R.C., 1985, The adequacy of the fossil record reconsidered. *In* J.C.W. Cope and P.W. Skelton (eds), *Evolutionary Case Histories from the Fossil Record, Special Papers in Palaeontology,* **33**: 7–16.

Paul, C.R.C., 1988, Extinction and survival in the echinoderms. *In* G.P. Larwood (ed.), *Extinction and Survival in the Fossil Record, Systematic Association Special Volume 34,* Clarendon Press, Oxford: 155–170.

Paul, C.R.C., Mitchell, S.F., Marshall, J.D., Leary, P.N., Gale, A.S., Duane, A.M. and

Ditchfield, P.W., 1994, Palaeoceanographic events in the middle Cenomanian of northwest Europe, *Cretaceous Research*, **15**: 707–738.

Schlanger, S.O., Arthur, M.A., Jenkyns, H.C. and Scholle, P.A., 1987, The Cenomanian–Turonian oceanic anoxic event: I. Stratigraphy and distribution of organic carbon-rich beds and the marine $\delta^{13}C$ excursion. *In* J. Brooks and A.J. Fleet (eds), *Marine Petroleum Source Rocks, Geological Society Special Publication*, **26**: 371–399.

Scholle, P.A. and Arthur, M.A., 1980, Carbon isotope fluctuations in Cretaceous pelagic limestones: potential stratigraphic and petroleum exploration tool, *American Association of Petroleum Geologists Bulletin*, **64**: 67–87.

Schuchert, C., 1904, On Siluric and Devonic Cystidea and *Camarocrinus*, *Smithsonian Miscellaneous Collections*, **47**: 201–272.

Sepkoski, J.J., Jr, 1982, A compendium of fossil marine families, *Milwaukee Public Museum Contributions to Biology and Geology*, **51**: 125 pp.

Sepkoski, J.J., Jr, 1986, Phanerozoic overview of mass extinctions. *In* D.M. Raup and D. Jablonski (eds), *Patterns and Processes in the History of Life, Dahlem Konferenzen: Life Science Research Report*, **36**: 277–295.

Signor, P.W. and Lipps, J.H., 1982, Sampling bias, gradual extinction patterns and catastrophes in the fossil record. *In* H.T. Silver and P.H. Schultz (eds), *Geological Implications of Impacts of Large Asteroids and Comets on the Earth, Geological Society of America Special Paper*, **190**: 291–296.

Smith, A.B., 1988, Patterns of diversification and extinction in early Palaeozoic echinoderms, *Palaeontology*, **31**: 799–828.

Vaziri, M.R., 1997, *Microfaunal changes across the Cenomanian–Turonian Boundary in England*, Unpublished Ph.D. thesis, University of Liverpool: 295 pp.

6
Fossil Soils and Completeness of the Rock and Fossil Records

Gregory J. Retallack

INTRODUCTION

> Forests may be as dense and lofty as those of Brazil, and may swarm with quadrupeds, birds and insects, yet at the end of thousands of years one layer of black mould a few inches thick may be the sole representative of those myriads of trees, leaves, flowers and fruits, those innumerable bones and skeletons of birds, quadrupeds, and reptiles . . .　　　　　　　　(Lyell, 1877, p. 21)

The contrast between the exuberance of life on land and the paucity of its fossil record is striking, and Lyell's pessimism concerning the completeness of the fossil record remains justified, as we shall see. However, there is usually more than just 'black mould' remaining from where forests once grew. There are remains of the soils in which they were rooted and drew sustenance. Paleosols can be regarded as trace fossils of ecosystems, just as footprints in sandstone are trace fossils of a passing dinosaur. Paleosols are the result of a variety of processes such as weathering, root penetration and burrowing acting over the time during which the ecosystem thrived. Viewed in this way, paleosols are records of past environments independent of associated fossils and sedimentary rocks. The fossil record of soils now allows estimates of the completeness of both the rock and fossil records (Retallack, 1984, 1986).

For example, a paleosol with a subsurface horizon enriched in clay skins and with stout woody root traces is an indication of trees and other forest biota, even if leaves and trunks of the trees are not preserved. In some cases fossil trees and leaf litters are found in paleosols (Retallack, 1976; Gastaldo,

The Adequacy of the Fossil Record. Edited by S.K. Donovan and C.R.C. Paul.
© 1998 John Wiley & Sons Ltd.

1986; DiMichele and DeMaris, 1987; Retallack and Dilcher, 1988; Taylor *et al.*, 1992; Taylor and Taylor, 1993), but the paucity of such fossiliferous paleosols is a measure of the incompleteness of the fossil record. These formerly forested paleosols also represent times of non-deposition or gaps in the sedimentary record. The duration of that hiatus in sedimentation can be estimated to an order of magnitude from the degree of the development of paleosols by comparison with well-dated Quaternary soils (Leeder, 1975; Retallack, 1986; Bown and Kraus, 1987, 1993; Bown and Larriestra, 1990; Kraus and Bown, 1993a). These gaps in sedimentation are a measure of the incompleteness of the sedimentary record.

Some years ago I addressed such issues of completeness of the rock and fossil record using Eocene and Oligocene paleosols in Badlands National Park, South Dakota (Retallack, 1984, 1986). Subsequent studies (Towe, 1987; Bown and Beard, 1990) and my own reports of more than a thousand Mesozoic and Cenozoic paleosols (Retallack, 1975, 1976, 1977, 1979, 1983a,b, 1984, 1986, 1994a; Retallack and Dilcher, 1981a,b; Retallack and Ryburn, 1982; Retallack *et al.*, 1995, 1996, 1997, 1998) emphasized again pH, Eh and time for development of paleosols as important constraints in preservation of fossils in paleosols. Further studies have also showed how paleosols can be used to find the big gaps in non-marine sedimentary successions, because they can be guides to sequence stratigraphic architecture (Kraus, 1992; Wright, 1992; Wright and Marriott, 1993).

COMPLETENESS OF THE FOSSIL RECORD

After an organism dies, its remains must survive a variety of destructive processes in order to be preserved as part of the fossil record. The case of an antelope dying on an East African savannah is an appropriate model for preservation of fossils in paleosols (Shipman, 1981). Initially, predators and scavengers of all kinds, ranging from lions and hyenas to blowflies and carrion beetles to fungi and bacteria, devour the flesh and disarticulate the skeleton. Bones are destroyed by trampling, and by cracking and splitting during weathering at the surface. As the bones are overgrown by plants, covered by dust, fall down cracks and burrows or are trampled into the soil, they are destroyed by a new set of processes related to soil formation. Fossil assemblages in paleosols are further modified during late diagenesis, exposure and collecting. This is a complex chain of events and the prospects of reconstructing each aspect of it are daunting.

Paleosols come to the rescue by offering information on three important taphonomic variables: Eh, pH (Figure 6.1) and time of soil formation (Figure 6.2). The redox status of paleosols can be determined from colour and mineral composition. Chemically oxidized paleosols are reddish-coloured

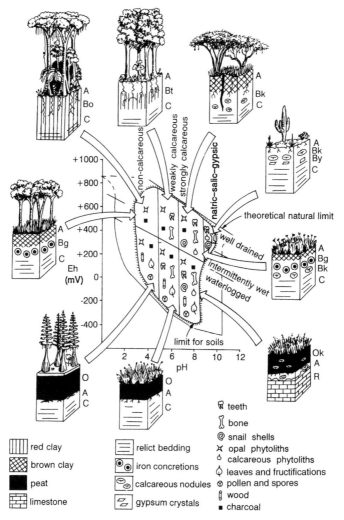

Figure 6.1   *Theoretical Eh–pH stability fields for common kinds of terrestrial fossils preserved in fossil soils and representative kinds of soils. Soil horizon shorthand is from Soil Survey Staff (1990). Scales on the soil columns are about 2 m and for forest vegetation about 10 m. (After Retallack, 1984)*

with ferric oxide minerals, and also have deeply penetrating root traces and burrows. Chemically reducing paleosols, on the other hand, are drab-coloured and have minerals such as siderite, as well as coal layers, aquatic fossils and tabular root systems (Retallack, 1976, 1990; Fastovsky and McSweeney, 1987; Besly and Turner, 1989; Blodgett *et al.*, 1993; PiPujol and Buurman, 1994).

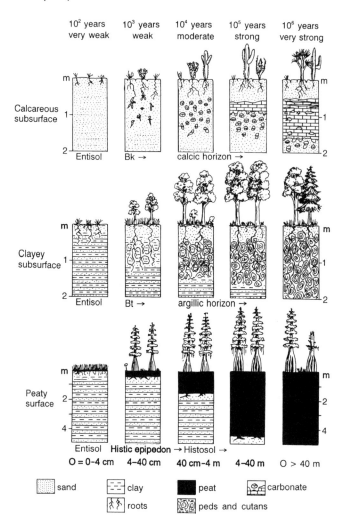

Figure 6.2 *Soil features used for estimating degree of development of paleosols. The scale for cartoons of ecosystem type differs from those on the soil columns. (Data from Retallack, 1990, 1991)*

Similarly, pH can be estimated from the mineral and chemical composition of paleosols. Unlike acidic paleosols, alkaline paleosols are calcareous and enriched in alkali and alkaline earth elements (Retallack, 1976, 1990; Percival, 1986; Hughes *et al.*, 1992). Surface environments can be geochemically characterized within a grid of Eh and pH (Krumbein and Garrels, 1952; Baas-Becking *et al.*, 1960), and so can the stability of different kinds of fossils in soils and paleosols (Figure 6.1).

Finally, time for development of paleosols can be estimated from the degree of expression of pedogenic structures compared with original structures such as sedimentary bedding (Bown and Kraus, 1987, 1993; Retallack, 1990, 1991; Kraus and Bown, 1993a). There are recognizable stages in the degree of subsurface enrichment in clay or carbonate and in the surface accumulation of peat that correspond to broad classes of development for humid, arid and waterlogged environments, respectively (Figure 6.2). The point here is that Eh, pH and time for development can be estimated from paleosols independent of their fossils. The role of these variables in fossil preservation in paleosols can thus be assessed from field studies of paleosols and the fossils in them.

A New Compilation

Much of the initial inspiration for this work came from a study of 87 paleosols of Eocene and Oligocene age in Badlands National Park, South Dakota (Retallack, 1984, 1986). Even at that time it was apparent that preservational preferences were statistical or probabilistic, rather than absolute, so that a larger data set was desirable. During the past decade I have been able to study several thousand paleosols of all geological ages. Those of early Palaeozoic and Precambrian age are not considered here because not all the different kinds of potential terrestrial fossils had evolved by then. Published reports of late Palaeozoic, Mesozoic and Cenozoic profiles now include 1478 paleosols, all studied within long stratigraphic sections using uniform field protocol (for example, Figure 6.3; Retallack, 1988a). The paleosol descriptions have been published elsewhere. In order of geological age these include late Permian and early Triassic paleosols of the Sydney Basin, Australia (Retallack, 1998), of southern Victoria Land, Antarctica (Retallack *et al.*, 1997) and at Graphite Peak, Antarctica (Retallack *et al.*, 1998); early Triassic paleosols near Sydney, Australia (Retallack, 1975, 1976, 1977); middle Triassic paleosols of New Zealand (Retallack, 1979, 1983a; Retallack and Ryburn, 1982); mid-Cretaceous paleosols of Kansas (Retallack and Dilcher, 1981a,b); late Cretaceous and Paleocene paleosols of Montana (Retallack, 1994a); early Eocene paleosols of Wyoming (Figure 6.3); late Eocene and early Oligocene paleosols of the Clarno and Painted Hills areas of Oregon (Bestland and Retallack, 1994a,b) and of Badlands National Park, South Dakota (Retallack, 1983b, 1984, 1986); and Miocene paleosols of Pakistan and Kenya (Retallack, 1991; Retallack *et al.*, 1995). These paleosols represent a variety of different kinds of biological and physical environments, including humid and arid palaeoclimates; swamps and deserts; high and low latitudes; forests and grassland; therapsid, dinosaur and mammal communities; uplands and lowlands; volcaniclastic and

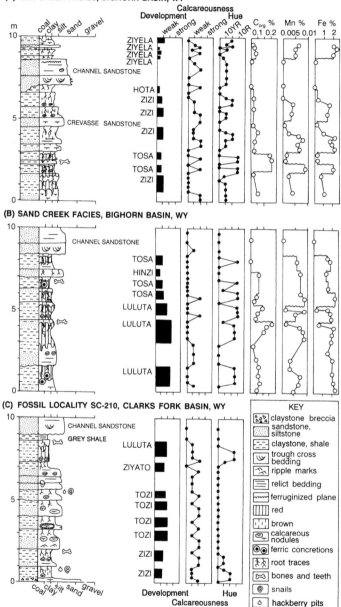

quartzofeldspathic parent materials; and young to ancient land surfaces.

These data are more representative than a pilot study (Retallack, 1984), but there is still room for improvement. The present data are biased toward paleosols of neutral Eh, acidic pH and weak development (Figure 6.4). A glaring deficiency is the under-representation of strongly developed paleosols (only 61 profiles) and very strongly developed ones (only three), when all other categories are represented by well in excess of 100 paleosol profiles. Phosphatic coprolites (three occurrences) and charcoal (six occurrences) also could be better represented. The calcareous phytoliths are all hackberry endocarps and represented by only 14 occurrences. This compilation thus represents a significant iteration of this kind of study, but not the last word.

Significantly, the results have changed little with escalation to 1478 from 87 paleosols (of Retallack, 1984). Bones, teeth, phosphatic coprolites, land snails and calcareous phytoliths are all significantly more abundant in calcareous than non-calcareous or pH-neutral paleosols (Figure 6.4). Oxidized rather than reducing, and moderately rather than very weakly developed, soils also favour their preservation, but these preferences are less well marked. Even though more frequently encountered in moderately developed paleosols, the quality and completeness of preservation of skeletons is usually better in weakly developed paleosols (Bown and Beard, 1990; Rose, 1990; Behrensmeyer *et al.*, 1995). Fossil plants and charcoal are clearly favoured by chemically reducing rather than oxidizing environments. Plant fossils also are preferred in acidic and weakly developed paleosols, whereas the limited data on charcoal indicate better representation in alkaline and moderately developed paleosols.

Eh/pH Control of Preservation

These new data support previous inferences from experimental and theoretical studies on preservation of fossils in soils (Figure 6.1). Original Eh and pH

Figure 6.3 (opposite)   *A comparison of different paleosol sequences within the Eocene Willwood Formation of Wyoming, showing recommended style of field logging. This includes estimates of soil development (from preservation of bedding), pH (reaction with applied dilute hydrochloric acid) and Eh (Munsell hue; see Retallack, 1988a, 1990). (A) Elk Creek facies west of Worland (of Bown and Kraus, 1981; NW¼ NE¼ NE¼ Sect. 4 T47N R94W Washakie Co.). (B) Sand Creek facies east of Worland (of Bown and Kraus, 1981; SE¼ NW¼ NW¼ Sect.1 T46N R92W Washakie Co.). (C) Fossil locality SC-210 west of Powell (of Winkler, 1983; NW¼ SW¼ NW¼ Sect. 25 T56N R102W, Park Co.). The chemical data include weight percentage free iron and manganese extracted by dithionite citrate and weight percentage carbon by colorimetry (Bown, 1979; Bown and Kraus, 1981). Pedotype names are from the Teton Dakota language (Buechel and Manhart, 1978) and mean scarlet (luluta), purple (tosa), orange-red (zizi), yellow (ziyela), cream to buckskin colour (hinzi), brownish grey (hota), green (ziyato) and light green (tozi)*

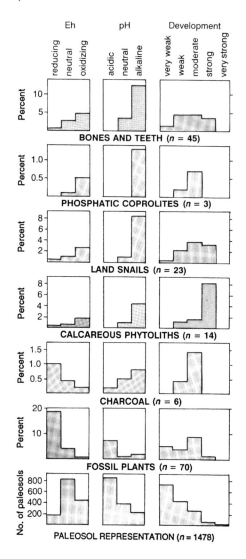

Figure 6.4 *A new compilation of the relative abundance of different kinds of fossils in paleosols of varied Eh, pH and time for development. Percentage values represent the proportion of all the studied paleosols of that category that yielded examples of a particular kind of fossil. (Data from Retallack, 1975, 1976, 1977, 1979, 1983a,b, 1984, 1986, 1991, 1994a,b; Retallack and Dilcher, 1981a,b; Retallack and Ryburn, 1982; Bestland and Retallack, 1994a,b; Retallack et al., 1995, 1997, 1998)*

are of greater significance for fossils in soils than in sediments because of the long time during which they are exposed to weathering, and because of the great range of Eh and pH conditions found in soils (Baas-Becking *et al.*, 1960).

Biogenic opal phytoliths and sponge spicules are stable in the widest range of Eh and pH, and accumulate in a variety of modern soils (Wilding and Drees, 1972; Wilding *et al.*, 1979). Their distribution may also be related to Eh because silica phytoliths become enriched as a concentrate in soils in which associated organic matter is destroyed by oxidation. Opal phytoliths and spicules are very weakly soluble at pH < 9, although appreciably more soluble than quartz, but both quartz and opal are dissolved at pH > 9 (Krauskopf, 1979).

The occurrence of calcareous phytoliths in soils has been little studied, although they have been obtained from prehistoric human coprolites (Bryant and Williams-Dean, 1975). The endocarps of hackberry, *Celtis occidentalis*, may contain 25–64% calcium carbonate by dry weight of the whole fruit and 2–7% silica (Yanovsky *et al.*, 1932). This may account for the favoured preservation of fossil hackberry endocarps in oxidized, calcareous paleosols of Tertiary age in North America and Kenya (Retallack, 1983b, 1991).

Aragonite (rarely calcite) shells of land snails are not found in acidic, non-calcareous soils, but are common in alkaline (pH 7.5–8.0) soils and paleosols (Evans, 1972; Pickford, 1986). Dissolution of shells in acidic soils proceeds rapidly once the organic periostracum decays (within about a year in non-waterlogged soils). Although the distribution of fossil land snails is to a certain extent due to differential preservation, many species have larger living populations in areas of calcareous soils.

Preservation of bones and teeth is also controlled more by pH than Eh, because these phosphatic fossils are prone to dissolution by acid (Chaplin, 1971; Pickford, 1986). In seasonally dry, subtropical Zimbabwe, bones in 700 year-old human graves were well preserved under alkaline (pH 6.2–7.9) termite mounds, but had been completely destroyed in adjacent acidic (pH 4.1–5.4) soils (Watson, 1967). In cool temperate, humid, north-western Europe, human bodies interred in acid, reducing peats for thousands of years (mostly from 100 BC to 500 AD) showed varying degrees of bone decalcification and in one case complete loss of bone within preserved skin and other soft tissues (Glob, 1969). Large bones have a lower surface to volume ratio than small bones and so persist preferentially in paleosols (Retallack, 1988b). This and their weaker mineralization may also explain the rarity of bones of young compared with old animals (Gordon and Buikstra, 1981; Carpenter, 1982). Teeth persist in more acid conditions and for a longer time than bones, and are found in Quaternary paleosols without associated bones (Evans, 1972). This may be because of the greater density and lesser permeability of teeth compared with bone (Shipman, 1981). Under extremely alkaline conditions (pH 9–10), the growth of salts may promote cracking of bone (Behrensmeyer, 1978). Even unpolluted rain is acidic (pH 5.6), so bones at the surface of the soil weather according to the following sequence: stage 1, superficial longitudinal cracking; stage 2, flaking of outer surface and cracking 1–1.5 mm deep; stage 3, exposed patches of fibrous bone; stage 4, very deep flaking and cracking; stage 5, bone falling apart in places and like a pile of splinters (Behrensmeyer, 1978).

Coprolites have varied original compositions, including fibre, pollen, spores, calcareous phytoliths, shells, scales, arthropod exoskeletons, bones and teeth. They can also be permineralized with calcite, siderite or silica (Bryant and Williams-Dean, 1975; Edwards and Folk, 1979). Organic matter of coprolites is preserved where aerobic bacterial decay is limited by freezing in permafrost, by strong acidity in peat bogs, by excessively reducing water-logged environments or by extreme desiccation in deserts (Heizer and Napton, 1969). Coprolites of birds and carnivorous mammals contain appreciable amounts of bone (Mellet, 1974) and other materials of similar phosphatic composition (Edwards and Folk, 1979). Avian guano is not obviously coprolitic because it is amorphous, but its composition becomes similar to that of bone as its organic matter decomposes or is leached. Fresh excrement of captive pelicans has only 4% P_2O_5, but this has been enriched to 19% in modern Peruvian accumulations of pelican guano, and 29% in old Peruvian guano (Hutchinson, 1950). Similar conditions and processes may account for the reduction in volume, shrivelling and then (after a few weeks or months) bleaching, cracking and powdering of the dung of domestic dogs (Edwards and Yatkola, 1974). Like bone, calcium phosphates are generally stable under alkaline conditions, but the degree of acidity at which they dissolve depends on their crystallinity and chemical composition (Lindsay and Vlek, 1977).

The preservation of organic matter, such as wood, leaf cuticle, plant fructifications, spores and pollen, is controlled more by Eh than by pH. This is mainly because of the activity of aerobic microbial heterotrophs, such as fungi and bacteria, rather than because of direct chemical oxidation (Elsik, 1971; Havinga, 1971; Scheihing and Pfefferkorn, 1984; Burnham, 1985). To a lesser extent, microbial activity also appears to be related to pH. Havinga (1971) observed greater microbial destruction of pollen and spores in the A horizon of a calcareous (3% $CaCO_3$), alkaline (pH 7.1), alluvial clay, than in the A horizon of an acidic (pH 4.6), podzolized sand. Others have observed that mildly alkaline soils (pH 6.5–8.1) are often barren of recognizable pollen and spores (Dimbleby, 1957), although in very alkaline soils (pH 9 or more) microbial activity may be inhibited and allow pollen preservation (Potter and Rowley, 1960). High alkalinity and a tendency to set like Portland cement may also explain the preservation of insects, millipedes, spiders, and the skin and tongues of lizards in carbonatite–nephelinite volcanic ash (Hay, 1986; Retallack, 1991; Retallack *et al.*, 1995).

Charcoal (or fusinite in the jargon of coal petrography; Diessel, 1992) can

Table 6.1 *Criteria for distinguishing between fossil charcoal and coalified wood fragments (after Cope and Chaloner, 1980)*

Fossil charcoal	Coalified wood
Equant shape	Usually elongate splinters
Sharply broken or rounded ends	Irregular or frayed ends
Black and opaque	Brown to black
Broken surface fibrous	Broken surface conchoidal
Broken surface showing cell structure	Broken surface structureless
No middle lamella between cell walls as seen under SEM	Middle lamella visible between crushed cells under SEM
Resistant to oxidation	Easily oxidized
Found in oxidized and gleyed paleosols	Found in gleyed paleosols
Glows on burning	Burns with bright flame

be distinguished from unburnt coalified wood by a variety of criteria, the best of which is the homogenization of adjacent cell walls visible under the scanning electron microscope (SEM) (Table 6.1). Charcoal is appreciably more resistant to microbial and chemical degradation, and has a longer residence time in soils, than other kinds of organic matter. Organic materials can be ranked in order of resistance to decay, with starch and fructose very easily degraded, but suberin, cutin, resin and amber less easily degraded, and lignin and sporopollenin decay-resistant (Tegelaar *et al.*, 1991). Nevertheless, charcoal is more prominent in oxidized than in reduced paleosols, where it is masked by associated uncharred organic material. Charcoal has been widely used for radiocarbon dating of late Quaternary paleosols, which appear to have formed under a wide range of Eh and pH conditions. A dedicated compilation of such data is in order, because pre-Quaternary records remain too sparse to be representative.

Time for Soil Development as a Factor

The destruction of potential fossils by processes controlled by Eh and pH in soils takes time. Fossils favoured for preservation should also increase in relative abundance within the soil through time. Apparent from this study is accumulation through time in paleosols of land snails, bones and teeth, and perhaps also of poorly represented calcareous phytoliths, phosphatic coprolites and charcoal (Figure 6.4). This is true also of organic matter accumulating within soils as peat. In contrast, there is declining representation of individual plant fossils in increasingly well-developed paleosols (Figure 6.4). Almost all non-peaty terrestrial ecosystems are oxidizing to some extent.

Without intervention of sedimentary cover characteristic of very weakly developed soils, individual fossil plants are bound to decay eventually.

Few studies document temporal trends in fossil preservation in soils and paleosols. In Zimbabwe, corrosion and flaking was notable in bones that were 20 years old, but 800 year-old graves no longer preserved skeletons (Watson, 1967). Accumulation of fossils in soils under favourable conditions also takes time. Surveys of bone accumulation in Amboseli National Park, Kenya (Behrensmeyer, 1982) indicated that thousands to tens of thousands of years are needed to accumulate a fossiliferous paleosol with 10–1000 bones per 100 m², representing the 10 major herbivores. Such processes give a time-averaged assemblage with a resolution of tens of thousands of years (Kowalewski, 1996).

Fossils freshly incorporated in the soil during burial may not have been affected by processes within the soils that tend to destroy fossils of that kind. Thus, fossil snails may be preserved in or on paleosols that were formerly acidic, although their remains are rare compared with those kinds of fossils favoured for preservation. Completely articulated skeletons are usually from animals overwhelmed by catastrophe in mudflows, stream channels or very weakly developed paleosols. In the Eocene Willwood Formation of Wyoming, there is greater representation of complete skeletons in weakly rather than strongly developed paleosols and prospecting has naturally focused on these deposits (Bown and Beard, 1990). Exceptionally preserved, articulated skeletons dominate museum displays, but are rare in paleosols (Retallack, 1988b, 1991).

COMPLETENESS OF THE ROCK RECORD

Ager (1973, p. 100) has well characterized the sedimentary record, which, 'like the life of a soldier, consists of long periods of boredom and short periods of terror'. The short periods of terror are events of sedimentation such as flooding, mudflows and volcanic ash falls. The periods of boredom are times of soil formation, and their duration can be estimated from the degree of development of the paleosols. Paleosols are thus important indicators of gaps in the sedimentary record of geological time. They can be helpful in assessing rates of evolution of mammals, and the duration and discreteness of biostratigraphic zones.

Palaeosols are often an obvious record of the completeness of stratigraphic sections. As soils develop, roots and burrows destroy bedding, schistosity or crystalline structure of parent sediments and rocks. Parent materials also are disrupted by cracking, frost heave and other processes. Weathering in the soil converts silt and sand grains of feldspar to clay and of magnetite to goethite. In this way a bedded silty sediment can be converted into a massive red claystone. Even at a glance over the spectacularly gullied cliffs of Badlands National Park, South Dakota, the purple-pink, clayey Chadron Formation can be seen to have much better developed soils, and thus a more incomplete record of geological time, than the silty and bedded Brule Formation. Similarly, the Elk Creek facies of the Willwood Formation of Wyoming is temporally more complete than the Sand Creek facies of that formation (Figure 6.3; Bown and Kraus, 1981; Retallack, 1984). The Dhok Pathan Formation of the Siwalik Group of Pakistan is temporally more complete (has more weakly developed paleosols; Retallack, 1991, 1995) than those mentioned above from Wyoming and South Dakota or the Chinji Formation of the Siwalik Group of Pakistan (Behrensmeyer *et al.*, 1995).

These general impressions can be quantified by drawing on information from a growing literature on times for development of Quaternary soils (Bockheim, 1980; Harden, 1990). Such studies aim to generate chronofunctions, which are mathematical relationships between soil variables such as clay enrichment and time for formation. Chronofunctions are based on carefully selected suites of soils for which the other variables of climate, vegetation, parent material and topographic setting are comparable. Such a group of soils is called a 'chronosequence'. Commonly these are soils on fluvial terraces at various elevations in incising river valleys, or on morainal ridges increasingly distant from retreating glaciers. Accurate dating of the surfaces is a major shortcoming of such studies, and includes radiocarbon dating and analysis of archaeological artefacts, fossil mammals and palaeomagnetism.

A qualitative scale of development also has been recognized (Birkeland, 1984; Retallack, 1990). This can be envisaged to proceed on three parallel tracks (Figure 6.2) representing humid well-drained soils (pedalfers of Marbut, 1935), aridland well-drained soils (pedocals of Marbut, 1935) and poorly drained soils (peat swamps in the geological sense). Development of pedalfers proceeds by obliteration of bedding or other original features of the parent material. With time, cracks are lined with clay that creates an increasingly clayey and thick subsurface horizon (Harden, 1982, 1990; Markewich *et al.*, 1990). For pedocals, the accumulation of carbonate proceeds through a series of stages from dispersed wisps to nodules and then massive to laminated layers (Gile *et al.*, 1966, 1980; Machette, 1985). In swamps, peat accumulates at rates of some 0.5–1 mm per year because slower rates allow oxidative decay of organic matter and faster rates create anoxic conditions that kill the roots of the trees supplying the organic debris (Falini, 1965; Moore and Bellamy, 1974). Because of the natural tempo imposed on soil formation by Pleistocene advances of ice sheets, moderately developed soils on the modern landscape are those that formed after the retreat of northern hemisphere ice some 15 000 years ago (Birkeland, 1984). These are pedalfers with clay enriched to about 8% more in subsurface than surface horizons of

Table 6.2 *Resolution and completeness of late Eocene and early Oligocene sequences of Badlands National Park, South Dakota and Painted Hills National Monument, Oregon, estimated from maximum and minimum times of paleosol development (see text) and from radiometric rates of sediment accumulation (following Sadler, 1981)*

		Chadronian		Orellan		Whitneyan	
		Painted Hills	Badlands	Painted Hills	Badlands	Painted Hills	Badlands
Thickness (m)		12	23	98	38	114	46
From paleosol minimum times	Time (years)	229 000	106 005	454 900	80 135	469 700	65 920
	Rate (m/1000 yrs)	0.052	0.213	0.215	0.479	0.243	0.701
	Resolution (years)	14 313	5889	3273	3339	2973	2535
	Completeness (% of 1000 yr intervals)	7	17	31	30	34	39
From paleosol maximum times	Time (years)	645 000	930 500	1 396 5000	648 500	1 385 500	307 000
	Rate (m/1000 yrs)	0.019	0.024	0.070	0.059	0.082	0.150
	Resolution (years)	40 313	51 694	10 047	27 020	8769	11 808
	Completeness (% of 1000 yr intervals)	2	2	10	4	11	9
From radiometric times	Time (years)	1 700 000	1 700 000	2 000 000	2 000 000	2 000 000	2 000 000
	Rate (m/1000 yrs)	0.007	0.013	0.049	0.019	0.057	0.023
	Resolution (years)	1 700 000	1 700 000	2 000 000	2 000 000	2 000 000	1 700 000
	Completeness (% of 1000 yr intervals)	0.9	1.6	6.1	2.4	7.1	2.9

the soil, pedocals with carbonate accumulation to the nodular stage and peat accumulated to a thickness of about 40 cm. These are the limits used by the US soil taxonomy for argillic and calcic horizons, and for Histosols (Soil Survey Staff, 1975, 1990). Weakly and strongly developed soils can be calibrated to orders of magnitude less or more time. This order of magnitude scale lacks precision, but has the advantage of wide applicability.

Such a yardstick of paleosol development can be used to estimate both resolution and completeness of stratigraphic sections. Resolution is a question of scale or the time interval in years for which a sequence can be considered to have a relatively complete record of events. Completeness, on the other hand, is a measure of the reliability of a sequence at a given resolution. A complete sequence at a resolution of 1000 years should have on average a bed, paleosol or other record of geological time for every interval of 1000 years. A sequence of paleosols can be regarded as complete at a resolution that is equivalent to the average time of formation of its individual paleosols. Completeness can be estimated for finer resolutions as a fraction or percentage of that resolution (Retallack, 1990).

A more complex way of estimating stratigraphic completeness is by comparing rates of sediment accumulation of a sequence with rates usual for that environment and time span, as estimated from a large compilation of rates (Sadler, 1981; Anders *et al.*, 1987). Rates of sediment accumulation can be distinguished from rates of rock accumulation (Kraus and Bown, 1986, 1993a), with rock accumulation converted to sediment accumulation by compensating for burial compaction using standard curves (Sclater and Christie, 1980; Baldwin and Butler, 1985). Such conversions are not necessary for using the compilations of Sadler (1981) and co-workers (Anders *et al.*, 1987), because their long-term rates were calculated for rock sections uncompensated for compaction. A shortcoming of this probabilistic approach using typical sediment accumulation rates is that it fails to demonstrate the exact position and duration of the gaps, as is potentially possible using paleosols (Retallack, 1994b).

A Tale of Two Sections

Whether a sequence of paleosols was the best for time or the worst for time can be illustrated by comparison of late Eocene and early Oligocene paleosols of Badlands National Park, South Dakota and Painted Hills National Monument, Oregon (Figure 6.5; Table 6.2). These two terrestrial sequences of paleosols can be correlated using mammalian biostratigraphy, palaeomagnetic stratigraphy and numerous $^{39}Ar/^{40}Ar$ radiometric dates (Prothero and Rensberger, 1985; Swisher and Prothero, 1990; Prothero and Swisher, 1992; Prothero, 1996; Tedford *et al.*, 1996; Bestland *et al.*, 1997), and these force

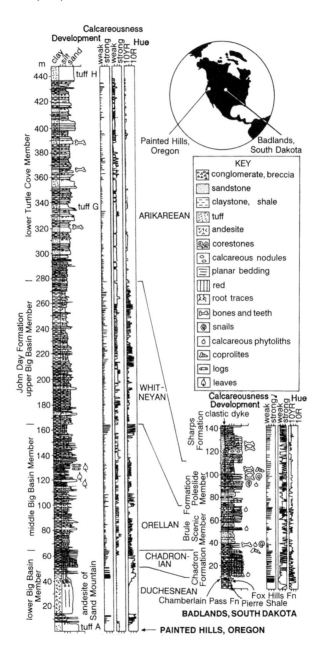

minor revision in previous estimates of completeness (Retallack, 1984, 1986). The Painted Hills sequence is thinner than the Badlands sequence for an interval correlated with the Chadronian North American Land Mammal 'Age', but thicker for the succeeding Orellan and Whitneyan North American Land Mammal 'Ages'. This indication of an inferior completeness of the Chadronian, but a superior completeness of the Orellan and Whitneyan in the Painted Hills, is confirmed by the nature of paleosols in each section. The Painted Hills have more weakly developed paleosols than the Badlands (generally shorter temporal resolution from paleosols in Table 6.2). In both areas resolution and completeness increase up-section, regardless of whether minimum or maximum rates of soil formation are used (Table 6.2). Palaeosols give a more detailed impression of these relative changes than radiometry.

These estimates help to resolve an important area of doubt remaining from earlier analysis of completeness of the Badlands section (Retallack, 1984). In that study it was shown that rates of sediment accumulation estimated from paleosols were much higher than rates estimated from radiometric data, as can be seen here (Table 6.2). Estimates of completeness and resolution thus remain more optimistic from paleosols than from radiometry. This discrepancy indicates that there are more gaps than meet the eye in these paleosol sections.

The hundreds of thousands of years of time missing are unlikely to be lost in underestimated soil development due to the paleosols reaching a steady state in age-diagnostic features, because this has not been found in recent analysis of adequately resolved chronosequence data used to estimate times of development of the paleosols (Harden, 1990). The missing time is only partly compensated by estimated maximal times of paleosol development, as recommended by Kraus and Bown (1986). Even with generous maximum times of soil formation, rates of sediment accumulation are still two to four times greater than rates estimated from radiometric data (Table 6.2).

Nor is much time likely to be lost in paleosol overprinting. By this is meant the obscuring of one soil by development of a later soil on the same material. Cases of overprinting have been detected in the Painted Hills where clasts or relict horizons of moderately to strongly developed pre-existing soils indicate an unusual degree of weathering in a paleosol (Bestland *et al.*, 1996). One moderately to strongly developed paleosol overprinting another is rare in my experience, but has also been noted elsewhere (Gardner *et al.*, 1988; Hughes *et al.*, 1992). The more usual case of a strongly developed soil

---

Figure 6.5 (opposite)   *A comparison of late Eocene and early Oligocene sequences of paleosols in Badlands National Park, South Dakota and Painted Hills National Monument, Oregon. (From Retallack, 1983b, and Bestland and Retallack, 1994b, with addition of 'Redcap paleosols' at the top of the Orellan-equivalent interval in the Painted Hills)* (hundreds of thousands of years) overprinting weakly developed ones (tens of years) involves too little time to be significant.

Instead, the time seems to have disappeared in major disconformities between packages of similar paleosols. The similarity of purple-pink clayey paleosols in the Chadron Formation and their strong contrast with brown silty paleosols in the Brule Formation is probably a reflection of a relatively short duration of accumulation of each formation compared with the gap in time between formations. Palaeovalleys and other erosional discontinuities at the base of the Chadronian and Duchesnean sequences have recently been mapped throughout the Badlands of South Dakota (Terry and Evans, 1994). Erosional gaps at the base of the Orellan and Whitneyan sequences in the Badlands of South Dakota have now been demonstrated by truncation of magnetic reversals along strike of these disconformity surfaces compared with sections further west (Prothero *et al.*, 1985; Prothero, 1996; Tedford *et al.*, 1996). Completeness of the record within members and formations in the Badlands may be superior to that within comparable units in the Painted Hills, but the Painted Hills overall provides the more complete record, because the disconformities between units are less profound.

This discovery has important implications for studies of mammalian evolution based on the fossil record of Badlands National Park. Because the sequence is a succession of high-resolution subsequences, separated by profound temporal gaps, it is not surprising that it has been used to support the concept of evolution by punctuated equilibrium (Heaton, 1993; Prothero *et al.*, 1998). Mammal fossils within the successive sedimentary units show little morphological change, interpreted as evolutionary stasis, and they appear and disappear abruptly near the boundaries of the units. Thus, it has become even clearer with this analysis than during initial studies (Retallack, 1984, 1986) that the Badlands of South Dakota are a poor choice for evolutionary studies. Even though the Painted Hills sequence has greater temporal completeness than the Badlands of South Dakota, paleosols in the Painted Hills are non-calcareous and unsuitable for bone preservation except at the very top of the sequence where mammal remains are fragmentary and poorly preserved. Better sequences for the study of mammalian evolution are in the Willwood Formation of the Bighorn and Clarks Fork Basins of Wyoming (Figure 6.3; Gingerich, 1976, 1980, 1985, 1987; Bown and Kraus, 1993; Gingerich and Gunnell, 1995), and in the Siwalik Group of Pakistan and India (Barry *et al.*, 1990, 1991; Retallack, 1991, 1995). These sequences were deposited under sustained high rates of sediment accumulation, and have long seccessions of weakly to moderately developed, calcareous paleosols. Nevertheless, it is doubtful that, even in these sections, major discontinuities can be assumed to be negligible, as implied by methods that allot geological time in proportion to the development of paleosols (Bown and Larriestra, 1990; Bown and Kraus, 1993; Kraus and Bown, 1993a). Such approaches assume a degree of completeness that is unrealistic. The degree of completeness of any section should be determined, not assumed.

Recent high-precision radiometric dating and magnetostratigraphic correlation show that the big temporal gaps in the Badlands sequence coincide with temporally lesser gaps in the Painted Hills, and these are synchronous with marine regressive lowstands in shallow marine rocks of late Eocene and early Oligocene age (Bestland *et al.*, 1997). This is a remarkable set of coincidences, considering that during the late Eocene and Oligocene the Badlands of South Dakota were at least 1600 km north of the Gulf coast of that time, and 1300 km east of the Painted Hills, which were separated from the former Pacific coast by a 300 km wide volcanic mountain range now preserved in the Western Cascades. It is difficult to imagine sea level having such long-distance effect into continental interior basins. Cratonic South Dakota and volcanic Oregon were also very different tectonic provinces and had dissimilar subsidence regimes. The common thread between these two sections is variation in climate, which has already been shown to be the principal control on sedimentation in the Badlands in a multifactorial comparison (Retallack, 1986).

Sequence Stratigraphy for Palaeosols

Possible explanations for the distribution of gaps in paleosol sequences can be gained by studying the pattern of development of paleosols (Figure 6.6), in an approach analogous to sequence stratigraphy of marginal marine rocks (Posamentier and Vail, 1988; Wilson, 1991; Emery and Myers, 1996). Two contrasting patterns of development of successive paleosols are evident in the Badlands and Painted Hills (Figure 6.5). The Painted Hills sequence has a sinuous pattern where development rises and falls in successive paleosols. One exception is the probable Chadronian portion of the Painted Hills sequence, which is like the Badlands sequence in showing a sawtooth pattern of development. Sawtooth patterns show a steady increase in development of successive paleosols until a sudden switch to weakly developed paleosols, then build again to better-developed paleosols higher in the next sequence. The sawtooth pattern shows ageing-upwards sequences of paleosols, whereas the sinuous to irregular pattern has alternating ageing-upwards and younging-upwards sequences. Other examples of sinuous patterns are in the Siwalik Group of Pakistan (Retallack, 1991, 1995), the Hiwegi Formation of Kenya (Retallack *et al.*, 1995) and the Fremouw Formation of Antarctica (Retallack *et al.*, 1997b). Other examples of the sawtooth pattern are in the Clarno Formation of Oregon (Retallack *et al.*, 1996), Chinle Group of Arizona (Kraus and Middleton, 1987), Beaufort Group of South Africa (Smith, 1990), Illawarra Coal Measures of Australia (Retallack, 1998) and Buckley Forma-

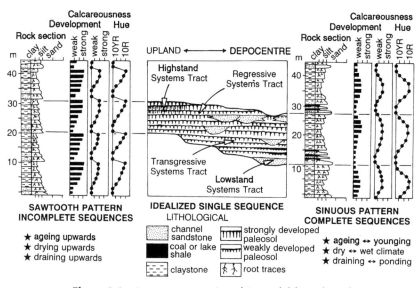

Figure 6.6   *A sequence stratigraphic model for paleosols*

tion of Antarctica (Retallack *et al.*, 1998). The sawtooth pattern is especially common in coal measures where clastic deposition declines up-sequence to allow accumulation of thick sequence-capping coal seams (Arditto, 1991; Hampson, 1995; Herbert, 1995; Holdgate *et al.*, 1995; Hampson *et al.*, 1996; Retallack, 1998).

The sinuous pattern is one of high sedimentation rates and high long-term completeness, whereas the sawtooth pattern is one of relatively low sedimentation rates and low long-term completeness. The sawtooth itself allows identification of the most profound gaps in the sequence. They are at the break between the most strongly and most weakly developed paleosols in the sequence. This can be true even if the most strongly developed paleosol is only moderately developed. Very strongly developed, thick paleosols with silcretes, laterites and bauxites show sedimentary hiatuses of millions of years, all by themselves (Retallack and German-Heins, 1994; Bestland *et al.*, 1996).

The causes of long-term sawtooth sequences of paleosols can be discerned from evidence within the paleosols themselves. A common explanation for such patterns is marine transgression to a stillstand (Kraus, 1992; Hampson, 1995; Herbert, 1995; Holdgate *et al.*, 1995; Hampson *et al.*, 1996). This may play a role in some coastal sequences (Retallack, 1979, 1983b, 1994a, 1998; Percival, 1986; Strasser, 1988). Marine transgression does not account for other sawtooth sections such as the Eocene–Oligocene sequences of South Dakota and Oregon (Figure 6.5), which were very remote from the sea.

Another sedimentary explanation for ageing-upwards sequences of paleosols is palaeochannel migration (Kraus, 1987; Smith, 1990; Kraus and Bown, 1993b). However, the Whitneyan, Orellan and Chadronian sequences described herein (Figure 6.5) each include numerous levels of palaeochannels, and are cycles of longer duration (millions of years; Table 6.2) than alluvial cycles (thousands of years). Both the Badlands and Painted Hills, as well as other paleosol sequences (Wright, 1992), include paleosols traceable laterally for distances of many kilometres, indicating major regional pauses in sedimentation.

Another common explanation is tectonic uplift of the source area (Kraus, 1992). Source-area uplift has been discounted in the Badlands of South Dakota for two out of three sequence boundaries by unchanging or decreased proportions of far-travelled clasts in lower compared with upper portions of the sawtooth developmental pattern of paleosols (Retallack, 1986). Uplift of the sedimentary basin is a plausible explanation for disconformities at the breaks in the sawtooth pattern, but not for the increased stability indicated by paleosols below the break. Declining subsidence up-section is also possible, but how would this be synchronized in 2-Myr pulses in such distant places as Oregon and South Dakota? Milankovitch-scale periodicity of 21 000, 41 000, 95 000, 123 000 and 413 000 years has been found in many paleosol sequences: in the Miocene of Pakistan (Retallack, 1991), Oligocene of Oregon (Bestland and Swisher, 1996), Cretaceous of Switzerland (Strasser, 1988), Triassic of United States (Olsen, 1986), and in the Pennsylvanian of United States (Cecil, 1990). This periodicity in paleosol sequences more likely has orbital or palaeoclimatic than tectonic explanations (Olsen, 1986). The role of tectonics can be envisaged as analogous to that of a tape recorder. If relative subsidence rates are high, then the sequence records palaeoenvironmental signals with high fidelity. If subsidence rates are low, then the recording is less faithful to the signal. High-fidelity recording is facilitated by active subsidence, but the message comes from elsewhere.

Yet another explanation is increased volcanic input in lower compared with upper parts of sawtooth developmental patterns. It is true that the weakly developed paleosols of the lower parts of sequences are separated by less-weathered volcaniclastic sediments in Oregon (Retallack *et al.*, 1996) and Kenya (Retallack, 1991). However, the really big eruptions, such as lava flows and ash-flow tuffs, are underlain and overlain by similar paleosols, leaving unaffected the sawtooth developmental pattern (Figure 6.5). Big eruptions can override the pattern, but small eruptions are either preserved or weathered away depending on their timing within the pattern. In the Badlands of South Dakota, volcanic ash either decreases or is unchanging across sequence boundaries despite an overall increase up-section (Retallack, 1986). The Milankovitch periodicities widely observed in paleosol sequences (Olsen, 1986; Strasser, 1988; Cecil, 1990; Retallack, 1991; Bestland and Swisher, 1996) also are unlikely to have a volcanic cause.

An explanation that does appear to have some merit is that the sawtooth pattern of paleosol development follows variation in degree of drainage of paleosols. The lower parts of paleosol sequences are commonly drab in colour, with minerals such as siderite and pyrite (Cecil *et al.*, 1985; Cecil, 1990; Bown and Kraus, 1993). Root traces in these paleosols are tabular rather than vertically penetrating and animal burrows are rare. These are all indications of poorly drained paleosols (Retallack, 1990; PiPujol and Buurman, 1994). In contrast, the upper parts of sequences are commonly red in colour with minerals such as goethite and haematite, and have evidence of clays and silts washed down into deeper parts of the profile. They also commonly have deeply penetrating root traces and common burrows of non-aquatic animals such as termites and mongooses. These are all indications of well-drained paleosols (Retallack, 1976, 1990; Fastovsky and McSweeney, 1987; Besly and Turner, 1989). Differences in drainage may be more subtle in coal-measure sequences formed in very waterlogged terrains, but can still be inferred from petrographic measures of declining vitrinite content up-section within coal seams (Smyth, 1970; Cook, 1975; Shibaoka and Smyth, 1975). Such trends could reflect declining subsidence rate up-section, but could also be due to filling to a new dynamic equilibrium. These two alternatives are difficult to distinguish on a local level, but regional correlations tend to support the second alternative. The draining-upwards paleosol sequences of the North American Chadronian and Orellan are better drained at any comparable level in Nebraska compared with South Dakota (Retallack, 1992) and in Oregon's Picture Gorge compared with the Painted Hills (Bestland and Retallack, 1994b). Because the better-drained sequence is also thinner and closer to the source area, this supports the concept of filling to a new equilibrium rather than differential subsidence.

A final possibility is the tendency of paleosols that are higher within a sawtooth pattern of paleosol development to indicate a drier palaeoclimate. This is especially evident in sequences of calcareous paleosols like those of South Dakota's Badlands, where the mean annual rainfall is reflected by the depth to the calcareous nodular horizon of the paleosols (Schultz and Stout, 1980; Thompson *et al.*, 1982; Retallack, 1986, 1994b). In such cases the calcareous nodular horizons become shallower and shallower in each successive paleosol up-section, reflecting an increasingly dry climate. Such climatic change could also explain in part the common observation of draining-upwards sequences of paleosols, such as coals grading up through drab clayey to red clayey paleosols (Cecil *et al.*, 1985; Cecil, 1990; Bestland *et al.*, 1997). Climate is also a plausible link to marine regression and transgression. Warmer, wetter, greenhouse climates with little polar and glacial ice alternating with glacial advances during cooler, drier, oxygen-rich climates have been a conspicuous feature of the Quaternary (Catt, 1986; Crowley, 1990). Drying-upwards sequences of paleosols may reflect such climatic variability on a variety of temporal scales ranging from millions to hundreds of years. Whatever the cause in any given section, it is common for paleosols to form sawtooth developmental sequences that age upwards, drain upwards and dry upwards. The abrupt changes in development represent discontinuities in sediment accumulation that are not apparent from sequences of high sediment accumulation rate in which development of successive paleosols form a sinuous pattern. The discontinuities of sawtooth developmental sequences are major gaps in the sedimentary record, more significant than gaps between individual paleosols or created by overprinted soil development.

CONCLUSIONS

Palaeosols constitute a record of the past that is complementary to the fossil record. Palaeosols are in a sense trace fossils of ecosystems, which can be compared with the record of ecosystems from fossils. Former Eh and pH can be inferred from paleosols, and these are important variables in the preservation of different kinds of fossils. Fossils of organic matter such as leaves, logs, pollen and charcoal, for example, are preserved preferentially in chemically reducing (low negative Eh) environments such as bogs and swamps. This is because they are readily decayed by fungi and other aerobic decomposers under oxidizing conditions. Fossils of aragonite and calcite, such as snail shells, and of phosphate, such as bones and teeth, are relatively resistant to decay, but prone to dissolution by acid. They are seldom found in acidic soils and paleosols, and preferentially accumulate in calcareous alkaline soils of dry regions.

Palaeosols also can reveal breaks in sedimentation by their degree of development. Weakly developed soils and paleosols show little change from their parent materials other than rooting. Strongly developed soils, on the other hand, show considerable subsurface accumulation of clay or carbonate, or surface accumulation of peat. Each paleosol is in itself a break in sedimentation, but there are common patterns of development of successive paleosols that can reveal longer breaks in sedimentation. Sawtooth patterns of development have ageing-upwards sequences of paleosols, and major discontinuities are commonly located between the most strongly developed and the least developed paleosols. Because many incomplete sequences of paleosols accumulate by this pattern of punctuated equilibrium, it is perhaps not surprising that this would be an inferred pattern of mammalian evolution in them. Sequences of paleosols best for evolutionary studies are those that accumulated at high long-term sedimentation rates, with relatively weakly developed paleosols showing sinuous patterns of paleosol development. Ageing-upwards sequences of paleosols commonly also are draining-upwards sequences and drying-upwards sequences, and in addition can be correlated with global changes in relative sea level. A common denominator of many of these changes is climate, which has a powerful influence in creating sequences of paleosols.

ACKNOWLEDGEMENTS

This research has been funded by two decades of NSF support, most recently by grants OPP 9315228 and SBR 9513175, as well as National Parks Service contract CX-9000-1-10009 and American Chemical Society grant 31270-AC-8. Colleagues especially influential in shaping my ideas about the subjects of this paper include Kay Behrensmeyer, Mary Kraus, Tom Bown, V. Paul Wright and Erick Bestland.

REFERENCES

Ager, D.V., 1973, *The Nature of the Stratigraphical Record* (1st edn), Wiley, New York: 114 pp.

Anders, M.H., Krueger, S.W. and Sadler, P.W., 1987, A new look at sedimentation rates and completeness of the stratigraphic record, *Journal of Geology*, **95**: 1–14.

Arditto, P.A., 1991, A sequence stratigraphic analysis of the late Permian succession in the southern coalfield, Sydney Basin, New South Wales, *Australian Journal of Earth Sciences*, **38**: 125–137.

Baas-Becking, L.G.M., Kaplan, I.R. and Moore, D., 1960, Limits to the natural environment in terms of pH and oxidation–reduction potential, *Journal of Geology*, **68**: 243–284.

Baldwin, B. and Butler, C.O., 1985, Compaction curves, *American Association of Petroleum Geologists Bulletin*, **69**: 622–626.

Barry, J.C., Flynn, L.J. and Pilbeam, D.R., 1990, Faunal diversity and turnover in a Miocene terrestrial sequence. In R. Ross and W. Allmon (eds), *Causes of Evolution: A Paleontological Perspective*, University of Chicago Press, Chicago: 381–421.

Barry, J.C., Morgan, M.E., Winkler, A.J., Flynn, L.J., Lindsay, E.H., Jacobs, L.L. and Pilbeam, D.R., 1991, Faunal interchange and Miocene terrestrial vertebrates of southern Asia, *Paleobiology*, **17**: 231–245.

Behrensmeyer, A.K., 1978, Taphonomic and ecological information from bone weathering, *Paleobiology*, **4**: 150–162.

Behrensmeyer, A.K., 1982, Time resolution in fluvial vertebrate assemblages, *Paleobiology*, **8**: 211–227.

Behrensmeyer, A.K., Willis, B.J. and Quade, J., 1995, Floodplains and paleosols of Pakistan Neogene and Wyoming Paleogene deposits: a comparative study, *Palaeogeography, Palaeoclimatology, Palaeoecology*, **115**: 37–60.

Besly, B.M. and Turner, P., 1989, Paleosols in Westphalian coal-bearing and red-bed sequences, central and northern England, *Palaeogeography, Palaeoclimatology, Palaeoecology*, **70**: 303–330.

Bestland, E.A. and Retallack, G.J., 1994a, Geology and paleoenvironments of the Clarno Unit, John Day Fossil Beds National Monument, Oregon, *National Parks Service Contract*, **CX-9000–1–10009**, 1600pp.

Bestland, E.A. and Retallack, G.J., 1994b, Geology and paleoenvironments of the Painted Hills Unit, John Day Fossil Beds National Monument, *National Parks Service Contract*, **CX-9000–1–10009**, 211pp.

Bestland, E.A. and Swisher, C.C., 1996, Milankovitch-scale climate cycles recorded in Oligocene paleosols from central Oregon, *Geological Society of America Abstracts with Programs*, **28** (5): 49.

Bestland, E.A., Retallack, G.J., Rice, A.E. and Mindszenty, A., 1996, Late Eocene detrital laterites in central Oregon: mass balance geochemistry, depositional setting and landscape evolution, *Geological Society of America Bulletin*, **108**: 285–302.

Bestland, E.A., Retallack, G.J. and Swisher, C.C., 1997, Stepwise climate change recorded in Eocene-Oligocene paleosol sequences from central Oregon, *Journal of Geology*, **105**: 153–172.

Birkeland, PW, 1984, *Soils and Geomorphology*, Oxford University Press, New York: 372 pp.

Blodgett, R.H., Crabaugh, J.P. and McBride, E.F., 1993, The color of red beds: a geologic perspective. *In* J.M. Bigham and E.J. Ciolkoscz (eds), *Soil Color, Special Publication of the Soil Science Society of America*, **31**: 127–159.

Bockheim, J.G., 1980, Solution and age of chronofunctions in studying soil development, *Geoderma*, **24**: 71–85.

Bown, T.M., 1979, Geology and mammalian paleontology of the Sand Creek Facies, lower Willwood Formation (Lower Eocene), Washakie County, Wyoming, *Memoir of the Geological Survey of Wyoming*, **2**: 151 pp.

Bown, T.M. and Beard, K.C., 1990, Systematic lateral variation in the distribution of fossil mammals in alluvial paleosols. *In* T.M. Bown and K.D. Rose (eds), *Dawn of the Age of Mammals in the Northern Part of the Rocky Mountain Interior, Geological Society of America Special Papers*, **243**: 135–151.

Bown, T.M. and Kraus, M.J., 1981, Lower Eocene alluvial paleosols (Willwood Formation, northwest Wyoming, U.S.A.) and their significance for paleoecology, paleoclimatology and basin analysis, *Palaeogeography, Palaeoclimatology, Palaeoecology*, **34**: 1–30.

Bown, T.M. and Kraus, M.J., 1987, Integration of channel and floodplain suites: I. Developmental sequence and lateral relations of alluvial paleosols, *Journal of Sedimentary Petrology*, **57**: 587–601.

Bown, T.M. and Kraus, M.J., 1993, Time stratigraphic reconstruction and integration of paleopedologic, sedimentologic and biotic events (Willwood Formation, Lower Eocene, northwest Wyoming, U.S.A.), *Palaios*, **8**: 68–80.

Bown, T.M. and Larriestra, C.N., 1990, Sedimentary paleoenvironments of fossil platyrrhine localities, Miocene Pinturas Formation, Santa Cruz Province, Argentina, *Journal of Human Evolution*, **19**: 87–119.

Bryant, V.M., Jr and Williams-Dean, G., 1975, The coprolites of man, *Scientific American*, **232** (1): 100–109.

Buechel, E. and Manhart, P., 1978, *A Dictionary of the Teton Dakota Sioux Language*, Red Cloud Indian School, Pine Ridge, South Dakota: 852 pp.

Burnham, R.J., 1985, Relationship between standing vegetation and leaf litter in a paratropical forest: implications for paleobotany, *Review of Paleobotany and Palynology*, **58**: 5–32.

Carpenter, K., 1982, Baby dinosaurs from the late Cretaceous Lance and Hell Creek Formations and description of a new species of theropod, *Contributions to Geology of the University of Wyoming*, **20**: 123–134.

Catt, J.A., 1986, *Soils and Quaternary Geology*, Clarendon Press, Oxford: 267 pp.

Cecil, C.B., 1990, Paleoclimate controls on stratigraphic repetition of chemical and siliciclastic rocks, *Geology*, **18**: 533–536.

Cecil, C.B., Stanton, R.W., Neuzil, S.G., Dulong, F.T., Ruppert, L.F. and Pierce, B.S., 1985, Paleoclimate controls on late Paleozoic sedimentation and peat formation in the central Appalachian Basin (U.S.A.), *International Journal of Coal Geology*, **5**: 195–230.

Chaplin, R.E., 1971, *The Study of Animal Bones from Archaeological Sites*, Seminar Press, London: 170 pp.

Cook, A.C., 1975, Spatial and temporal variation of the type and rank of Australian coals. *In* A.C. Cook (ed.), *Australian Black Coal: Its Occurrence, Mining, Preparation and Use*, Australasian Institute of Mining and Metallurgy, Illawarra Branch, Wollongong: 63–84.

Cope, M.J. and Chaloner, W.G., 1980, Fossil charcoal as evidence of past atmospheric composition, *Nature*, **283**: 647–648.

Crowley, T.J., 1990, Are there satisfactory geologic analogs for a future greenhouse warming? *Journal of Climate*, **3**: 383–422.

Diessel, C.F.K., 1992, *Coal-bearing Depositional Sequences*, Springer, Berlin: 721 pp.

Dimbleby, G.W., 1957, Pollen analysis of terrestrial soils, *New Phytologist*, **56**: 12–28.

DiMichele, W.A. and DeMaris, P.J., 1987, Structure and dynamics of a Pennsylvanian-age *Lepidodendron* forest: colonizers of a disturbed swamp habitat in the Herrin (no. 6) coal of Illinois, *Palaios*, **2**: 146–157.

Edwards, P. and Folk, R.L., 1979, Coprolites. *In* R.W. Fairbridge and D. Jablonski (eds), *The Encyclopedia of Paleontology*, Dowden, Hutchinson & Ross, Stroudsburg, Pennsylvania: 224–225.

Edwards, P. and Yatkola, D., 1974, Coprolites of the White River (Oligocene) carnivorous mammals: origin and paleoecologic and sedimentologic significance, *Contributions to Geology of the University of Wyoming*, **13**: 67–73.

Elsik, W.C., 1971, Microbiological degradation of sporopollenin. *In* J. Brooks, P.R. Grant, M. Muir, P. Gijzel and G. Shaw (eds), *Sporopollenin*, Academic Press, New York: 480–511.

Emery, D. and Myers, K. (eds), 1996, *Sequence Stratigraphy*, Blackwell, Oxford: 254 pp.

Evans, J.G., 1972, *Land Snails in Archaeology*, Seminar Press, London: 436pp.

Falini, F., 1965, On the formation of coal deposits of lacustrine origin, *Geological Society of America Bulletin*, **76**: 1317–1346.

Fastovsky, D.E. and McSweeney, K., 1987, Paleosols spanning the Cretaceous–Paleogene transition, eastern Montana and western North Dakota, *Geological Society of America Bulletin*, **99**: 66–77.

Gardner, T.W., Williams, E.G. and Holbrook, PW, 1988, Pedogenesis of some Pennsylvanian underclays: ground-water, topographic and tectonic controls. *In* J. Reinhardt and W.R. Sigleo (eds), *Paleosols and Weathering through Geologic Time: Principles and Applications*, Geological Society of America Special Paper, **216**: 81–101.

Gastaldo, R.A., 1986, Implications on the paleoecology of autochthonous lycopods in clastic sedimentary environments of the early Pennsylvanian of Alabama, *Palaeogeography, Palaeoclimatology, Palaeoecology*, **53**: 191–212.

Gile, L.H., Peterson, F.F. and Grossman, R.B., 1966, Morphological and genetic sequences of carbonate accumulation in desert soils, *Soil Science*, **101**: 347–360.

Gile, L.H., Hawley, J.W. and Grossman, R.B., 1980, Soils and geomorphology in the Basin and Range area of southern New Mexico: guidebook to the Desert Project, *Memoir of the New Mexico Bureau of Mines and Mineral Resources*, **39**: 222 pp.

Gingerich, P.D., 1976, Paleontology and phylogeny: patterns of evolution at the species level, *American Journal of Science*, **276**: 1–28.

Gingerich, P.D., 1980, Evolutionary patterns in early Cenozoic mammals, *Annual Reviews of Earth and Planetary Science*, **8**: 407–424.

Gingerich, P.D., 1985, Species and the fossil record: concepts, trends and transitions, *Paleobiology*, **11**: 27–41.

Gingerich, P.D., 1987, Evolution and the fossil record: patterns, rates and processes, *Canadian Journal of Zoology*, **65**: 1053–1060.

Gingerich, P.D. and Gunnell, G.F., 1995, Rates of evolution in Paleocene–Eocene mammals of the Clarks Fork Basin, Wyoming, and a comparison with Neogene Siwalik Group, Pakistan, *Palaeogeography, Palaeoclimatology, Palaeoecology*, **115**: 225–247.

Glob, P.V., 1969, *The Bog People: Iron Age Man Preserved* (trans. by R. Bruce-Mitford), Cornell University Press, Ithaca, New York: 200 pp.

Gordon, C.C. and Buikstra, J.E.. 1981, Soil pH, bone preservation and sampling bias at mortuary sites, *American Antiquity*, **46**: 566–571.

Hampson, G., 1995, Discrimination of regionally extensive coals in the Upper Carboniferous of the Pennine Basin, U.K., using high resolution sequence stratigraphic concepts. *In* M.K.G. Whateley and D.A. Spears (eds), *European Coal Geology, Special Paper of the Geological Society of London*, **82**: 79–97.

Hampson, G.J., Elliott, T. and Flint, S.S., 1996, Critical application of high resolution sequence stratigraphic concepts to the Rough Rock Group (Upper Carboniferous) of northern England. *In* J.A. Howell and J.F. Aitken (eds), *High Resolution Sequence Stratigraphy: Innovation and Applications, Special Publication of the Geological Society of London*, **104**: 221–246.

Harden, J.W., 1982, A quantitative index of soil development from field descriptions: examples from a chronosequence in central California, *Geoderma*, **28**: 1–28.

Harden, J.W., 1990, Soil development on stable landforms and implications for landscape studies, *Geomorphology*, **3**: 391–398.

Havinga, A.J., 1971, An experimental investigation into the decay of pollen and spores in various soil types. *In* J. Brooks, P.R. Grant, M. Muir, P. van Gijzel and G. Shaw (eds), *Sporopollenin*, Academic Press, London: 446–479.

Hay, R.L., 1986, Role of tephra in preservation of fossils in Cenozoic deposits of East Africa. *In* L.E. Frostick, R.W. Renaut, I. Reid and J.J. Tiercelin (eds), *Sedimentation in the East African Rifts, Special Publication of the Geological Society of London*, **25**: 339–345.

Heaton, T.H., 1993, The Oligocene rodent *Ischyromys* of the Great Plains: replacement mistaken for anagenesis, *Journal of Paleontology*, **67**: 297–308.

Heizer, R.F. and Napton, L.K., 1969, Biological and cultural evidence from prehistoric human coprolites, *Science*, **165**: 563–568.

Herbert, C., 1995, Sequence stratigraphy of the late Permian Coal Measures in the Sydney Basin, *Australian Journal of Earth Sciences*, **42**: 391–405.

Holdgate, G.R., Kershaw, A.P. and Sluter, I.R.K., 1995, Sequence stratigraphic analysis and the origins of Tertiary brown coal lithotypes, Latrobe Valley, Gippsland Basin, Australia, *International Journal of Coal Geology*, **28**: 249–275.

Hughes, R.E., DeMaris, P.J. and White, W.A., 1992, Underclays and related paleosols associated with coals. *In* I.P. Martini and W. Chesworth (eds), *Weathering, Soils and Paleosols*, Elsevier, Amsterdam: 501–523.

Hutchinson, G.E., 1950, Survey of contemporary knowledge of biogeochemistry: 3. The biogeochemistry of vertebrate excretion, *Bulletin of the American Museum of Natural History*, **96**: 554 pp.

Kowalewski, M., 1996, Time-averaging, overcompleteness and the geological record, *Journal of Geology*, **104**: 317–326.

Kraus, M.J., 1987, Integration of channel and floodplain suites: II. Vertical relations of alluvial paleosols, *Journal of Sedimentary Petrology*, **57**: 602–612.

Kraus, M.J., 1992, Mesozoic and Tertiary paleosols. *In* I.P. Martini and W. Chesworth (eds), *Weathering, Soils and Paleosols*, Elsevier, Amsterdam: 525–542.

Kraus, M.J. and Bown, T.M., 1986, Paleosols and time resolution in alluvial stratigraphy. *In* V.P. Wright (ed.), *Paleosols: Their Origin, Classification and Interpretation*, Blackwells, Oxford: 180–207.

Kraus, M.J. and Bown, T.M., 1993a, Short term sediment accumulation rates determined from Eocene alluvial paleosols, *Geology*, **21**: 743–746.

Kraus, M.J. and Bown, T.M., 1993b, Paleosols and sandbody prediction in alluvial sequences. *In* C. North and D.T. Prosser (eds), *Characterization of Fluvial and Aeolian Reservoirs, Special Publication of the Geological Society of London*, **73**: 23–51.

Kraus, M.J. and Middleton, L.T., 1987, Dissected paleotopography and base level changes in a Triassic fluvial sequence, *Geology*, **15**: 18–21.

Krauskopf, K.B., 1979, *Introduction to Geochemistry*, McGraw-Hill, New York: 721 pp.

Krumbein, W.C. and Garrels, R.M., 1952, Origin and classification of chemical sediments in terms of pH and oxidation–reduction potential, *Journal of Geology*, **60**: 1–33.

Leeder, M.R., 1975, Pedogenic carbonates and flood sediment accumulation rates: a quantitative model for arid-zone lithofacies, *Geological Magazine*, **112**: 257–270.

Lindsay, W.L. and Vlek, P.L.G., 1977, Phosphate minerals. *In* J.B. Dixon and S.B. Weed (eds), *Minerals in Soil Environments*, Soil Science Society of America, Madison, Wisconsin: 639–672.

Lyell, C., 1877, *Principles of Geology, 1* (11th edn), Appleton, New York: 671 pp.

Machette, M.N., 1985, Calcic soils of the southwestern United States. *In* D.L. Weide (ed.), *Soils and Quaternary Geology of the Southwestern United States, Geological Society of America Special Paper*, **203**: 1–21.

Marbut, C.F., 1935, *Atlas of American Agriculture, Part III, Soils of the United States*, Government Printer, Washington: 98 pp.

Markewich, H.W., Pavich, M.J. and Buell, G.R., 1990, Contrasting soils and landscapes of the Piedmont and coastal Plain, eastern United States, *Geomorphology*, **3**: 417– 447.

Mellett, J.S., 1974, Scatological origin of microvertebrate accumulations, *Science*, **185**: 349–350.

Moore, P.D. and Bellamy, D.J., 1974, *Peatlands*, Springer, New York: 221 pp.

Olsen, P.E., 1986, A 40 million-year lake record of early Mesozoic orbital climatic forcing, *Science*, **234**: 842–848.

Percival, C.J., 1986, Paleosols containing an albic horizon: examples from the Upper Carboniferous of northern England. *In* V.P. Wright (ed.), *Paleosols: Their Recognition and Interpretation*, Blackwell, Oxford: 87–111.

Pickford, M., 1986, Sedimentation and fossil preservation in the Nyanza Rift System. *In* L.E. Frostick, R.W. Renaut, I. Reid and J.J. Tiercelin (eds), *Sedimentation in the East African Rifts, Special Publication of the Geological Society of London*, **25**: 345–362.

PiPujol, M.D. and Buurman, P., 1994, The distinction between ground-water gley and surface-water gley phenomena in Tertiary paleosols of the Ebro Basin, NE Spain, *Palaeogeography, Palaeoclimatology, Palaeoecology*, **110**: 103–113.

Posamentier, H.W. and Vail, P.R., 1988, Eustatic controls for clastic deposition: II. Sequence and systems tract models. *In* C.K. Wilgus, B.S. Hastings, C.G. St C. Kendall, H.W. Posamentier, C.A. Ross, and J.C. Van Wagoner (eds), *Sea Level*

Changes: An Integrated Approach, Special Publication of the Society of Economic Paleontologists and Mineralogists, **42**: 125–154.

Potter, L.D. and Rowley, J., 1960, Pollen rain and vegetation, San Augustin Plains, New Mexico, *Botanical Gazette*, **122**: 1–25.

Prothero, D.R., 1996, Magnetic stratigraphy of the White River Group in the High Plains. *In* D.R. Prothero and R.J. Emry (eds), *The Terrestrial Eocene–Oligocene Transition in North America*, Cambridge University Press, New York: 262–277.

Prothero, D.R. and Rensberger, J.M., 1985, Magnetostratigraphy of the John Day Formation, Oregon, and the North American Oligocene–Miocene boundary, *Newsletters in Stratigraphy*, **15**: 59–70.

Prothero, D.R. and Swisher, C.C., 1992, Magnetostratigraphy and geochronology of the terrestrial Eocene–Oligocene transition in North America. *In* D.R. Prothero and W.A. Berggren (eds), *Eocene–Oligocene Climatic and Biotic Evolution*, Princeton University Press, Princeton: 46–73.

Prothero, D.R., Denham, C.R. and Farmer, H.G., 1985, Magnetostratigraphy of the White River Group and its implications for Oligocene geochronology, *Palaeogeography, Palaeoclimatology, Palaeoecology*, **42**: 151–156.

Prothero, D.R., Heaton, T.H. and Stanley, S.M., 1998, Speciation and faunal stability in mammals during the Eocene–Oligocene climatic crash. *In* C. Janis, K.M. Scott and L. Jacobs (eds), *Tertiary Mammals of North America*, Cambridge University Press, New York.

Retallack, G.J., 1975, The life and times of a Triassic lycopod, *Alcheringa*, **1**: 3–29.

Retallack, G.J., 1976, Triassic paleosols of the upper Narrabeen Group of New South Wales: I. Features of the paleosols, *Journal of the Geological Society of Australia*, **23**: 383–399.

Retallack, G.J., 1977, Triassic paleosols of the upper Narrabeen Group of New South Wales: II. Classification and reconstructions, *Journal of the Geological Society of Australia*, **24**: 19–35.

Retallack, G.J., 1979, Middle Triassic coastal plain deposits in Tank Gully, Canterbury, New Zealand, *Journal of the Royal Society of New Zealand*, **9**: 397–414.

Retallack, G.J., 1983a, Middle Triassic estuarine deposits near Benmore Dam, southern Canterbury and northern Otago, New Zealand, *Journal of the Royal Society of New Zealand*, **13**: 107–127.

Retallack, G.J., 1983b, Late Eocene and Oligocene paleosols from Badlands National Park, *Geological Society of America Special Paper*, **193**: 82 pp.

Retallack, G.J., 1984, Completeness of the rock and fossil record: estimates using fossil soils, *Paleobiology*, **10**: 59–78.

Retallack, G.J., 1986, Fossil soils as grounds for interpreting long-term controls on ancient rivers, *Journal of Sedimentary Petrology*, **56**: 1–18.

Retallack, G.J., 1988a, Field recognition of paleosols. *In* J. Reinhardt and W.R. Sigleo (eds), *Paleosols and Weathering through Geologic Time: Principles and Applications*, *Geological Society of America Special Paper*, **216**: 1–20.

Retallack, G.J., 1988b, Down-to-earth approaches to vertebrate paleontology, *Palaios*, **3**: 335–344.

Retallack, G.J., 1990, *Soils of the Past*, Unwin-Hyman, London: 520 pp.

Retallack, G.J., 1991, *Miocene Paleosols and Ape Habitats in Pakistan and Kenya*, Oxford University Press, New York: 346 pp.

Retallack, G.J., 1992, Paleosols and changes in climate and vegetation across the Eocene/Oligocene boundary. *In* D.R. Prothero and W.A. Berggren (eds), *Eocene–Oligocene Climatic and Biotic Evolution*, Princeton University Press, Princeton: 382–398.

Retallack, G.J., 1994a, A pedotype approach to latest Cretaceous and earliest Tertiary paleosols in eastern Montana, *Geological Society of America Bulletin*, **106**: 1377–1397

Retallack, G.J., 1994b, The environmental factor approach to the interpretation of paleosols. *In* R. Amundson, J. Harden and M. Singer (eds), *Factors of Soil Formation, Special Paper of the Soil Science Society of America*, **33**: 31–64.

Retallack, G.J., 1995, Paleosols of the Siwalik Group as a 15 Ma record of South Asian paleoclimate. *In* S. Wadia, R. Korisettar and V.S. Kale (eds), *Quaternary Environments and Geoarchaeology of India: Essays in Honour of S.N. Rajaguru, Memoir of the Geological Society of India*, **32**: 36–51.

Retallack, G.J., 1997, Post-apocalyptic greenhouse revealed by Triassic paleosols in the Sydney Basin, Australia, *Geological Society of America Bulletin*, **110**.

Retallack, G.J. and Dilcher, D.L., 1981a, A coastal hypothesis for the dispersal and rise to dominance of flowering plants. *In* K.J. Niklas (eds), *Paleobotany, Paleoecology and Evolution*, 2, Praeger Press, New York: 27–77.

Retallack, G.J. and Dilcher, D.L., 1981b, Early angiosperm reproduction: *Prisca reynoldsii* gen. et sp. nov. from mid-Cretaceous coastal deposits in Kansas, U.S.A., *Palaeontographica*, **B179**: 103–137.

Retallack, G.J. and Dilcher, D.L., 1988, Reconstructions of selected seed ferns, *Annals of the Missouri Botanical Gardens*, **75**: 1010–1057.

Retallack, G.J. and German-Heins, J., 1994, Evidence from paleosols for the geological antiquity of rain forest, *Science*, **265**: 499–502.

Retallack, G.J. and Ryburn, R.J., 1982, Middle Triassic deltaic deposits in Long Gully, near Otematata, north Otago, New Zealand, *Journal of the Royal Society of New Zealand*, **12**: 207–227.

Retallack, G.J., Bestland, E.A. and Dugas, D.P., 1995, Miocene paleosols and habitats of *Proconsul* on Rusinga Island, Kenya, *Journal of Human Evolution*, **29**: 53–91.

Retallack, G.J., Bestland, E.A. and Fremd, T.J., 1996, Reconstructions of Eocene and Oligocene plants and animals of central Oregon, *Oregon Geology*, **58**: 51–69.

Retallack, G.J., Krull, E.S. and Robinson, S.E., 1997, Permian and Triassic paleosols and paleoenvironments of southern Victoria Land, Antarctica, *US Antarctic Journal*, **30**: 33–36.

Retallack, G.J., Krull, E.S. and Robinson, S.E., 1998, Permian and Triassic paleosols and paleoenvironments of the central Transantarctic Mountains, Antarctica, *US Antarctic Journal*.

Rose, K.D., 1990, Postcranial skeletal remains and adaptations in early Eocene mammals from the Willwood Formation, Bighorn Basin, Wyoming. *In* T.M. Bown and K.D. Rose (eds), *Dawn of the Age of Mammals in the Rocky Mountain Interior, North America, Geological Society of America Special Paper*, **243**: 107–123.

Sadler, P.M., 1981, Sediment accumulation rates and the completeness of stratigraphic sections, *Journal of Geology*, **89**: 569–584.

Scheihing, MH and Pfefferkorn, H.M., 1984, The taphonomy of land plants in the Orinoco Delta: a model for the incorporation of plant parts in clastic sediments of late Carboniferous age in Euramerica, *Review of Palaeobotany and Palynology*, **41**: 25–40.

Schultz, C.B. and Stout, M., 1980, Ancient soils and climatic changes in the central Great Plains, *Transactions of the Nebraska Academy of Science*, **8**: 187–205.

Sclater, J.G. and Christie, A.F., 1980, Continental stretching as an explanation of the post-mid-Cretaceous subsidence of the central North Sea Basin, *Journal of Geophysical Research*, **85**: 3711–3739.

Shibaoka, M. and Smyth, M., 1975, Coal petrography and the formation of coal seams in some Australian sedimentary basins, *Economic Geology*, **70**: 1463–1473.

Shipman, P., 1981, *Life History of a Fossil*, Harvard University Press, Cambridge: 222 pp.

Smith, R.M.H., 1990, Alluvial paleosols and pedofacies sequences in the Permian lower Beaufort of the southwestern Karoo Basin, South Africa, *Journal of Sedimentary Petrology*, **60**: 258–262.

Smyth, M., 1970, Type seam sequences for some Permian Australian coals, *Proceedings of the Australasian Institute of Mining and Metallurgy*, **233**: 7–15.

Soil Survey Staff, 1975, *Soil Taxonomy, Handbook of the U.S. Department of Agriculture*, **436**: 754 pp.

Soil Survey Staff, 1990, *Keys to Soil Taxonomy, Technical Monograph, Soil Management Support Services, Blacksburg, Virginia*, **19**: 422 pp.

Strasser, A., 1988, Shallowing upward sequence in Purbeckian peritidal carbonates (lowermost Cretaceous, Swiss and French Jura Mountains), *Sedimentology*, **35**: 367–383.

Swisher, C.C. and Prothero, D.R., 1990, Single crystal $^{40}Ar/^{39}Ar$ dating of the Eocene–Oligocene transition in North America, *Science*, **249**: 760–762.

Taylor, E.L. and Taylor, T.N., 1993, Fossil tree rings and paleoclimate from the Triassic of Antarctica. *In* S.G. Lucas and M. Morales (eds), *The Non-marine Triassic, Bulletin of the New Mexico Museum of Natural History and Science*, **3**: 453–455.

Taylor, E.L., Taylor, T.N. and Cuneo, N.R., 1992, The present is not a key to the past: a polar forest from the Permian of Antarctica, *Science*, **257**: 1675–1677.

Tedford, R.H., Swinehart, J.B., Swisher, C.C., Prothero, D.R., King, S.A. and Tierney, J.E., 1996, The Whitneyan–Arikareean transition in the High Plains. *In* D.R. Prothero and R.J. Emry (eds), *The Terrestrial Eocene–Oligocene Transition in North America*, Cambridge University Press, New York: 312–334.

Tegelaar, E.W., Kerp, H., Visscher, H., Schenck, P.A. and DeLeeuw, J.W., 1991, Bias of the paleobotanical record as a consequence of variation in the chemical composition of higher vascular plant cuticles, *Paleobiology*, **17**: 133–144.

Terry, D.O. and Evans, J.E., 1994, Pedogenesis and paleoclimatic implications of the Chamberlain Pass Formation, basal White River Group, Badlands National Park, South Dakota, *Palaeogeography, Palaeoclimatology, Palaeoecology*, **110**: 197–215.

Thompson, G.R., Fields, P.W. and Alt, A.P., 1982, Land-based evidence for Tertiary climatic variations in the northern Rockies, *Geology*, **10**: 412–417.

Towe, K.M., 1987, Fossil preservation. *In* R.S. Boardman, A.H. Cheetham and A.J. Rowell (eds), *Fossil Invertebrates*, Blackwell Scientific, Palo Alto: 36–41.

Watson, J.P., 1967, A termite mound in an Iron Age burial ground in Rhodesia, *Journal of Ecology*, **55**: 663–669.

Wilding, L.P. and Drees, L.R., 1972, Biogenic opal in Ohio soils, *Proceedings of the Soil Science Society of America*, **35**: 1004–1010.

Wilding, L.P., Hallmark, C.T. and Smeck, N.E., 1979, Dissolution and stability of biogenic opal, *Proceedings of the Soil Science Society of America*, **43**: 800–802.

Wilson, R.C.L., 1991, Sequence stratigraphy: an introduction, *Geoscientist*, **1**: 13–23.

Winkler, D.A., 1983, Paleoecology of an early Eocene mammalian fauna from paleosols in the Clarks Fork Basin, northwestern Wyoming, *Palaeogeography, Palaeoclimatology, Palaeoecology*, **43**: 261–298.

Wright, V.P., 1992, Paleopedology: stratigraphic relationships and empirical models. *In* I.P. Martini and W. Chesworth (eds), *Weathering, Soils and Paleosols*, Elsevier, Amsterdam: 475–499.

Wright, V.P. and Marriott, S.B., 1993, The sequence stratigraphy of fluvial depositional systems: the role of floodplain sediment storage, *Sedimentary Geology*, **86**: 203–210.

Yanovsky, E., Nelson, E.K. and Kingsbury, R.M., 1932, Berries rich in calcium, *Science*, **75**: 565–566.

7
Phylogenetic Analyses and the Quality of the Fossil Record

Peter J. Wagner

INTRODUCTION

Reconstructing phylogeny has long been of interest to palaeontologists. This has become especially true in recent years, as several workers have stressed the importance of considering evolutionary relationships when evaluating palaeontological data (for example, Smith, 1994). Early workers apparently relied extensively on stratigraphic data when estimating phylogenies, both for establishing models of morphological evolution and for estimating relationships. However, many recent workers, especially those from the cladistic school, consider stratigraphic data useless for assessing phylogeny. Some have gone so far as to suggest that estimated phylogenies represent the best assessment of the quality of the fossil record (see, for example, Smith, 1988, 1994; Norell, 1992; Norell and Novacek, 1992a,b; Benton and Storrs, 1994; Benton, 1995). Whether stratigraphic data have a role in evaluating phylogeny (and, if so, how such data are best utilized) or whether stratigraphy should be evaluated in the light of parsimony analyses are important issues to palaeontologists. This paper will review: (1) the expected relationship between stratigraphic data and estimates of phylogeny; (2) whether estimates of phylogeny (especially those derived from parsimony) meaningfully assess the quality of the fossil record; and (3) whether stratigraphic data can improve estimates of phylogeny.

The issues addressed herein are important for two reasons. First, several workers have assessed the quality of the fossil record based on cladograms. The quality of the fossil record obviously is important when assessing several

The Adequacy of the Fossil Record. Edited by S.K. Donovan and C.R.C. Paul.
© 1998 John Wiley & Sons Ltd.

evolutionary issues, especially diversity patterns (for example, Raup, 1975; Signor, 1985; Smith, 1988; Miller and Foote, 1996), so cladograms are even more useful than commonly portrayed if they offer a useful tool for evaluating the record. Second, estimates of phylogeny have been used to test macroevolutionary hypotheses about rates of morphological evolution (for example, Wills *et al.*, 1994; Anstey and Pachut, 1995; Wagner, 1995b, 1997), to discern patterns of speciation (Jackson and Cheetham, 1994; Wagner and Erwin, 1995) and to calculate rates of origination and/or extinction (for example, Sanderson and Bharathan, 1993; Purvis *et al.*, 1995). Accurate estimates of phylogeny clearly are important for both issues. Unfortunately, minimum-step parsimony, the preferred method of cladists, is not foolproof – numerous simulation studies have shown that parsimony often fails under even unrealistically simple models of character evolution (for example, Rohlf *et al.*, 1990; Huelsenbeck and Hillis, 1993; Kim, 1993; Mooers *et al.*, 1995). We should be wary of an evaluator of stratigraphic sampling that is itself an error-prone estimate. We also should consider whether stratigraphic data can improve parsimony estimates of phylogeny, and thus improve our insights into evolutionary patterns and processes.

The major concern for both issues addressed herein is the accuracy of phylogenetic estimates. Thus, these arguments are irrelevant to one particular branch of cladistics, pattern cladistics, which focuses exclusively on classification based on parsimonious hierarchies (Patterson and Rosen, 1977; Nelson, 1979; Harvey and Pagel, 1991; Frost and Kluge, 1994; Norell, 1996).

EVALUATING THE FOSSIL RECORD WITH CLADOGRAMS

Smith's (1988) classic study of early Palaeozoic echinoderms suggested that an apparent mass extinction among supposedly primitive echinoderms at the end of the Cambrian followed by the radiation of more derived taxa in the late early Ordovician might be an artefact of poor sampling in particular intervals and misleading taxonomy. A phylogeny (estimated by parsimony) suggested that many of the late Cambrian echinoderms had either descendants or close relatives in the Arenig (early Ordovician). If true, then many more species survived the Cambrian than previously thought and the Tremadoc (earliest Ordovician) must have had numerous unsampled echinoderm species. Thus, Smith's results had implications not simply for macroevolution, but also for sampling.

Smith's basic approach has an intuitive appeal and has been adopted by several workers. Estimated phylogenies can imply gaps in the fossil record, so accurate estimates of phylogeny also assess completeness. Given a set of observed ranges (Figure 7.1A), cladograms usually suggest unsampled an-

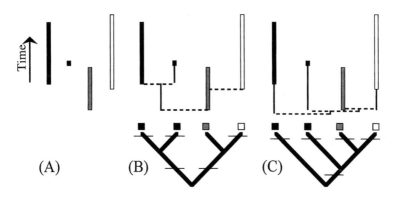

Figure 7.1   *(A) Temporal/stratigraphic ranges of four species as observed from the fossil record. (B) An estimated phylogeny, suggesting a ghost lineage (Norell, 1992) linking the dark-grey species to the black species and a ghost taxon (Norell, 1993) linking that clade to the light clade. Ghost lineages/taxa are shown by thin vertical black lines. Because the light-grey species is considered ancestral to the white species, no ghost lineage is required (Norell, 1992). (C) Another estimated phylogeny for the same species, which suggests many more ghost lineages. The light-grey species no longer is considered ancestral to the white species, implying that some precursors of the latter species at least existed since the first appearance of the former species. The dark-grey species is considered a sister taxon of the light clade, meaning that it, too, must have had unsampled precursors at least as old as the light-grey species.*

cestors linking observed taxa and frequently suggest that observed species diverged prior to their first known appearance (Figure 7.1B,C). The amount of incongruence is dependent on an estimated phylogeny – Figure 7.1B posits far fewer gaps because the tree is far more congruent with observed ranges than is Figure 7.1C, and even posits sampled ancestors. Of course, there can be only one phylogeny and we are forced to use estimates of that phylogeny. If Figure 7.1B is closer to the truth, but Figure 7.1C is a more parsimonious solution, then our incorrect estimate of phylogeny will yield an incorrect estimate of incompleteness. Therefore, the reader should bear in mind that cladograms can evaluate the quality of the fossil record only if the cladogram is reasonably correct (Norell and Novacek, 1992b).

Several metrics for measuring the congruence between phylogeny and stratigraphy have been proposed, each of which describes something slightly different. Hitchin and Benton (1997) recently contrasted these metrics, so I discuss them only in terms of their adequacy for evaluating the quality of the fossil record. In particular, I will focus on how well these metrics serve as a sampling metric, that is, an evaluation of the proportion of species that are sampled and/or the probability of sampling a taxon over a given interval.

Gauthier *et al.* (1988) attempted to quantify congruence by comparing order in which clades appear in the fossil record with the order in which clades were posited to appear by a cladogram (Figure 7.2). This approach considers the fossil record congruent with stratigraphy when the immediate outgroup of each clade (for example, taxon a relative to clade b–e in Figure 7.2A) is as old as, or older than, that clade. The technique was further refined by Norell and Novacek (1992a,b), who used the test statistic from Spearman's rank correlation test (a non-parametric analogue of correlation tests; Sokal and Rohlf, 1981, p. 607) to evaluate the congruence. A perfect correlation yields $\rho = 1.0$ (for example, Figure 7.2B; note that taxon e is omitted from the calculation as there is a necessary tie in clade rank for the final node). Poor fits (for example, Figure 7.2C) yield a lower ρ, with $\rho = 0$ indicating that there is no association between clade rank and first appearance, and $\rho = -1.0$ indicating complete incongruence.

Examining rank correlations can be done only after the tree has been reduced to a completely pectinate (or unbalanced) topology. However, trees of any size are expected to be more symmetrical (or balanced) and thus include numerous subclades (for example, Figure 7.3). The problem for rank correlation is that cladograms predict divergence times only vertically; that is, positing a j–k clade in Figure 7.3 makes a prediction about when taxon i diverged, but not about the other 'node 4' taxa, clades g–h, d–e and b–c. Thus, a rank within any one subclade cannot be equated with the same rank in a different subclade.

Huelsenbeck (1994) noted the problems cited above and proposed an alternative metric, the Stratigraphic Consistency Index (SCI), which describes the proportion of nodes that are congruent with stratigraphy. Congruency here uses the same definition, with congruent nodes being those whose sister taxon is the same age or older in the fossil record.

$$\text{SCI} = \frac{\text{Number of consistent nodes}}{\text{Number of nodes} - 1} \tag{1}$$

If the tree is fully bifurcating, then the denominator is the number of taxa −2. The basal node (for example, node 1 in Figure 7.3) cannot be examined without adding additional outgroup taxa.

Congruence metrics suffer drawbacks. Siddall (1996) suggested that the SCI is biased toward favouring more pectinate trees, reasons for which are discussed below. A more critical problem is that congruence metrics really do not measure what they are claimed to measure. Congruence metrics assume that a clade should appear at the same time as or after its sister taxon. However, we really expect it to appear after its sister taxon only if the sister taxon is paraphyletic; otherwise, sister clades should appear within the lifetime of a common ancestor. This usually will be within the same stratigraphic interval. Thus, if two sister clades have different first appearances

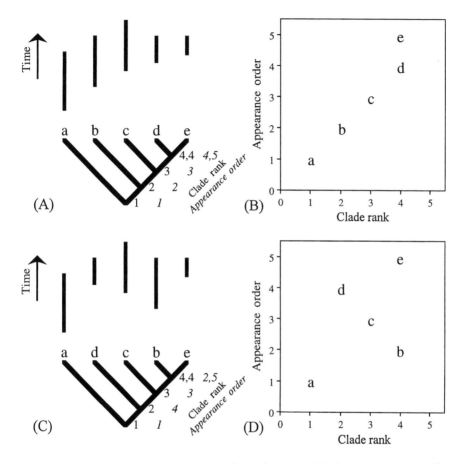

Figure 7.2 *Congruence between stratigraphy and estimated phylogeny, as measured by Gauthier et al. (1988) and Norell and Novacek (1992a,b). Clade ranks (that is, the node to which a taxon is linked, ranked relative to the base of the tree) are contrasted with order of appearance. If the immediate outgroup of a clade (for example, taxon a relative to the remainder of either tree) is older than that clade, then the node is considered congruent. (A) One estimate of phylogeny for five taxa that is highly congruent with stratigraphy. Note that this completely congruent tree still implies gaps in sampling. (B) Association between clade rank and appearance order. Note that Spearman's rank correlation tests omit taxon e, allowing a perfect fit (Norell and Novacek, 1992a,b). (C) Another estimate for the same five taxa. (D) A correspondingly much poorer association between clade rank and appearance order. Note that this test can be applied only to pectinate topologies.*

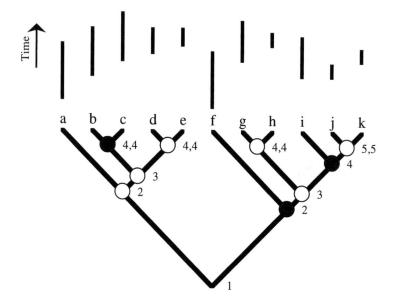

Figure 7.3   *Congruence as measured by the Stratigraphic Consistency Index (SCI) (Huelsenbeck, 1994). Numbers by the nodes give rank order of appearances. White nodes are considered consistent because the sister taxon of that clade appears prior to (or at the same time as) the clade defined by that node. For example, the d–e clade is considered consistent with stratigraphy because its sister taxon (the b–c clade) appears earlier in the fossil record. Black nodes are considered inconsistent because the sister taxon appears later. In this case, the b–c clade is considered inconsistent because it precedes its sister taxon in the record. As shown by that particular example, the b–c clade and d–e clade will both be considered consistent only if both appear at the same time. Note also that even congruent nodes often require ghost taxa. For example, a ghost taxon is needed to link the d–e clade phylogenetically to the b–c clade*

(invoking a ghost lineage), one is considered congruent with stratigraphy and the other is not. In Figure 7.3, node g–h is considered congruent because its sister taxon, clade i–k, appears earlier but node i–k is considered incongruent. There is, of course, some probability that both will be first sampled from the same interval, which is at least equal to R^2, where R is the probability of sampling a taxon per interval.

What type of distribution do we expect for the SCI? If all taxa are monophyletic and evolution proceeds in a bifurcating manner, then we expect R^2 to be the proportion of nodes to have both members sampled in their first stratigraphic unit. (If taxa within a clade cannot be summarized by a single preservation rate, then the proportion will be the average of $R_i * R_j$ for all sister pairs i and j.) We expect half of the remaining nodes (that is, $[1 - R^2]/2$) to be congruent because they appear after their sister taxa. Thus, the expected SCI,

$$E[SCI] = \left(R^2 + \frac{1 - R^2}{2} \right) \pm \text{binomial error} \qquad (2)$$

Thus, as R goes to 1.0, the SCI should go to 1.0; as R goes to zero, the SCI should go to 0.5.

Given such a distribution, what does a particular SCI mean? An overly high SCI (that is, SCI > E[SCI] + 2SD) implies that too many immediate outgroups appear prior to their derived sister taxon. This is consistent with those taxa being paraphyletic and thus 'ancestral' to their sister taxa. An overly low SCI implies that the cladogram is incorrect, and thus that parsimony is an inappropriate model for estimating the phylogeny of that clade. Note that the same relationship is expected between R and the congruence metric of Gauthier *et al.* (1988).

The other problem with congruence metrics is highlighted by the fact that two sister clades will produce one incongruency and one congruency whenever they do not appear at the same time. However, this is not a problem when a clade's sister taxon is a species, as species are not considered inconsistent. Thus, as Siddall (1996) originally noted, the SCI usually can be greatest for pectinate trees, as that is the easiest way to make all clades appear consistent with stratigraphy.

Norell's (1992) Z-statistic attempted to evaluate how much of an inferred history is actually sampled, based on ghost lineages (that is, extensions of first appearances interpolated from a cladogram):

$$Z = 1 - \frac{\text{Average ghost lineage/taxon duration}}{\text{Age of the clade}} \qquad (3)$$

The ghost taxon needed to link clade g–h with clade i–k now detracts from estimated congruence. A ghost taxon is necessary for clade g–k if taxon f cannot be considered ancestral to that clade; if taxon f can be so considered, then no stratigraphic gap need be invoked. Thus, Norell's statistic is specific to particular estimates of phylogeny and different interpretations of the same cladogram can yield different Z-statistics (Figure 7.4).

Unfortunately, the Z-statistic does not measure the proportion of a clade's history that is unsampled; to do this, the denominator should be the summed duration of the ranges within a clade rather than the duration of the clade itself. The Relative Completeness Index (RCI) comes closer to this goal, describing the proportion of the temporal distributions (as inferred from an estimate of phylogeny) that is represented by observed fossils (Benton and Storrs, 1994). (Note that the RCI has nothing to do with the Rescaled Consistency Index (also abbreviated RCI), which measures homoplasy invoked by a phylogenetic estimate; Farris, 1989.) The metric is measured as:

$$RCI = \frac{\Sigma \ \text{Minimum gap implied by species x}}{\Sigma \ \text{Observed range of species x}} \tag{4}$$

Worst-case scenarios, where species are known only by single horizons and thus have no true ranges, yield an RCI of $-\infty$. If there are no inferred gaps, then the RCI = 1.0. Note that as 'observed' gaps (that is, those within a taxon's range) are not included when calculating the RCI, it measures only the implications of an estimated phylogeny.

Gaps and ranges in previous studies have been measured in millions of years (for example, Benton and Storrs, 1994; Benton, 1995). The RCI also can be described using Fisher's (1991, 1994) stratigraphic debt concept, in which gaps and durations are described by the number of intervals in which taxa are sampled. In this case:

$$RCI = 1 - \frac{\Sigma \ \text{Number of intervals that species x invokes a ghost lineage/taxon}}{\Sigma \ \text{Duration of species x (in intervals)}} \tag{4a}$$

This might be preferable in cases where time scales are uncertain or the error on the estimates is great (for example, when evaluating Palaeozoic taxa).

Like the Z-statistic, the RCI is affected by ancestor–descendant estimates, which reduce minimum implied stratigraphic gaps (or stratigraphic debt). Thus, the same cladogram can yield two different RCIs, depending on whether one allows (and data support) ancestor–descendant relationships.

Because RCI ranges from $-\infty$ to $+1.0$, it does not offer an assessment of the proportion of species sampled or the probability of sampling a species per time/stratigraphic unit. Again, this could be done by adjusting the denominator so that it reflects the sum of ranges inferred by the phylogeny (that is, observed ranges plus ghost lineages/taxa). This phylogenetic sampling metric gives the proportion of a clade's reconstructed history that is represented by ghost lineages and ghost taxa (see Smith and Littlewood, 1994):

$$\text{Implied gappiness} = \frac{\text{Sum of inferred gaps}}{\text{Sum of observed ranges} + \text{Sum of inferred gaps}} \tag{5}$$

where inferred ranges are the observed ranges plus ghost lineage extensions implied by a phylogeny. If all species are known from single occurrences (and have no ranges themselves), then the denominator is equal to the numerator, implying that there is nearly an infinitesimal chance of sampling a species. If stratigraphic units are used (as when calculating Fisher's stratigraphic debt), then the metric estimates the probability of sampling a taxon

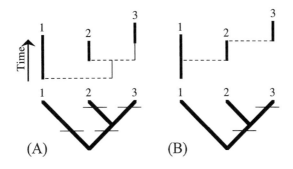

Figure 7.4 *The effect of ancestor–descendant estimates on inferred stratigraphic gaps, using different phylogenies derived from the same cladogram. Metrics that use such gaps (for example, the Z-statistic, Relative Completeness Index and Fisher's stratigraphic debt) therefore are metrics of particular phylogenies rather than of cladograms. (A) All three taxa are diagnosed by autapomorphies (thin bars), meaning that none appears to be ancestral to the others. Ghost lineages and taxa (thin vertical black lines) are invoked, adding to the stratigraphic gaps. (B) An alternative phylogeny in which species 1 and 2 are identical to hypothesized ancestors. No stratigraphic gaps are invoked in this case*

per stratigraphic unit. However, note that this estimate will be an over-estimate unless gaps within observed taxon ranges are added to the numerator. Even then, there is no estimate of 'gaps' between last appearances and true extinctions (that is, the upper end of the Signor–Lipps effect; Signor and Lipps, 1982).

The 'gappiness' sampling metric can be used to estimate the proportion of taxa sampled if the typical duration of those taxa can be estimated (see Foote and Raup, 1996). However, the overestimate of the probability of sampling a taxon should lead to an overestimate of the proportion of taxa sampled. Alternatively, one can examine the number of taxa hypothesized by a clado-gram that actually are observed. Parsimony reconstructs the character states of estimated ancestors; if those reconstructions are accurate and if sampling is good, then we expect that some of those reconstructions should be matched by observed species (Foote, 1996). If none of the estimated ancestors is observed, then the cladogram suggests that we have very poor sampling, too poor, in fact, actually to estimate the proportion of species sampled. Note that only hypothesized taxa are important – if one includes observed and hypothesized taxa, then one would always conclude that more than 50% of taxa were sampled as there are at most $n - 1$ ancestors reconstructed for n taxa. Note also that this phylogenetic sampling metric is dependent not just on an accurate cladogram, but on an accurate phylogeny. We expect some ancestral species to be autapomorphic under parsimony optimization (Alroy, 1995; Wagner, 1995a, 1996) and we expect incorrect cladograms to posit non-existent species. Thus, whereas the 'gappiness' metric above should overestimate sampling, this phylogenetic sampling metric should chronically underestimate the quality of the record.

The incompleteness of the fossil record has long been known (Darwin, 1859) and palaeontologists began confronting the issue long before the advent of cladistics. Thus, there are methods for evaluating incompleteness that do not use estimated phylogenies. Non-phylogenetic metrics do not depend on accurate phylogenetic estimates, making them very appealing. Foote and Raup (1996) assessed completeness using the distribution of observed stratigraphic ranges. Foote and Raup found that the probability of sampling a taxon from one time/stratigraphic unit can be estimated from the proportions of taxa sampled from one, two and three stratigraphic/time units. The average duration of a taxon can be estimated from the same data (that is, the slope of a log-transformed histogram of the frequencies of taxa that are known from 1...n intervals; Van Valen, 1973). An estimate of the probability of sampling a taxon over the average taxon duration yields the probability of sampling any one taxon and, hence, the proportion of taxa we expect to have sampled.

One concern regarding any statistical analysis in palaeontology is that phylogenetic autocorrelation (that is, the non-independence of species within a phylogeny) can produce seemingly significant associations among taxa within a clade (Raup *et al.*, 1973; Felsenstein, 1985b). However, the Foote–Raup metric actually depends on phylogenetic autocorrelation. Extrapolating from observed distributions to infer total numbers of taxa assumes that the unsampled taxa have similar potentials for being preserved and sampled as do the unsampled taxa. Fortunately, factors affecting preservation and sampling potential (for example, skeleton types, ecological and geographical ranges) should not be distributed independently across phylogeny. This is especially true for unsampled portions of observed taxa (that is, ghost lineages), but also true for their unsampled ancestors (that is, ghost taxa) unless factors decreasing the probability of sampling a species increase the probability of that species leaving descendants. However, no model of speciation makes this prediction (Wagner and Erwin, 1995) and several models, especially those focusing on peripheral isolates (for example, Wright, 1931, 1932; Mayr, 1963; Eldredge and Gould, 1972), predict that factors favouring speciation should also favour sampling. This is important because the phylogenetic sampling metrics discussed above offer estimates only on ghost lineages and ghost taxa.

The fact that the Foote–Raup metric applies only to taxa such as those in the sampling pool means that it should overestimate sampling. For example, if a clade of snails includes a subclade that lost the shell, then the Foote–Raup metric would apply only to the shelled members of the clade. Differences in ecological and/or geographical distributions should have analogous effects. However, phylogenetic sampling metrics will apply only to the same taxa and thus suffer the same deficiency. Thus, both phylogenetic and non-phylogenetic sampling metrics apply to the branches, but not the twigs. The advantage of non-phylogenetic estimates of the probability of sampling a taxon per time/stratigraphic interval is that they incorporate gaps within observed ranges and the upper end of the Signor–Lipps effect. Phylogenetic methods can do the former only if observed gaps are tallied in an independent analysis. Given that phylogenetic sampling metrics should become less reliable as estimated phylogenies become more inaccurate, non-phylogenetic sampling metrics clearly are preferable for evaluating the quality of fossil data.

EVALUATING CLADOGRAMS WITH THE FOSSIL RECORD

Fossil data are potentially useful for phylogenetic analyses because they add extinct species to (otherwise) neontological analyses, offer the potential to root trees and allow evaluations of parsimony estimates. The importance of the first issue has been dealt with in detail elsewhere (Gauthier *et al.*, 1988; Donoghue *et al.*, 1989; Novacek, 1992; Eernisse and Kluge, 1993; Wills *et al.*, 1995) and requires only a brief summary. Like any other method of estimation, sampling affects the precision and accuracy of parsimony analyses (Lecointre *et al.*, 1993; Mooers, 1995). Solely neontological studies lack all extinct forms, which often is a major sampling deficiency. Despite the incomplete nature of fossils (even complete fossils preserve only portions of anatomy and no molecular data), they improve the accuracy of parsimony estimates by improving taxon sampling. The absence of ancestors (or at least close relatives of ancestors) appears to hinder parsimony, in which the ability to reconstruct ancestral characters decreases with increased divergence time (Huelsenbeck, 1991b; Maddison, 1995). Many extant species might be ancestral to other extant species, but as divergence times become greater, the probability of ancestral morphologies surviving both extinction and anagenetic change becomes increasingly smaller (Foote, 1996). Many taxa diverged from their closest living relatives many millions of years ago, introducing long-branch problems (see Felsenstein, 1978). Compensation for this sampling loss makes fossil taxa useful.

Fossil data offer the potential to root trees and determine character polarity. Currently, the preferred method is to use one or more outgroup taxa, that is, taxa assumed to be related closely to the clade of interest (the ingroup; see, for example, Maddison *et al.*, 1984). The outgroup method assumes that characters shared between the outgroup and ingroup are primitive to both; this sets the stage for polarizing characters within the ingroup. Primitive characters must evolve prior to more derived characters, so the null expectation is that primitive characters should appear in older strata than do their derived counterparts. However, using fossil data to polarize characters has been disparaged, either because of the numerous seemingly apocryphal examples where features known to be derived precede those known to be primitive (Harvey and Pagel, 1991) or because of the opinion that fossil homologies can be interpreted only in the context of extant taxa (Nelson, 1978; Patterson, 1994). Nevertheless, stratigraphy appears to play a tacit role when palaeobiologists choose outgroups – my own personal observation is that the outgroup species of many (if not most) major analyses are species contemporaneous with or slightly older than the oldest ingroup species.

Systematists often have a good reason for choosing particular outgroups. However, there frequently might be situations where the nearest relatives of a clade are disputed or are completely unknown. Choosing an outgroup then becomes an unwarranted assumption unless extensive separate analyses are performed (Wagner, 1997, 1998). In such cases, Huelsenbeck (1994) suggested that trees be rooted with stratigraphic data. The agreement of any parsimony tree with stratigraphy depends on how it is rooted; if there is no compelling outgroup, then the rooting(s) with the minimum incongruence with stratigraphy should be favoured. Huelsenbeck used the SCI metric, but one could use the Rescaled Consistency Index or stratigraphic debt in the same manner. This use of stratigraphic data is fully consistent with the suggestion that systematists use stratigraphic data as ancillary evidence (Harper, 1976; Smith, 1994). Stratigraphic rooting should be applied when there is uncertainty about outgroups and almost certainly is preferable to the use of hypothetical ancestors employed by many studies.

The appropriate use of stratigraphic data in refining estimates of phylogenetic relationships is controversial. As noted above, stratigraphic data probably played a role in early assessments of phylogeny. The pre-numerical studies are justifiably criticized for being difficult if not impossible to repeat. Interestingly, the criticism is more valid for the use of stratigraphy than for the use of morphology: whereas workers at least outlined important characters uniting taxa, stratigraphic gaps seemed to be invoked at convenience. Workers seemingly tried to link roughly contemporaneous taxa, but, if a long stratigraphic gap was required to support a particular idea, then it was done so without any evaluation of that gap.

With the advent of numerical techniques, phylogenetic analyses became far easier to repeat and evaluate because morphological characters were explicitly coded. Workers differing in opinions about homologies or models of morphological evolution then could critically recode and reanalyse estimates by other workers. (This property is usually attributed to cladistics alone, but it is equally true of phenetics.) The first techniques explicitly to utilize stratigraphic data were phenetic (Gingerich, 1979). Cladistic approaches have been slower to develop, probably because cladists contended that stratigraphic data were insufficient to support trees even a single step longer than the most parsimonious (see, for example, Nelson, 1978; Smith, 1994). Fundamentally, the contention has been that only hierarchical information (for example, character data) can describe phylogenetic patterns, whereas linear information (for example, stratigraphic data) cannot (for example, Rieppel and Grande, 1994). At best, stratigraphy was used as ancillary evidence – if (as usually happens) there are multiple equally parsimonious trees, then the tree that is least incongruent with stratigraphy should be favoured (Smith, 1994; Suter, 1994; Littlewood and Smith, 1995).

Advances in cladistic methodology have rendered as suspect the view that only maximally parsimonious trees should be considered valid approximators of phylogeny. These advances have occurred on two fronts. One concerns whether character data offer appreciably better support for one topology than another. Systematists now commonly use bootstrapping of characters (Felsenstein, 1985a; Sanderson, 1989, 1995) and analyses of particular branch lengths (Davis, 1993; Bremer, 1994) to assess this concern. It also is common to assess whether character data even contain the type of signal expected if phylogenetic autocorrelation is the primary source of structure (Archie, 1989; Faith, 1991; Huelsenbeck, 1991a; Alroy, 1994) and protocols exist for assessing which portions of cladograms are supported by that structure (Hillis and Huelsenbeck, 1992). Finally, advocates of congruence methods have noted that the optimal tree might not be the shortest one in any particular data set (see, for example, Swofford, 1991; Kim, 1993).

The second front involves simulation studies (see above). Spawned by the need to assess the numerous methods proposed for examining molecular data (including parsimony), simulations show that even using very simple evolutionary models, small amounts of homoplastic noise can disrupt hierarchical information, leading parsimony to incorrect solutions. Smith (1994) dismissed the relevance of these simulations to morphological data because simulations cannot hope to emulate the (presumably) complicated patterns of morphological evolution. However, the performance of parsimony typically worsens when underlying models of change are complicated (Huelsenbeck and Hillis, 1993; Kuhner and Felsenstein, 1994), as complicated models violate one of the key assumptions of unweighted parsimony (that is, that all characters are equally likely to change). This can be addressed through *a priori* character weighting, but weighting schemes usually are difficult to justify statistically.

The tendency for true trees to be longer than parsimony estimates can be demonstrated with simulations. Table 7.1 shows how parsimony underestimates true evolution under different sampling regimes and different rates of character state change, based on simulations using an MBL model of evolution (Raup *et al.*, 1973) in which simulations were run until six species

Table 7.1   *Underestimate of true tree length by parsimony and stratocladistics, based on simulations with 50 binary characters*

	Underestimate of true tree length					
	Species sampled (%)					
$f(\Delta/Br)$	95	89	77	61	42	21
Parsimony						
0.02	0.445	0.498	0.648	0.990	1.635	2.278
0.04	0.788	1.105	1.300	1.983	2.825	4.323
0.06	1.433	1.545	2.323	3.413	5.233	7.413
0.08	2.268	2.598	3.150	4.760	7.835	11.643
Stratocladistics						
0.02	0.170	0.160	0.120	0.178	0.608	1.283
0.04	0.313	0.503	0.573	1.105	1.638	3.510
0.06	0.573	0.683	1.390	2.323	4.078	6.488
0.08	1.090	1.520	1.938	3.503	6.588	10.673

$f(\Delta/Br)$ gives the frequency of character state change per branch on the real phylogeny. If sampling is 75%, then four times as many changes are expected over three sampled branches; if sampling is 50%, then twice as many changes are expected per sampled branch. Each cell represents the results of 1000 simulations under the sampling and evolutionary parameters specified, using an MBL model of cladogenesis and character evolution (see Raup et al., 1973). Note that stratocladistics underestimates true morphological evolution nearly as greatly as does parsimony.

were sampled. Note that the estimate error increases with both increasing rates of morphological evolution and decreasing intensities of sampling.

If systematists should consider a range of trees in addition to the most parsimonious ones, then expanding the ancillary role of stratigraphy becomes very appealing. The logical basis for this is straightforward. First, we cannot assume that morphological data contain only hierarchical information; instead, we must assume that both hierarchical and non-hierarchical information are present. Second, we expect the hierarchical signal to be congruent with linear stratigraphic data. Third, homoplasies should be distributed throughout a tree; random connections within the true phylogeny often will link disparately aged taxa, so we expect non-hierarchical homoplasies to be incongruent with stratigraphy. Thus, apparent hierarchy that deviates from linear patterns is evidence that the apparently synapomorphic characters are in fact convergent.

Although there have been few examples, trees with excellent congruence with stratigraphic data often are only a few steps longer than the maximally parsimonious trees. Huelsenbeck (1994) found that an 'optimal' tree (see below) was only two steps longer than the maximally parsimonious (20 vs. 18 steps of a maximum of 33). Wagner (1995a) found a tree positing no statistically significant gaps that was only seven steps longer than the most parsimonious (202 vs. 195 of a maximum of 618 steps). In an extensive analysis of

29 mammal data sets, Clyde and Fisher (1997) found that retention indices (RIs, which contrast the difference between the observed and maximum possible lengths with the difference between the minimum and maximum possible lengths; Farris, 1989) decreased only slightly to provide much higher stratigraphic RIs (median morphological RIs: 0.80 down to 0.75; median stratigraphic RIs: 0.58 up to 0.79).

Three methods incorporating stratigraphic data into cladistic analyses have been proposed: reweighting with stratigraphic consistency (Huelsenbeck, 1994), incorporating coded stratigraphic characters (stratocladistics; Fisher, 1991, 1994), and sieving trees with 95% confidence intervals (Wagner, 1995a; see also Jackson and Cheetham, 1994). The difference between reweighting and stratocladistics is similar to that between partition–congruence methods and total-evidence parsimony. In partition–congruence approaches, the optimal tree is found by analysing morphological and molecular data sets separately and then producing a consensus of the results. This treats each data type as separate but equal. Reweighting with stratigraphic data does the same; as proposed by Huelsenbeck (1994), the morphological length of a tree is rescaled by the SCI of that tree so that:

$$\text{Modified tree length} = \frac{\text{Raw tree length}}{\text{SCI}} \tag{6}$$

As the modified length is inversely proportional to congruence and directly proportional to the number of steps, trees that are highly inconsistent with stratigraphy and/or invoke many steps will have large modified lengths.

Stratocladistics is much more similar to total evidence in that stratigraphic data are analysed along with morphological data. The length of any tree is:

$$\text{Total parsimony debt} = \text{Number of steps} + \text{Stratigraphic debt} \tag{7}$$

Stratigraphic debt is reduced by making earlier-appearing sister species ancestral to their later-appearing sister taxa, so the total parsimony debt is dependent on a particular phylogeny (Figure 7.4). Immediate reversals (which leave ancestral species with autapomorphies under parsimony optimizations) are invoked when the reduction in stratigraphic debt equals the number of reversed characters because that satisfies a secondary parsimony criterion common to graph theory, thus reducing the number of necessary links (Alroy, 1995; see also Remane, 1985). Methodologically, this is important because appreciable proportions of ancestors will have autapomorphies under parsimony optimization even with the simplest models of character state evolution (Alroy, 1995; Wagner, 1995a, 1996). The ability to recognize these cases is required for phylogenetic methods to be accurate. Explicit phylogenies (including ancestor–descendant estimates where required) are necessary for evolutionary biologists because ancestor–descendant estimates are far more powerful than sister-taxon statements for testing a wide variety of macroevolutionary hypotheses. (Arguments about assumptions inherent to ancestor–descendant estimates stem from pattern cladistics philosophy and thus do not pertain to tests of macroevolutionary hypotheses.)

Coding stratigraphic characters represents a potential problem for stratocladistics, although the same obviously can be said of morphological characters. However, whereas there will always be some subjectivity in the coding of morphological characters, one can use statistical methods to justify the designation of stratigraphic intervals. One possible approach is to estimate the probability of sampling a taxon per given stratigraphic interval (Foote and Raup, 1996). By comparing this with the frequency of character state change per branch from the parsimony tree (and assuming that parsimony is not underestimating that frequency too severely), one can find intervals for which the probability of a gap is approximately equal to the probability of an additional character state change. This type of approach might also circumvent the fact that the number of characters will affect the influence of stratigraphy – that is, the more characters present, the greater number of homoplasies expected per branch and the greater amount of stratigraphic debt necessary to overcome those homoplasies. This is analogous to the concern that molecular data will swamp out morphological data in total-evidence analyses. Conventional wisdom would counter that we have more confidence in results when we have more characters; however, greater numbers of characters can yield greater support for incorrect topologies when models of evolution deviate strongly from the models assumed by the reconstruction methods (Felsenstein, 1978).

Wagner (1995a) proposed a method in which statistical analyses of gaps are used to find the shortest tree that posits no statistically significant gaps (see also Strauss and Sadler, 1989; Marshall, 1990; Jackson and Cheetham, 1994). Like stratocladistics, the method will reverse putative autapomorphies if the apparent first appearance of a sister species is significantly earlier than the first appearance of its sister taxon. Unlike stratocladistics, the number of autapomorphies reversed does not play a role. However, reversing numerous autapomorphies greatly increases the chance that the tree will not be the shortest one found. Sieving trees with confidence intervals also differs from stratocladistics in that gaps are assessed on a case-by-case basis. Thus, the same gap in different parts of the phylogeny will yield different results if the gap involves well-sampled species in some cases and poorly sampled species in others. A final difference is that gaps in different sampling realms can be incorporated to utilize taphonomic controls (Bottjer and Jablonski, 1988) and to accommodate difference in biogeography or ecology (Wagner, 1995a, fig. 2).

A potential problem with the confidence-interval approach concerns multiple independent samples, especially in analyses of large numbers of species. Suppose there are 20 gaps, each of which equals the 95% confidence intervals of the species involved. As there is a 1 in 20 chance of observing such a gap, we expect that one of those gaps should be real (Rice, 1989) and rejecting all of them will probably yield a Type I error. Interestingly, there is an even bigger concern with Type II errors. Suppose we have 10 gaps each of which equals the 90% confidence intervals of the species involved. Using the traditional $\alpha = 0.05$, we would accept all of these gaps. However, there is only a 1 in 10 chance of any one of those gaps being real; this is analogous to observing 10 straight heads on a coin weighted so that it should yield tails 9 times in 10. In fact, we expect that at most only 3 of those flips will yield heads. By focusing only on individual gaps, the confidence-intervals approach cannot assess whether the entire field of gaps deviates significantly from the observed record.

The tacit implication of arguments against using stratigraphic data is that their inclusion should yield overly long trees. However, preliminary simulations suggest that these methods tend to underestimate true tree lengths. Stratocladistics tends to come closer than the other methods, but it still produces trees that are nearly as short as parsimony trees and only a little closer to the true length (Table 7.1). If simulation results such as those shown in Table 7.1 are typical, then present methods incorporating stratigraphic data into phylogeny reconstruction underestimate rates of morphological evolution. Alternative methods therefore need to be considered. Huelsenbeck and Rannala (1997) have proposed a maximum likelihood method that incorporates stratigraphic data into phylogenetic analyses. Under maximum likelihood, one evaluates the likelihood of a particular hypothesis (in this case, an estimated phylogeny) given observed patterns (in this case, the stratigraphic patterns of first and last appearances). Because the fossil record is imperfect, likelihood should offer best support for estimates that include gaps in the fossil record. However, likelihood should offer poor support for phylogenies with a large proportion of nearly significant gaps. In addition, likelihood tests can be modelled to take into account variation in sampling probabilities owing to difference in taphonomy and biogeographical and/or ecological distributions. Thus, likelihood might offer advantages over the methods described above.

The method proposed by Huelsenbeck and Rannala (1997) applies best to a series of anagenetically evolving lineages sampled over a period without any faunal turnover. Workers dispute the ubiquity of both anagenesis (see, for example, Eldredge and Gould, 1972; Bookstein, 1987; Lynch, 1989) and consistent turnover (see, for example, Vrba, 1993; Morris *et al.*, 1995; Jablonski and Sepkoski, 1996). However, methods accommodating alternative scenarios can be derived from Huelsenbeck and Rannala's basic framework. Two types of previously proposed sampling metrics could be used in likelihood tests: those estimating the probability of sampling a taxon per stratigraphic interval (for example, Foote and Raup, 1996) and those estimating the distribution of stratigraphic gaps within taxon ranges (for example, Strauss and Sadler, 1989; Marshall, 1994). If using the former type of metric (as would be appropriate for a supraspecific analysis), then one could calculate the stratigraphic debt inherent to a series of trees. If one is examining species with horizon data, then one would assess the likelihood of observing a particular collection of range extensions, given the known sampling densities of the individual species. Using the approach initially described by Strauss and Sadler (1989), one could use a Dirichlet distribution to calculate the exact probability of various gaps within a lineage. The most likely gap between true origin and first appearance should be approximately equal to the average gap within the observed range, with larger and smaller gaps being less likely. For each tree, one would simulate sampling using the probabilities calculated from the observed data. The frequency of simulations yielding as many stratigraphic gaps as implied by the hypothesized tree then gives the likelihood of that hypothesized tree, based on the observed stratigraphic data.

Morphological evolution also can be incorporated into likelihood tests, although it is difficult to justify models of morphological evolution. This highlights the great irony of the debate presented here, namely that stratigraphic data actually are easier to evaluate quantitatively than are morphological data. However, one can use the parsimony estimate as a null model to estimate per-branch frequencies of character change, and then simulate morphological change under those frequencies. The likelihood of two independent hypotheses (in this case, tree length and stratigraphic debt) co-occurring is directly proportional to the product of the likelihoods of observing either (Edwards, 1992). Thus:

$$L[\text{stratigraphic debt} = Z \ \& \ \text{tree length} = Y \mid f\{\text{sampling}\} \ \& \ f\{\text{change}\}] \propto$$
$$L[\text{stratigraphic debt} = Z \mid f\{\text{sampling}\}] \times L[\text{tree length} = Y \mid f\{\text{change}\}] \qquad (8)$$

where inferred stratigraphic debt and tree lengths are a property of a particular estimate and f[sampling] and f[change] are commonly calculated from observed data. The support for alternative phylogenies then can be evaluated using log likelihood ratio tests (Huelsenbeck and Rannala, 1997; see, for example, Sokal and Rohlf, 1981, p. 695).

CONCLUSIONS

Cladograms probably are not appropriate for evaluating the quality of the fossil record. Phylogenetic sampling metrics depend on the accuracy not only of cladograms, but also of phylogenies, whereas non-phylogenetic methods do not. The biases of non-phylogenetic sampling metrics (that is, underestimating difficult-to-sample species) are also biases of phylogenetic sampling metrics. As trees highly consistent with stratigraphy are often only slightly longer than parsimony trees, and as parsimony frequently under-estimates tree lengths in simulations using the simplest models of evolution, the repeated claim that stratigraphic data cannot offer information about phylogenetic relationships also needs to be reconsidered. Of course, the informativeness of the fossil record will depend on the group being studied – statements based on the record of marine molluscs obviously do not apply to the record of dinosaurs (or vice versa). Moreover, much work remains to be done on determining exactly which approaches are most likely to yield correct results under particular circumstances. Careful evaluation of mor-phological data and parsimony analyses will always be important first steps toward elucidating phylogenetic patterns. Given the need for accurate as-sessments of phylogeny, they probably should not represent the only steps.

ACKNOWLEDGEMENTS

I thank the editors and J. Huelsenbeck for critical comments and discussion.

REFERENCES

Alroy, J., 1994, Four permutation tests for the presence of phylogenetic structure, *Systematic Biology*, **43**: 430–437.

Alroy, J., 1995, Continuous track analysis: a new phylogenetic and biogeographic method, *Systematic Biology*, **44**: 153–172.

Anstey, R.L. and Pachut, J.L., 1995, Phylogeny, diversity history and speciation in Paleozoic bryozoans. *In* D.H. Erwin and R.L. Anstey (eds), *New Approaches to Speciation in the Fossil Record*, Columbia University Press, New York: 239–284.

Archie, J.W., 1989, A randomization test for phylogenetic information in systematic data, *Systematic Zoology*, **38**: 239–252.

Benton, M.J., 1995, Testing the time axis of phylogenies, *Philosophical Transactions of the Royal Society of London*, **B349**: 5–10.

Benton, M.J. and Storrs, G.W., 1994, Testing the quality of the fossil record: paleon-tological knowledge is improving, *Geology*, **22**: 111–114.

Bookstein, F.L., 1987, Random walk and the existence of evolutionary rates, *Paleobiol-ogy*, **13**: 446–464.

Bottjer, D.J. and Jablonski, D., 1988, Palaeoenvironmental patterns in the evolution of post-Paleozoic benthic marine invertebrates, *Palaios*, **3**: 540–560.

Bremer, K., 1994, Branch support and tree stability, *Cladistics*, **10**: 295–304.

Clyde, W.C. and Fisher, D.C., 1997, Comparing the fit of stratigraphic and morphologic data in phylogenetic analysis, *Paleobiology*, **23**: 1–19.

Darwin, C., 1859, *The Origin of Species*, John Murray, London.

Davis, J.I., 1993, Character removal as a means for assessing stability of clades, *Cladistics*, **9**: 201–210.

Donoghue, M.J., Doyle, J.A., Gauthier, J., Kluge, A.G. and Rowe, T., 1989, The importance of fossils in phylogeny reconstruction, *Annual Review of Ecology and Systematics*, **20**: 431–460.

Edwards, A.W.F., 1992, *Likelihood* (expanded edn), Johns Hopkins University Press, Baltimore: 275 pp.

Eernisse, D.J. and Kluge, A.G., 1993, Taxonomic congruence versus total evidence, and amniote phylogeny inferred from fossils, molecules, and morphology, *Molecular Biology and Evolution*, **10**: 1170–1195.

Eldredge, N. and Gould, S.J., 1972, Punctuated equilibria: an alternative to phyletic gradualism. *In* T.J.M. Schopf (ed.), *Models in Paleobiology*, Freeman, San Francisco: 82–115.

Faith, D.P., 1991, Cladistic permutation tests for monophyly and nonmonophyly, *Systematic Zoology*, **40**: 366–375.

Farris, J.S., 1989, The retention index and the rescaled consistency index, *Cladistics*, **5**: 417–419.

Felsenstein, J., 1978, Cases in which parsimony or compatibility methods will be positively misleading, *Systematic Zoology*, **27**: 401–410.

Felsenstein, J., 1985a, Confidence limits on phylogenies: an approach using the bootstrap, *Evolution*, **39**: 783–791.

Felsenstein, J., 1985b, Phylogenies and the comparative method, *American Naturalist*, **125**: 1–15.

Fisher, D.C., 1991, Phylogenetic analysis and its implication in evolutionary paleobiology. *In* N.L. Gilinsky and P.W. Signor (eds), *Analytical Paleobiology*, Paleontological Society, Knoxville, Tennessee: 103–122.

Fisher, D.C., 1994, Stratocladistics: morphological and temporal patterns and their relation to phylogenetic process. *In* L. Grande and O. Rieppel (eds), *Interpreting the Hierarchy of Nature – From Systematic Patterns to Evolutionary Theories*, Academic Press, Orlando: 133–171.

Foote, M., 1996, On the probability of ancestors in the fossil record, *Paleobiology*, **22**: 141–151.

Foote, M. and Raup, D.M., 1996, Fossil preservation and the stratigraphic ranges of taxa, *Paleobiology*, **22**: 121–140.

Frost, D.R. and Kluge, A.G., 1994, A consideration of epistemology in systematic biology, with special reference to species, *Cladistics*, **10**: 259–294.

Gauthier, J., Kluge, A.G. and Rowe, T., 1988, Amniote phylogeny and the importance of fossils, *Cladistics*, **4**: 105–209.

Gingerich, P.D., 1979, The stratophenetic approach to phylogeny reconstruction in vertebrate paleontology. *In* J. Cracraft and N. Eldredge (eds), *Phylogenetic Analysis and Paleontology*, Columbia University Press, New York: 41–77.

Harper, C.W., Jr, 1976, Phylogenetic inference in paleontology, *Journal of Paleontology*, **50**: 180–193.

Harvey, P.H. and Pagel, M.D., 1991, *The Comparative Method in Evolutionary Biology*, Oxford University Press, Oxford: 239 pp.

Hillis, D.M. and Huelsenbeck, J.P., 1992, Signal, noise, and reliability in molecular phylogenetic analysis, *Journal of Heredity*, **83**: 189–195.

Hitchin, R. and Benton, M.J., 1997, Congruence between parsimony and stratigraphy: comparisons of three indices, *Paleobiology*, **23**: 20–32.

Huelsenbeck, J.P., 1991a, Tree-length distribution skewness: an indicator of phylogenetic information, *Systematic Zoology*, **40**: 257–270.

Huelsenbeck, J.P., 1991b, When are fossils better than extant taxa in phylogenetic analysis? *Systematic Zoology*, **40**: 458–469.

Huelsenbeck, J.P., 1994, Comparing the stratigraphic record to estimates of phylogeny, *Paleobiology*, **20**: 470–483.

Huelsenbeck, J.P. and Hillis, D.M., 1993, Success of phylogenetic methods in the four-taxon case, *Systematic Biology*, **42**: 247–264.

Huelsenbeck, J.P. and Rannala, B., 1997, Maximum likelihood estimation of topology and node times using stratigraphic data, *Paleobiology*, **23**: 174–180.

Jablonski, D. and Sepkoski, J.J., Jr, 1996, Paleobiology, community ecology, and scales of ecological pattern, *Ecology*, **77**: 1367–1368.

Jackson, J.B.C. and Cheetham, A.H., 1994, Phylogeny reconstruction and the tempo of speciation in cheilostome Bryozoa, *Paleobiology*, **20**: 407–423.

Kim, J., 1993, Improving the accuracy of phylogenetic estimation by combining different methods, *Systematic Biology*, **42**: 331–340.

Kuhner, M.K. and Felsenstein, J., 1994, A simulation comparison of phylogeny algorithms under equal and unequal evolutionary rates, *Molecular Biology and Evolution*, **11**: 459–468.

Lecointre, G., Philippe, H., Ván Lé, H.L. and Le Guyader, H., 1993, Species sampling has a major impact on phylogenetic inference, *Molecular Phylogenetics and Evolution*, **2**: 205–224.

Littlewood, D.T.J. and Smith, A.B., 1995, A combined morphological and molecular phylogeny for sea urchins (Echinoidea: Echinodermata), *Philosophical Transactions of the Royal Society of London*, **B347**: 213–234.

Lynch, J.D., 1989, The gauge of speciation: on the frequencies of modes of speciation. *In* D. Otte and J.A. Endler (eds), *Speciation and its Consequences*, Sinauer, Sunderland, Massachusetts: 527–553.

Maddison, W.P., 1995, Calculating the probability distributions of ancestral states reconstructed by parsimony on phylogenetic trees, *Systematic Biology*, **44**: 474–481.

Maddison, W.P., Donoghue, M.J. and Maddison, D.R., 1984, Outgroup analysis and parsimony, *Systematic Zoology*, **33**: 83–103.

Marshall, C.R., 1990, Confidence intervals on stratigraphic ranges, *Paleobiology*, **16**: 1–10.

Marshall, C.R., 1994, Confidence intervals on stratigraphic ranges: partial relaxation of the assumption of randomly distributed fossil horizons, *Paleobiology*, **20**: 459–469.

Mayr, E., 1963, *Animal Species and Evolution*, Harvard University Press, Cambridge, Massachusetts: 250 pp.

Miller, A.I. and Foote, M., 1996, Calibrating the Ordovician radiation of marine life: implications for Phanerozoic diversity trends, *Paleobiology*, **22**: 304–309.

Mooers, A. Ø., 1995, Tree balance and tree completeness, *Evolution*, **49**: 379–384.

Mooers, A. Ø., Page, R.D.M., Purvis, A. and Harvey, P.H., 1995, Phylogenetic noise leads to unbalanced cladistic tree reconstructions, *Systematic Biology*, **44**: 332–342.

Morris, P.J., Ivany, L.C., Schopf, K.M. and Brett, C.E., 1995, The challenge of paleoecological stasis: reassessing sources of evolutionary stability, *Proceedings of the National Academy of Sciences, USA*, **92**: 11269–11273.

Nelson, G.J., 1978, Ontogeny, phylogeny, paleontology and the biogenetic law, *Systematic Zoology*, **31**: 461–478.

Nelson, G.J., 1979, Cladistic analyses and synthesis: principles and definitions, with a historical note on Adamson's *Families des Plantes* (1763–1764), *Systematic Zoology*, **28**: 1–21.

Norell, M.A., 1992, Taxic origin and temporal diversity: the effect of phylogeny. *In* M.J. Novacek and Q.D. Wheeler (eds), *Extinction and Phylogeny*, Columbia University Press, New York: 89–118.

Norell, M.A., 1993, Tree-based approaches to understanding history: comments on ranks, rules, and the quality of the fossil record, *American Journal of Science*, **293–A**: 407–417.

Norell, M.A., 1996, Ghost taxa, ancestors and assumptions: a comment on Wagner, *Paleobiology*, **22**: 454–455.

Norell, M.A. and Novacek, M.J., 1992a, Congruence between superpositional and phylogenetic patterns: comparing cladistic patterns with fossil records, *Cladistics*, **8**: 319–337.

Norell, M.A. and Novacek, M.J., 1992b, The fossil record and evolution: comparing cladistic and paleontologic evidence for vertebrate history, *Science*, **255**: 1690–1693.

Novacek, M.J., 1992, Fossils, topologies, missing data, and the higher level phylogeny of eutherian mammals, *Systematic Biology*, **41**: 58–73.

Patterson, C., 1994, Bony fishes. *In* D.R. Prothero and R.M. Schoch (eds), *Major Features of Vertebrate Evolution*, Paleontological Society, Knoxville, Tennessee: 57–84.

Patterson, C. and Rosen, D.E., 1977, Review of ichthyodectiform and other Mesozoic teleost fishes and the theory and practice of classifying fossils, *Bulletin of the American Museum of Natural History*, **158**: 81–172.

Purvis, A., Nee, S. and Harvey, P.H., 1995, Macroevolutionary inferences from primate phylogeny, *Proceedings of the Royal Society of London*, **B260**: 329–333.

Raup, D.M., 1975, Taxonomic diversity estimation using rarefaction, *Paleobiology*, **1**: 333–342.

Raup, D.M., Gould, S.J., Schopf, T.J.M. and Simberloff, D.S., 1973, Stochastic models of phylogeny and the evolution of diversity, *Journal of Geology*, **81**: 525–542.

Remane, J., 1985, Der Artbegriff in Zoologie, Phylogenetic und Biostratigraphie, *Paläontological Zeitschrift*, **59**: 171–182.

Rice, W.R., 1989, Analysing tables of statistical data, *Evolution*, **43**: 223–225.

Rieppel, O. and Grande, L., 1994, Summary and comments on systematic pattern and evolutionary process. *In* O. Rieppel and L. Grande (eds), *Interpreting the Hierarchy of Nature – From Systematic Patterns to Evolutionary Theories*, Academic Press, Orlando: 133–171.

Rohlf, F.J., Chang, W.S., Sokal, R.R. and Kim, J., 1990, Accuracy of estimated phylogenies: effects of tree topology and evolutionary model, *Evolution*, **44**: 1671–1684.

Sanderson, M.J., 1989, Confidence limits on phylogenies: the bootstrap revisited, *Cladistics*, **5**: 113–129.

Sanderson, M.J., 1995, Objections to bootstrapping phylogenies: a critique, *Systematic Biology*, **44**: 299–320.

Sanderson, M.J. and Bharathan, G., 1993, Does cladistic information affect inferences about branching rates? *Systematic Biology*, **42**: 1–17.

Siddall, M.E., 1996, Stratigraphic consistency and the shape of things, *Systematic Biology*, **45**: 111–115.

Signor, P.W., 1985, Real and apparent trends in species richness through time. *In* J.W.

Valentine (ed.), *Phanerozoic Diversity Patterns: Profiles in Macroevolution*, Princeton University Press, Princeton, New Jersey: 129–150.

Signor, P.W. and Lipps, J.H., 1982, Sampling bias, gradual extinction patterns and catastrophes in the fossil record, *Geological Society of America Special Paper*, **190**: 291–296.

Smith, A.B., 1988, Patterns of diversification and extinction in early Palaeozoic echinoderms, *Palaeontology*, **31**: 799–828.

Smith, A.B., 1994, *Systematics and the Fossil Record – Documenting Evolutionary Patterns*, Blackwell Scientific, Oxford: 223 pp.

Smith, A.B. and Littlewood, D.T.J., 1994, Paleontological data and molecular phylogenetic analysis, *Paleobiology*, **20**: 259–273.

Sokal, R.R. and Rohlf, F.J., 1981, *Biometry* (2nd edn), W.H. Freeman, New York: 859 pp.

Strauss, D. and Sadler, P.M., 1989, Classical confidence intervals and Bayesian probability estimates for ends of local taxon ranges, *Mathematical Geology*, **21**: 411–427.

Suter, S.J., 1994, Cladistic analysis of cassiduloid echinoids: trying to see the phylogeny for the trees, *Biological Journal of the Linnean Society*, **53**: 31–72.

Swofford, D.L., 1991, When are phylogeny estimates from molecular and morphological data incongruent? *In* M.M. Miyamoto and J. Cracraft (eds), *Phylogenetic Analysis of DNA Sequences*, Oxford University Press, New York: 295–333.

Van Valen, L., 1973, A new evolutionary law, *Evolutionary Theory*, **1**: 1–30.

Vrba, E.S., 1993, Turnover-pulses, the Red Queen, and related topics, *American Journal of Science*, **293–A**: 418–452.

Wagner, P.J., 1995a, Stratigraphic tests of cladistic hypotheses, *Paleobiology*, **21**: 153–178.

Wagner, P.J., 1995b, Testing evolutionary constraint hypotheses with early Paleozoic gastropods, *Paleobiology*, **21**: 248–272.

Wagner, P.J., 1996, Ghost taxa, ancestors, assumptions and expectations: a reply to Norell, *Paleobiology*, **22**: 456–460.

Wagner, P.J., 1997, Patterns of morphological diversification among the Rostroconchia, *Paleobiology*, **23**: 115–150.

Wagner, P.J., 1998, Phylogenetics of the earliest gastropods, *Smithsonian Contributions to Paleobiology*.

Wagner, P.J. and Erwin, D.H., 1995, Phylogenetic tests of speciation hypotheses. *In* D.H. Erwin and R.L. Anstey (eds), *New Approaches to Studying Speciation in the Fossil Record*, Columbia University Press, New York: 87–122.

Wills, M.A., Briggs, D.E.G. and Fortey, R.A., 1994, Disparity as an evolutionary index: a comparison of Cambrian and Recent arthropods, *Paleobiology*, **20**: 93–131.

Wills, M.A., Briggs, D.E.G., Fortey, R.A. and Wilkinson, M., 1995, The significance of fossils in understanding arthropod evolution, *Verhandelung Deutsch Zoologische Gesammung*, **88**: 203–215.

Wright, S., 1931, Evolution in Mendelian populations, *Genetics*, **16**: 97–159.

Wright, S., 1932, The roles of mutation, inbreeding, crossbreeding and selection in evolution, *Proceedings of the Sixth International Congress of Genetics*, **1**: 356–366.

8
'Taxonomic Barriers' and Other Distortions within the Fossil Record

Carl F. Koch

INTRODUCTION

The distribution of fossil species in time and space provides important information about the past. The areal distributions of species contribute greatly to our understanding of their life history, life habits and environmental tolerances. Temporal distributions document the history of life and provide an essential component for biostratigraphic determinations.

The history of the biosphere in general is of interest for many reasons. For example, we would like to know about patterns of diversity in time and space, about regions of high or low endemism, and about regions that served as a source of species to repopulate other regions or replace one fauna with another. We would simply like to understand what happened and when.

Many factors confound palaeontologists attempting to unravel these mysteries. These factors include the fact that the published fossil record is biased in favour of biostratigraphically important taxa, because these groups have been studied in greater detail and by more workers than other taxa. Other complications result from the fact that North America and Europe have, in general, been studied more intensely than other areas, and that the reported distributions of fossils are subject to the vagaries of sedimentary rock distribution and the preservational quality inherent to the various sediment types. Also important is the fact that a particular taxon may be known by different names on different continents, in different basins or at different

The Adequacy of the Fossil Record. Edited by S.K. Donovan and C.R.C. Paul.
© 1998 John Wiley & Sons Ltd.

times. These taxonomic barriers act to shorten the taxon's distribution metrics and to elevate species-richness values.

For each of these confounding factors there exists a cautionary procedure that diminishes their effects on analytical results. The first step is to accept and recognize these distortions in the fossil record and the second step is to deal with them.

The seriousness of the effect that distortions have on palaeontological studies varies depending on the scope and nature of the study. This chapter examines some of the more obvious distortions of the published fossil record so that future studies might consider and ameliorate their effects.

All of the examples presented are taken from work on Cretaceous larger invertebrates of North America and Europe. Similar problems almost certainly occur for other fossil groups, other continents and other time periods, with the possible exception of smaller forms such as nannofossils, for which materials can be conveniently shared among workers on the various continents.

THE PUBLISHED FOSSIL RECORD

The published fossil record consists of illustrations, descriptions and occurrence information produced by natural historians (mostly palaeontologists) over the last 500 or so years. However, most of the record currently used has been produced since the early 1800s, when scientists such as Blumenbach, Schlotheim, Cuvier, Brongniart, Sowerby and others began in earnest to illustrate and describe fossil species (see Laudan, 1987, for a review). In general, the early descriptions were very short by modern standards, but they defined many currently used specific names. Some early illustrations are simply inadequate (for example, Morton, 1834). Other early illustrations were very carefully executed (such as d'Orbigny, 1844–1847), but sometimes were more beautiful than accurate.

The published fossil record will always include the information of the earliest workers, despite some obvious weaknesses (for example, selective collecting, particularly during the reconnaissance of new areas). We cannot purge the published fossil record of this information, nor should we. Rather, care should be taken to refrain from contrasting such data with those which have met a higher standard.

BIAS IN FAVOUR OF BIOSTRATIGRAPHICALLY IMPORTANT TAXA

Newell (1959) noted that 'stratigraphers now outnumber palaeontologists five or six to one in the Palaeontological Society'. Therefore, the published fossil record owes much to studies which had as their primary function the solving of biostratigraphic problems. It is reasonable that the published record may be biased in favour of biostratigraphically important taxa because they have been studied by more persons and in greater taxonomic detail than fossils that are seldom used in solving stratigraphic problems.

An example of such bias was documented in Koch (1978). In this study, data were gathered on the presence, location and abundance of larger invertebrates – mostly molluscs from the upper part of the Cenomanian (*Sciponoceras gracile* Biozone, Upper Cretaceous) in the western interior of North America, and this new information was compared with the published fossil record at the time of the study. These new data resulted from: (1) thoroughly examining 32 bulk collections made by Dr Erle G. Kauffman between 1964 and 1975 (housed at the Smithsonian Institution); (2) perusing the unpublished files of the US Geological Survey (USGS), and subsequently examining over 300 USGS collections in Washington, DC, and Denver, Colorado; and (3) undertaking 12 personal field collections to spot check previous collections and to verify questionable reports. These new data were organized into 110 location files such that collections made within a 16 km radius (=10 miles) were combined. This resulted in 1050 occurrences. An occurrence in this case is defined as the presence of a species in a location file. Previously published data recorded 203 occurrences from 16 locations.

The 13 previously published reports recorded 65 species for their 203 occurrences. The new data gathered increased the number of taxa to 170 species, as a result of a fivefold increase in the number of occurrences. A comparison of the published with the newly gathered data showed that the number of ammonite taxa increased by a factor of 1.4, but bivalve and gastropod taxa increased in number by factors of 3.3 and 3.2, respectively; that is, the proportion of bivalve taxa increased from 40% to 56% and the proportion of ammonite taxa decreased from 28% to 18%. This demonstrated that the biostratigraphically important ammonites had previously been more thoroughly studied and reported.

Many previously published reports had as their purpose the solving of biostratigraphic problems and thus much of the collecting upon which they were based was selective. The newly gathered data included more than 44 bulk collections. Examination of bulk collections for a given time or place invariably results in the recognition of additional taxa. Selective collecting tends to favour the more noticeable specimens.

RARE SPECIES AND AMOUNT OF MATERIAL STUDIED

Palaeontologists have long recognized that the number of taxa found in a biogeographical area, time period or collection increases as more material is

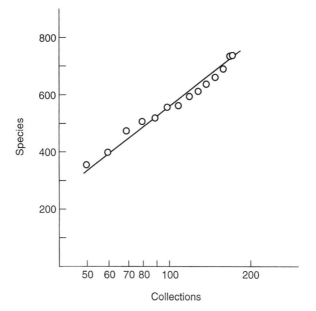

Figure 8.1   *Increase in the number of species with additional collections for larger invertebrates of mid-Maastrichtian sedimentary rocks of the US eastern Gulf coastal plain*

analysed (for example, Durham 1967; Raup 1972; Sheehan 1977; and many others). For studies of species richness in periods of epoch size or larger, sediment volume, sampling intensity, etc. (see Signor, 1982, for review) have been used to estimate the amount of material studied such that the effect of inequities of sample size can be considered.

This phenomenon is illustrated by three examples. One is presented in Figure 8.1, which shows that the number of species found increased with increased number of collections for mid-Maastrichtian macrofossils of the US Gulf coastal plain. The total number of specimens counted and identified was 45 000. Jackson *et al.* (1993, fig. 1) showed similar trends for four mollusc data sets of late Miocene to Pleistocene age. They plotted the number of specimens versus number of subgenera. The largest data set analysed was about 20 000 specimens and the smallest about 4000.

Interestingly, in none of the five data sets used herein (Figure 8.1), or of the four data sets of Jackson *et al.* (1993, fig. I), does the trend deviate significantly from a straight line as large numbers of specimens are analysed. This means that the rate at which new species or new subgenera are added to the data continues to be high even when large numbers of specimens are analysed.

On a much larger scale, Abdel-Gawad (1986) reported 14 new species among the Maastrichtian non-cephalopod mollusc specimens of the middle Vistula Valley, central Poland. Evidently the well-studied European sedimentary rocks continue to yield new species as more material is analysed.

The increase in species richness with increased sampling intensity results because most species are rare and occur infrequently. For example, Koch (1987) found that more than half of the 576 species encountered among 539 collections of sedimentary rocks of mid-Maastrichtian age in the US Atlantic and Gulf coastal plain occurred very uncommonly (in less than 1% of the collections). Similar numbers are found for nine other large (1000–20 000 occurrences) data sets of Cretaceous molluscs and Recent Benthic foraminifera. All 10 data sets were discussed in Koch (1987).

Genera and subgenera are often very rare. For the data set above (576 species in 539 collections), 120 bivalve genera are present. Of these, 30 (25%) are represented by one or two specimens, 73 (61%) are represented by only one species, and 36 (41%) occur in less than 1% of these collections. Similar patterns of rare genera and subgenera apparently exist among the late Miocene to Pleistocene specimens of Jackson et al. (1993).

Koch and Lundquist (1990) compared the numerical relationships between specimens, species, genera and families given for Pleistocene gastropods and bivalves of the Californian Province (Valentine, 1989) with the data for mid-Maastrichtian molluscs of the US Atlantic and Gulf coastal plain, and found these relationships to be nearly identical. That is, the percentages of genera represented by only one species are 57.8% (Pleistocene) and 60.9% (late Cretaceous); the percentages of families represented by only one species are 24.0% and 23.0%; and the percentages of families represented by a single genus are 46.5% and 41.0%, respectively. In addition, the frequency distributions of the number of genera represented by 1, 2, 3, etc. species are the same. The patterns for numbers of species per family and numbers of genera per family are also not significantly different. The evidence points to the fact that most species are rare and that a large proportion of genera are rare, at least as far back as the late Cretaceous.

DIFFERENCES IN STUDY INTENSITY

Historically, the fossils of Europe and North America have been studied for more time and more intensely than those from elsewhere. Most palaeontologists would agree that the fauna of these two areas are better known than those of the rest of the world. Raup and Jablonski (1993), in a study of end-Cretaceous extinctions, compiled data on Maastrichtian bivalve genera world-wide. They counted the number of bivalve genera that have been reported for each 10° square and displayed the numbers on a Maastrichtian palaeogeographical map. Table 8.1 summarizes their data for the purpose of evaluating the differences in study intensity throughout the world, on the assumption that more genera result from more study. Temperate North

Table 8.1   *Distribution of Maastrichtian bivalve genera by large geographical area (data from Raup and Jablonski, 1993)*

Geographical area	Generic occurrences	Percentage of total (%)
Northern hemisphere	1740	80
Southern hemisphere	446	20
North temperate (30–60°N)		
North America	612	28
Europe	485	22
Others	77	3
Tethyan (0–30°N)		
Central and South America	101	5
Europe, Africa and Middle East	377	17
Other	58	3

America and boreal Europe (30–60°N lat) account for 50% of the generic occurrences, with values that are about the same for the two regions, 28% and 22%. The Tethyan region (0–30°N lat) contains 25% of the generic occurrences; however, the amount of data from east of the Atlantic is more than three times the data from west of this ocean (17% compared with 5%). The amount of data from the northern hemisphere is four times greater than that from the southern hemisphere.

Studies comparing the faunal similarity or species richness between regions must take into account the effect that large numbers of rare species might have on the results. Koch (1987) gave calculations and curves designed to predict the effects of rarity of species on measured temporal and geographical distribution of species. For example, two large data sets taken from the same region should have only 70–80% of the species in common. Thus, if we compared the Upper Cretaceous fauna of North America with the coeval fauna of Europe we would expect only 70 or 80% of the species to be the same, even if the faunas of the two regions were identical. This fact requires that species common to two regions must be less than 70% before we can say that faunal differences exist between the two regions. When one region is better known than another, then the predicted percentage of common species may be very much smaller, even if the faunas are in fact exactly the same. Koch (1987) presented the mathematics needed to predict the amount of difference between two regions that would be observed if the faunal make-up of the large and the small data sets were in fact the same. For the convenience of workers who have difficulty with equations, he provided 'user-friendly' plots. The plot that is best for comparison of a large data set with a smaller one is given in Figure 8.2. For example, in the Raup and Jablonski (1993) study, data for the northern hemisphere are about four times

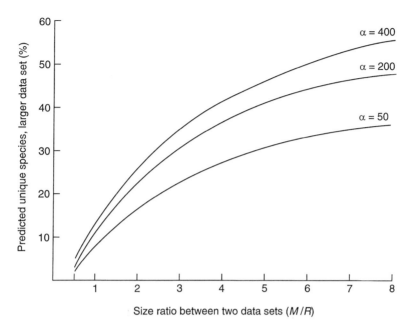

Figure 8.2 *Predicted percentage of unique species in the larger of two data sets as a function of the size ratio between the two. α is the constant of Fisher's log series distribution (Koch, 1987). All calculations used 10 000 occurrences (=M) as the size of the larger data set; R = number of occurrences in the smaller data set. (From Koch, 1987)*

the amount for the southern hemisphere. Figure 8.2 predicts that 35% of the species found in the northern hemisphere will not be found among the southern hemisphere specimens for no reason other than that so many species are rare. Figure 8.2 also predicts that a comparison of the eastern Tethys fauna with the western Tethys fauna would result in about 28% of the eastern Tethys species not being found in the smaller western Tethys data set, because many rare species of the eastern Tethys have a low probability of discovery in the smaller western Tethys data set.

EFFECTS OF SEDIMENTARY ROCK TYPE AND PRESERVATION

Most benthic fossils are sedimentary facies fossils; that is, they only occur in sedimentary rocks which represent the palaeoenvironment that fulfilled the life requirements of the organisms. Further, some sedimentary rocks do not protect fossil hard parts from chemical dissolution and, thus, some sedimentary rock types seldom preserve remains of their fauna.

An example that illustrates well the effect that sedimentary rock type can

Table 8.2   *Results of comparing North American bivalve species with European specimens as a function of frequency of occurrence among 502 North American collections and the expected results for 61 European collections. Species and specimens are larger invertebrates of Upper Cretaceous sedimentary rocks in eastern USA and western Europe (from Koch, 1995)*

Frequency of occurrence, North America	No. of species	Observed in 61 European collections	Expected in 61 European collections
< 10	64	12	20
10–20	26	16	22
21–40	29	17	28
41–90	26	17	26
> 90	19	16	19
Total	164	78 (48%)	115 (70%)

have on comparisons of one region with another is given in Koch and Morgan (1988) and Koch (1995). Koch (1995) found that when bivalve specimens from the Campanian and Maastrichtian in Europe are compared with species documented for similar sedimentary rocks of the Atlantic and Gulf coastal plain of North America, 78 species are recognized as being the same on both continents. The North American data set results from examination of 502 bulk collections and the European data set is based on examination of specimens from 61 locations and housed in 12 European museums.

A sampling-effect model (Koch, 1987), which accounts for the disparity in data set size, predicts that 115 rather than 78 of the 164 North American bivalve species should be found, if the two data sets are, in fact, samples of the same fauna (Table 8.2). The difference of 37 species is mainly attributed to species–sediment interaction and the fact that 15% of the North American collections are from limestones, whereas 77% of the European collections are from similar carbonate rocks. When the sampling-effect model uses frequency of species occurrence for the 74 North American limestone collections, the observed similarity values are close to those predicted. The model predicts 80–84 species; 75 species were actually observed (Table 8.3). It can be concluded that species–sediment relationships are strong for these bivalve species and that the sampling-effect model provides a good prediction if proper attention is given to the sedimentary make-up of the two data sets.

Differences in preservational quality can also affect comparison between two areas. For example, many bivalves and most gastropods have aragonitic hard parts and are seldom recognized in collections in which aragonite is not preserved. Koch and Sohl (1983) classified collections into six types based on preservational quality. *Type I* (aragonite and calcite well preserved) represents collections in which both the aragonitic- and calcitic-shelled species are

Table 8.3 *Results of comparing North American bivalve species with European speci-mens as a function of frequency of occurrence among 74 North American limestone collections and the expected results for 47 European collections. Species and specimens are larger invertebrates of Upper Cretaceous sedimentary rocks in eastern USA and western Europe (from Koch, 1995)*

Frequency of occurrence, North America	No. of species	Observed in 47 European collections	Expected in 47 European collections
0	70	14	8–12
1	32	14	15
2, 3	20	14	15
4–7	15	10	15
8–15	13	11	13
> 15	14	12	14
Total	164	*75 (46%)	80–84 (49–51%)

*Three additional species occur in the European collections, but only in siliciclastics.

fully preserved. Such preservation normally takes place in the silty, fine-grained, quartzose sandstone facies that contain sufficient clay to seal the shells from dissolution by groundwater. *Type II* (calcite, aragonite traces and moulds) represents those collections from which most of the aragonite has been leached, leaving only films of aragonitic shell material or moulds, but in which calcite shells are preserved except for their aragonitic layers (that is, inner shell layer of pectens, etc.). In the *Type III* (calcite, aragonite-replaced and moulds) collections, calcite shells are preserved, but aragonitic shells may be partially or totally replaced by other minerals. Preserved material in category *Type IV* (calcite and moulds) consists of calcitic shells and phos-phatic internal moulds or external moulds in the matrix. In *Type V* (moulds only) collections, even calcite is lost and only moulds remain. In collections of internal phosphatic moulds, high diversity is sometimes maintained by the entombment of smaller elements of the fauna within the mould infilling. These may be recovered by breaking the moulds. *Type VI* (calcite only) collections contain only calcitic-shelled species. No trace of aragonitic shells is seen either as phosphatic internal moulds or as impressions in the matrix.

Koch and Sohl (1983) applied this classification scheme to 83 fossil collec-tions of Maastrichtian age from the US eastern Gulf coastal plain. Results are shown in Table 8.4. Notice that Type I collections (aragonite and calcite well preserved) yield both more specimens and more species than any other type.

The effects of preservational quality can confound palaeoecological stu-dies of endemism, eurytopy versus stenotopy, species longevity and other investigations of the distribution of organisms in space and time. These effects are especially significant for studies based on presence/absence data,

Table 8.4   *Average number of taxa recovered for the six preservational categories and average number of specimens for each category. Specimens are large invertebrates of mid-Maastrichtian sedimentary rocks of the US eastern Gulf coastal plain (from Koch and Sohl, 1983)*

Preserve Type	No. of Collections	Average no. of specimens	No. of taxa
I	13	1206	74
II	12	219	39
III	11	296	49
IV	41	216	40
V	3	474	52
VI	3	157	18
Total	89	389	0 30 60 90 120 150 180

|- - - - ☐☐☐☐—I— ▬▬▬ - - - - -|
min   −1 SD avg. +1 SD   max

such as published faunal lists and selectively collected samples. In such studies, collections of good preservational quality may be interpreted to represent strata of high diversity, taxa having the more durable hard parts will appear to be widespread and long-ranging, and taxa that are preserved only in collections of the best preservational quality will appear stenotopic, endemic and rare.

Faunal elements that are recovered only in collections in which aragonite shells are well preserved present a serious problem for palaeoecologists because no solution for the sampling problem is currently known. These taxa may have been original constituents of the biotope at a locality where they have been subsequently lost during diagenesis. Palaeoecologists must be aware of this phenomenon and guard against misapplication of their data. For now, the best suggestion is that palaeontologists studying the distribution of organisms in time and space use collections of equivalent preservation. Use of published data for which preservational quality is unknown should be especially avoided.

In a comparison of latest Cretaceous bivalve species from the US Atlantic and Gulf coastal plain with coeval specimens from European sedimentary rocks (Koch, 1996), 68% of calcitic-shelled, but only about 33% of the aragonitic-shelled, US bivalve species were represented among the European specimens. This resulted from differences in preservational quality between the two areas.

Preservation of aragonitic-shelled fossils in Europe is most often in the form of internal moulds because the aragonite has been leached (Type IV). For example, Abdel-Gawad (1986, p. 52) noted that 'in the Late Cretaceous chalk and opoka facies of the North European Province, there is a preferential preservation of calcitic shells', and he classified his studied specimens as within Type IV (calcite and moulds). Most of the specimens in European collections observed by the author are Type VI (calcite only). In contrast, aragonitic-shelled fossils with original aragonite well preserved are common in the US collections from the Owl Creek Formation (Mississippi and Tennessee), Severn Formation (Maryland), Providence Sands (Georgia and Alabama) and the Corsicana Formation (Texas) (Koch, 1996).

The published fossil record sometimes reports facies and preservational information along with fossil occurrences. In such situations, workers can limit comparisons to collections with like sedimentary rock type and/or preservational quality. Without such information, results and conclusions are less likely to be reliable.

ON 'TAXONOMIC BARRIERS'

Taxa are often called by more than one name because many palaeontologists have restricted their interest to one basin, geographical area or time period. We cannot determine the true distribution of such taxa because of this 'taxonomic barrier'. Palaeontologists have long known of this phenomenon and have mentioned it in their writings.

Sohl (1964, p. 161) stated that 'the apparent differences in species between the Texas and the Tennessee and Mississippi areas are artifacts of too stringent a taxonomy and an evident belief on the part of some workers that gastropods had very narrow dispersal limits'. Ager (1987, p. 74) wrote that 'the limits of the old Austro-Hungarian Empire are defined almost as clearly by fossils as they are by the distribution of Hapsburg-Yellow government buildings . . . because the specialists who ranged through that multiplicity of lands and languages also took their fossils back to the great repositories in the capital city of Vienna'. Further, he noted the 'imperial' distribution of his youth in which fossil distributions can be traced from the British Isles to the far reaches of the British Empire.

In his presidential address to the Palaeontological Society (Sohl, 1987, p. 1106), Sohl noted 'the tendency among earlier workers to overemphasize the dissimilarities between Cretaceous and Tertiary gastropod faunas . . . because past workers had a tendency to either work on Cretaceous or on Tertiary assemblages but seldom both'. Erwin (1991, p. 518), in a discussion of the end-Permian mass extinction, observed that 'many palaeontologists have specialized on one side of the boundary or the other, producing two sets of names, one Paleozoic, the other Mesozoic, for the same families, genera and species'.

That 'taxonomic barriers' exist between geographical areas and between time periods is not in dispute. The questions are: what are the magnitudes of the barriers, and to what extent do such barriers affect the outcome of various palaeontological studies?

Comparison of Late Cretaceous Bivalve Species on Both Sides of the North Atlantic

Life was in some ways much simpler for palaeontologists 150 years ago. Forbes (1845) compared some specimens collected in the USA by Charles Lyell with European specimens at his disposal and found that four bivalve species are 'common to the American and European Cretaceous Systems'. Five years later Alcide d'Orbigny (1850) published his well-known *Prodrome de Paleontologie Stratigraphique Universelle des Animaux Mollusques et Rayonnés*, in which he divided the world's fossiliferous strata into 27 stages based on the species content of each stage. In the process he listed 11 of his 488 bivalve species of the 22nd stage (Senonien) as occurring in both North America and Europe. Since the days of Forbes, Lyell and d'Orbigny, fewer than a dozen temperate zone, late Cretaceous bivalves have been added to the list of species common to the two continents.

There exist several reasons why so few species have been added to the list. For example, both the number of known late Cretaceous bivalves and the number of publications about late Cretaceous bivalves increased markedly during the last half of the 19th century. For instance, d'Orbigny dealt with less than 1500 Cretaceous bivalve species in 1850, but Stoliczka (1871) listed about 9750 species. As more species became known, the taxonomists' problems of assigning specimens to known species or describing new taxa became more time-consuming. In addition, synonymizing species based on text descriptions and illustrations such as practised by some earlier workers diminished greatly. Taxonomic rigour required that actual specimens, type or otherwise, should be compared before assigning specimens to an individual species.

As the plates and figures illustrating various fossil species of one continent became readily available elsewhere, some palaeontologists noted the close similarities between their local species and species on the other side of the Atlantic Ocean, but stopped short of formally synonymizing the several species without examination of type specimens. However, some did document their observations. For example, Wade (1926, p. 72) considered two species, each of which is known from only one continent, and wrote 'perhaps on a comparison of real specimens they would be found identical'. Stephenson (1941) cited one European species as 'scarcely distinguishable' from one of his North American species. More recently, Dhondt (1987, p. 51) wrote about some German fossils, 'I would like to suggest that the Vaals specimens are almost identical' with the North American species. In other words, from time to time palaeontologists have noted species in common between North America and Europe, but attention to rigorous taxonomic practice prevented them from synonymizing species based on only perusal of figures and plates.

In order to measure the extent of the 'taxonomic barrier' between North America and Europe, illustrations of North American bivalve species were taken to museums in Europe known to house bivalve specimens collected from European Upper Cretaceous strata representing boreal marine environments. The North American illustrations included 1500 photographs of bivalve specimens collected from middle Maastrichtian-age sediments in the US Atlantic and Gulf coastal plain by USGS palaeontologists between the late-1800s and 1986. These photographs were collected into loose-leaf binders along with about 2000 figures and plates of species copied from published reports and monographs. Combined, these illustrations pictured 240 bivalve species, including many of the type specimens, and illustrated intraspecific variation present within the North American species concept.

These bivalve species had been extensively studied between 1977 and 1986 as part of a project led by Norman F. Sohl of the USGS and the author (Koch and Sohl, 1983; Sohl and Koch, 1983, 1984, 1987; Koch, 1987). In all, 95 000 bivalve specimens were examined and assigned to 240 species. The relative abundance of each species was recorded, as were size, habitat, feeding type, preservational potential and sedimentary rock specificity. This information is helpful in understanding the distribution of these species in time and space.

The 12 museums visited by the author in the summer and autumn of 1987 were the British Museum (Natural History), London; Sedgwick Museum, Cambridge; University Museum, Oxford; Norwich Castle Museum, Norwich; Musée National d'Histoire Naturelle, Paris; Université Pierre and Marie Curie, Paris; Université Claude Bernard, Lyon; Université Paul-Sabatier, Toulouse; La Cité à Fossiles, Musée de Paleontologie, Sarlat, western France; Institut Royal des Sciences Naturelles de Belgique, Brussels; Geologisk Museum, Copenhagen; and Museum für Naturkunde, Berlin. Materials examined were collected from seven western European countries: England, Denmark, France, Sweden, Germany, Belgium and The Netherlands.

The nearly impossible task of relating European specific names to the North American names was not attempted; instead the European specimens were examined without regard to the name on the label as long as the specimen was collected from sediments of the relevant area and age. The specimens were compared with the appropriate illustrations among the 3500 illustrations of North American specimens. When necessary, text descriptions of the North American species were consulted. If the European specimen fitted within the species concept of the North American species, then we could state that the particular North American species ranged into western and northern Europe during the late Cretaceous.

Not surprisingly, many of the European specimens fitted easily into the comparable North American species concepts. At least 54 bivalve species of the North American Upper Cretaceous Atlantic and Gulf coastal plain ranged into Europe. Another 24 species were noted as probably conspecific and will almost certainly be added to the list when more material is examined.

Many practical considerations come into play when we evaluate these results. As discussed in a previous section, many of the North American species are rare and the probability of finding them again anywhere is very low. Also very significant (as discussed above) is the considerable disparity in fossil preservation and inequities in the sedimentary rock types from which fossil collections have been made between the two continents. Not previously discussed is the fact that some of the North American bivalve species are not defined well enough to allow comparisons. For example, poorly defined listings such as 'Inoceramus sp.' or 'radiolitid indeterminate' tell us that these types of bivalves lived in the US coastal plain, but we cannot say which inoceramid or radiolitid species it was. Other species are listed in terms of known species (for example, *Nemodon* cf. *N. grandis*, *Pycnodonte* aff. *P. belli*) or new species (such as *Nemodon* n. sp. large, *Tellina* n. sp. smooth). Species that fall into one of these three categories were removed from consideration, leaving a total of 164 species.

About 78 of 164 late Cretaceous bivalve species of the US Atlantic and Gulf coastal plain were found to be represented among specimens collected in northern and western Europe, and thus the known geographical range of many species can be extended several thousand kilometres eastward. Previously published information would indicate that fewer than 20 temperate zone bivalve species were common to both continents during the late Cretaceous. This disparity is attributed to the 'taxonomic barrier' that results when species are given different names on different continents. The species found to have existed on both continents are proportionately more often calcitic than aragonitic-shelled forms. Not surprisingly, species that occur frequently among the North American collections are also proportionately much better represented on the list of species common to the two areas than the rarer forms. The effects of these factors imply that the true number of species that ranged through both continents is masked in the published fossil record. The number of species found to be common to North America and Europe will almost certainly increase with further study.

Revision of the *Nucula percrassa* Group in the Upper Cretaceous of the US Mid-Atlantic and Gulf Coastal Plains

Wingard and Sohl (1990) employed discriminant and qualitative analyses of *Nucula* specimens from the Lower Campanian through Maastrichtian units of the Mid-Atlantic and Gulf coastal plain to resolve questions about their taxonomic, geographical and stratigraphic relationship. The results indicated that for this place and time there are no genetically based morphological features that can distinguish between the five species names commonly used for *Nucula percrassa* group specimens. Wingard and Sohl synonymised the five species under the name *Nucula percrassa* Conrad. The authors stated that 'this study illustrates instances of nomenclature biases based on assumed stratigraphic or geographic isolation of populations or on topological differences. The existence of such biases can have profound effects on the calculation of rates of evolution and extinction. For the Nuculoida, the bias was toward shorter species duration and increased species diversity' (Wingard and Sohl, 1990, p. D1).

Taxonomy of Upper Cretaceous and Lower Tertiary Crassatellidae in the Eastern United States

To test the accuracy of the published record, Wingard (1993) conducted a quantitative and qualitative analysis of the Crassatellidae (Bivalvia) of the US Mid-Atlantic and Gulf coastal plain for the Upper Cretaceous and Lower Tertiary. She found that 38 specific names have been used in publication, but that only eight valid recognizable species exist in the Crassatellidae within the limits of the study. She stated that 'calculations of evolutionary and palaeobiologic significance based on poorly defined, overly split fossil groups, such as the Crassatellidae are biased in the following ways: (1) rates of evolution and extinction are higher; (2) faunal turnover at mass extinctions appears more catastrophic; (3) species diversity is high; (4) average species durations are shortened; and (5) geographic ranges are restricted' (Wingard, 1993, p. 1).

Wingard further concluded that faunal changes among the Crassatellidae at the Cretaceous/Tertiary boundary did occur, but they were not catastrophic. The number of species remained nearly constant on each side of the boundary. The published record indicates that 20 species went extinct at or near the boundary and 14 originated shortly after the boundary. Wingard's analyses indicate that the numbers are five and three, respectively.

Bivalve Genera Across the Cretaceous/Tertiary Boundary

Recent studies of bivalves of the uppermost Cretaceous strata along the US Atlantic and Gulf coastal plains assigned specimens to 15 genera previously listed in the *Treatise on Invertebrate Paleontology* as originating in the early Cenozoic (Koch, 1994). Work on the very uppermost Cretaceous hard ground in Denmark recognized 10 additional genera previously thought to be restricted to the Cenozoic that, in fact, have their origin in the Cretaceous (Heinberg, 1979). Some of this information may have resulted because the newer studies were more detailed and used advanced preparation techniques, but the remainder can be attributed to temporal bias of previous work. These two studies indicate that no more than 45% of the late Cretaceous bivalve genera went extinct at the Cretaceous/Tertiary boundary in contrast to the 62% suggested by the published fossil record.

CONCLUSIONS

Some of the distortions of the published fossil record have been demonstrated. They are:

1. Bias in favour of biostratigraphically important fossils
2. Effects of rare species and amount of material studied
3. Effects of differences in study intensity
4. Effects of sedimentary rock type and preservation
5. 'Taxonomic barriers' between basins, continents and eras

In addition to these, other biases have been noted. For example, Valentine (1989, p. 83) suggested that 'sampling of the Pleistocene record has been biased towards shallow-water assemblages. Fragile and minute forms are probably under-represented in the record.' My list of distortions is obviously incomplete, but most of the major known biases have been discussed.

These problems can be dealt with in some situations. For example, difference in study intensity can be compensated for using the methods of Koch (1987). 'Taxonomic barriers' do not affect results within a basin, continent or era, but come into play when we look for patterns of dispersal, or diversity on a world-wide scale, or relationships of geographical range and species longevity, etc. Each researcher must decide whether these distortions are important considerations or not for his or her particular study.

If the examples and discussion given in this paper help raise the consciousness of palaeontologists to practical problems such that more cautious and considered analyses are forthcoming, and, if the paper causes future workers to take on the challenge to overcome these distortions, then it has achieved two of its goals. Business as usual is not a reasonable choice.

REFERENCES

Abdel-Gawad, G.I., 1986, Maastrichtian non-cephalopod mollusks (Scaphopoda, Gastropoda and Bivalvia) of the middle Vistula Valley, central Poland, *Acta Geologica Polonica*, **36**: 69–224.

Ager, D.V., 1987, The politics of palaeontology, *Geology Today*, **3**: 74–75.

Dhondt, A.V., 1987, Bivalves from the Hochmoos Formation (Gasau-Group, Oberosterreich, Austria), *Annals Naturhistorische Museum Wien*, **88A**: 41–101.

Durham, J.W., 1967, The incompleteness of our knowledge of the fossil record, *Journal of Paleontology*, **51**: 559–564.

Erwin, D.H., 1991, The mother of mass extinctions, *Palaios*, **6**: 517–518.

Forbes, E., 1845, On the fossil shells collected by Mr. Lyell from the Cretaceous formations of New Jersey, *Quarterly Journal of the Geological Society of London*, **1**: 61–62.

Heinberg, C., 1979, Bivalves from the latest Maastrichtian of Stevns Klint and their stratigraphic affinities. *In* T. Birkelund and R.G. Bromley (eds), *Cretaceous–Tertiary Boundary Events Symposium*, University of Copenhagen, Copenhagen: 58–64.

Jackson, J.B.C., Jung, P., Coates, A.G. and Collins, L.S., 1993, Diversity and extinction of tropical American mollusks and the emergence of the Isthmus of Panama, *Science*, **260**: 1624–1626.

Koch, C.F., 1978, Bias in the published fossil record, *Paleobiology*, **4**: 367–372.

Koch, C.F., 1987, The prediction of sample size effects on temporal and geographic distribution patterns of species, *Paleobiology*, **13**: 100–107.

Koch, C.F., 1994, Are mass extinctions massive? – A K/T boundary example, *Geological Society of America Abstracts with Programs*, **26** (4): 22.

Koch, C.F., 1995, Sampling effects, species–sediment relationships and observed geographic distribution: an uppermost Cretaceous bivalve example, *Geobios*, **28**: 237–241.

Koch, C.F., 1996, Bivalve species' distribution in uppermost Cretaceous boreal marine beds, Europe and North America and the implied taxonomic problems, *Proceedings of the Fourth International Cretaceous Symposium, Hamburg, Germany, Mitteilungen aus dem Geologisch-Paläontologischen Institut de Universität Hamburg, Special Issue*: 75–81.

Koch, C.F. and Lundquist, J.J., 1990, Hierarchal fabric similarities of a late Cretaceous and Pleistocene mollusc data set, *Geological Society of America Abstracts with Programs*, **22** (7): A266.

Koch, C.F. and Morgan, J.P., 1988, On the expected distribution of species ranges, *Paleobiology*, **14**: 126–138.

Koch, C.F. and Sohl, N.F., 1983, Preservational effects in paleoecology studies: Cretaceous mollusc examples, *Paleobiology*, **9**: 26–34.

Laudan, R., 1987, *From Mineralogy to Geology, the Foundations of a Science 1650–1830*, University of Chicago Press, Chicago: 278 pp.

Marshall, C.R., 1991, Estimation of taxonomic ranges from the fossil record. *In* N.F. Gilinsky and P.W. Signor (eds), *Analytical Paleobiology, Paleontological Society Short Courses in Paleontology, 4*, University of Tennessee, Knoxville: 19–38.

Morton, S.G., 1834, *Synopsis of Organic Remains of the Cretaceous Groups of the United States*, Philadelphia, Pennsylvania: 84 pp.

Newell, N.D., 1959, Adequacy of fossil record, *Journal of Paleontology*, **33**: 488–499.

Orbigny, A. d', 1844–1847, *Paleontologie Francaise. Description des Mollusques et Rayonnés Fossiles. Terrain Crétacés: 3. Lamellibranches*, Paris: 807 pp.

Orbigny, A. d', 1850, *Prodrome de Paléontologie Stratigraphique Universelle des Animaux Mollusques et Rayonnés*, 2, Victor Masson, Paris: 428 pp.

Raup, D.M., 1972, Taxonomic diversity during the Phanerozoic, *Science*, **177**: 1065–1071.

Raup, D.M. and Jablonski, D., 1993, Geography of the end-Cretaceous marine bivalve extinctions, *Science*, **260**: 971–973.

Sheehan, P.M., 1977, Species diversity in the Phanerozoic: a reflection of labor by systematists? *Paleobiology*, **3**: 325–328.

Signor, P.W., III, 1982, Species richness in the Phanerozoic: compensating for sampling bias, *Geology*, **10**: 625–628.

Sohl, N.F., 1964, Neogastropoda, Opisthobranchia, and Basommatophora from the Ripley, Owl Creek, Prairie Bluff Formations, *US Geological Survey Professional Paper*, **331–B**: 153–344.

Sohl, N.F., 1987, Cretaceous gastropods: contrast between Tethys provinces and the temperate provinces, *Journal of Paleontology*, **61**: 1085–1111.

Sohl, N.F. and Koch, C.F., 1983, Upper Cretaceous (Maestrichtian) molluscs from the *Haustator bilira* assemblage zone in the east Gulf Coastal Plain, *US Geological Survey Open File Report*, **83–451**: 339 pp.

Sohl, N.F. and Koch, C.F., 1984, Upper Cretaceous (Maestrichtian) Mollusca from the *Haustator bilira* assemblage zone in the west Gulf Coastal Plain, *US Geological Survey Open File Report*, **84–687**: 271 pp.

Sohl, N.F. and Koch, C.F., 1987, Upper Cretaceous (Maestrichtian) larger invertebrates from the *Haustator bilira* assemblage zone in the Atlantic Coastal Plain with further data from the East Gulf, *US Geological Survey Open File Report*, **87–0194**: 172 pp.

Stephenson, L.W., 1941, The larger invertebrate fossils of the Navarro Group of Texas, *University of Texas Publications*, **4101**: 641 pp.

Stoliczka, F.W., 1871, Cretaceous fauna of southern India, III, Series VI, The Pelecypoda, *Memoirs of the Geological Survey of India, Palaeontologia Indica*: 537 pp.

Valentine, J.W., 1989, How good is the fossil record? Clues from the Californian Pleistocene, *Paleobiology*, **15**: 83–94.

Wade, B., 1926, The fauna of the Ripley Formation on Coon Creek, Tennessee, *US Geological Survey Professional Paper*, **137**: 271 pp.

Wingard, G.L., 1993, A detailed taxonomy of Upper Cretaceous and Lower Tertiary Crassatellidae in the Eastern United States: an example of the nature of extinction at the boundary, *US Geological Survey Professional Paper*, **1535**: 131 pp.

Wingard, G.L. and Sohl, N.F., 1990, Revision of the *Nucula percrassa* Conrad, 1853 group in the Upper Cretaceous of the Gulf and the mid-Atlantic Coastal Plains: an example of bias in the nomenclature, *US Geological Survey Bulletin*, **1881**: D1–D25.

9
Patterns of Occurrence of Benthic Foraminifera in Time and Space

Stephen J. Culver and Martin A. Buzas

INTRODUCTION

Microfossil workers have an advantage over other palaeontologists by virtue of the great abundance of their items of study. The astronomical number of nannofossils in chalk and of planktic foraminifera in deep-sea oozes comes to mind, but benthic microfossil groups are also represented in the modern and fossil biosphere by incredibly high densities and abundances. For example, larger foraminifera occur in obvious rock-forming quantities in the limestone used to build the pyramids of Egypt, while the density of living benthic foraminifera can exceed $10^6\,\mathrm{m}^{-2}$ with biomass estimates of over $10\,\mathrm{g\,m}^{-2}$ (Saidova, 1967; Wefer and Lutze, 1976; Buzas, 1978). The benthic foraminifera are also extremely diverse (morphologically and taxonomically), with thousands of morphospecies (Figure 9.1) belonging to some 3620 fossil and Recent genera (Loeblich and Tappan, 1987).

These demographic characteristics alone make benthic foraminifera attractive subjects for study, but their invaluable utility for solving geological problems via biostratigraphy and palaeoenvironmental reconstructions means that there is a vast published database available on their distribution through time and space. A portion of this huge database has been harnessed by the authors over the past two decades to address various biogeographical and palaeobiological questions. In this chapter we bring together and briefly review the findings of these studies, while also presenting data previously unpublished in the palaeontological literature. Via our review, we demonstrate that the fossil record of the benthic foraminifera, when viewed in the

The Adequacy of the Fossil Record. Edited by S.K. Donovan and C.R.C. Paul.
© 1998 John Wiley & Sons Ltd.

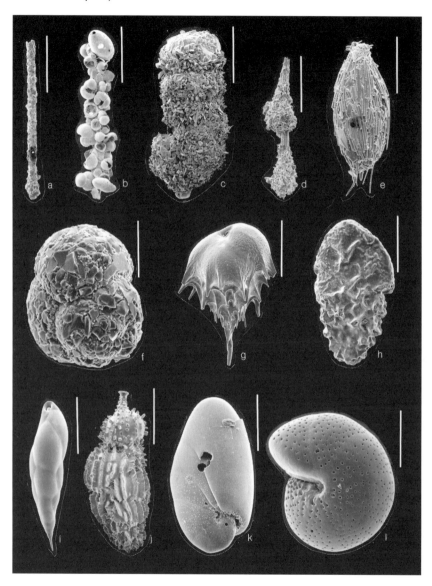

context of modern distributional data, exhibits meaningful patterns and is, indeed, a more than adequate resource in palaeobiological investigations.

THE DATA

Three large data sets have been compiled to help us understand patterns of benthic foraminifera in the fossil record. To provide a framework for comparison, we first documented and interpreted modern distributional data for all modern benthic foraminiferal species from the continental margins (intertidal zone to abyssal plain) around North and Central America (Data Set 1) (Figure 9.2). Data from several hundred publications resulted in 61369 records of 2329 species from 2673 localities. The number of species recorded in the literature was reduced by about one-third by extensive taxonomic work using the collections of the National Museum of Natural History, Smithsonian Institution, Washington, DC; the Natural History Museum, London; and several other smaller collections. The original and synonymized data were published in Culver and Buzas (1980, 1981, 1982, 1985, 1986, 1987).

A subset of these data (from the Atlantic continental margin) was chosen for extension into the fossil record. This second, large-scale data set documents the world-wide stratigraphic record and, hence, the ranges (partial species durations) of the over 800 species that occur on the Atlantic margin today (Data Set 2). This was achieved through a search of thousands of papers, a job made possible only by the Cushman/Todd card catalogue, in the Todd Library at the Smithsonian Institution, that contains details of virtually all records of all species that were referred to in publications from the 1920s to 1960s. Lithostratigraphic information was standardized to current chronostratigraphic placement of all units. Taxonomic standardization

---

Figure 9.1 (opposite)   *The wide variety of morphology and the degree of morphological complexity is only hinted at by these few examples of modern benthic foraminifera from the continental slope and rise off New Jersey, USA.* (a) Hyperammina cylindrica Parr, *finely agglutinated, unilocular; scale bar 1200 μm.* (b) Rhabdammina agglutissima *Hofker, agglutinated (other foraminiferal tests), unilocular; scale bar 1360 μm.* (c) Ammobaculites *sp., finely agglutinated, planispiral to uniserial; scale bar 500 μm.* (d) Reophax distans Brady, *finely agglutinated, uniserial; scale bar 500 μm.* (e) Technitella *sp. cf.* T. legumen Norman, *agglutinated (sponge spicules), unilocular; scale bar 176 μm.* (f) Ammoglobigerina globigeriniformis *(Parker and Jones), coarsely agglutinated, trochospiral; scale bar 50 μm.* (g) Bulimina mexicana Cushman, *calcareous, coarsely perforate, triserial, spinose; scale bar 176 μm.* (h) Bolivina *sp., calcareous, coarsely perforate, biserial, reticulate; scale bar 86 μm.* (i) Eubuliminella exilis *(Brady), calcareous, finely perforate, triserial; scale bar 231 μm.* (j) Uvigerina peregrina dirupta Todd, *calcareous, finely perforate, triserial, spinose and costate; scale bar 300 μm.* (k) Nonionella turgida Williamson, *calcareous, finely perforate, trochospiral; scale bar 100 μm.* (l) Melonis pompilioides *(Fichtel and Moll), calcareous, coarsely perforate, planispiral; scale bar 150 μm*

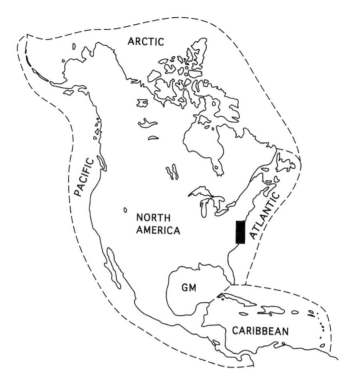

Figure 9.2   *Five geographical regions around North and Central America from which benthic foraminiferal data have been collected. Black rectangle indicates the Salisbury–Albemarle embayment(s). GM, Gulf of Mexico*

of the data was addressed as for the modern database; the dangers of not dealing with taxonomic inconsistencies in compiled data were illustrated by Culver *et al.* (1987).

Data Set 2 dealt with partial species durations and with distributions on a large (provincial) scale. It provoked some questions of community dynamics through time that could only be addressed with a third large data set centred on a particular geographical area and incorporating full species durations (Data Set 3). We chose the Salisbury–Albemarle embayment(s) that were centred on the mid-Atlantic shelf of the USA during the Cenozoic (Figure 9.2). Foraminiferal census data from six Cenozoic formations were combined, and subjected to extensive and exhaustive species-level taxonomic standardization. A total of 357 species was recognized as valid. For each of these, again using the literature and the Cushman/Todd card catalogue, we documented the world-wide geographical distribution through time, as well as the time and location of first and last reliable occurrences. The age of each of 142 formations where the 357 species were recorded was taken from the

US Geological Survey stratigraphic CD-Rom file. The age of origination for each species was defined as the midpoint of the lowermost planktic microfossil zone occurring in the oldest formation where the species was found. The age of last occurrences was similarly defined as the midpoint of the uppermost planktic microfossil zone occurring in the youngest formation where the species was found.

These three data sets (one of modern data and two of fossil data) have allowed us to begin to understand the nature and reliability of observed distributional patterns in time and space. They have permitted us to characterize and attempt to explain patterns of ubiquity, endemism, duration, diversity, origination, extinction and community dynamics for benthic foraminifera over the past 55 Myr.

MODERN FORAMINIFERA

The Statistical Distribution of Benthic Foraminiferal Species Occurrence (Data Set 1)

Naturalists have long known that, in any taxonomic group, only a few species are abundant and the majority are rare, that is, that counts of the number of individuals of each species fit the log series (Fisher *et al.*, 1943) or the log normal (Preston, 1948) distributions. This pattern has been demonstrated for benthic foraminifera (see, for example, Buzas *et al.*, 1977; also see any substantial foraminiferal data set). During compilation of Data Set 1 it became apparent that most available data did not contain counts of species. Rather, the data often consisted of species lists for particular localities. In other words, the data were of species occurrence and not of species abundance. Buzas *et al.* (1982) found that the number of species occurring at 1, 2, . . . *n* localities conforms to a log series distribution. Once the constants x (a number close to 1) and α (a measure of species diversity independent of the number of localities) have been evaluated by the equations

$$S = -\alpha \ln (1 - x)$$

and

$$N = \alpha x / (1 - x)$$

where the number of occurrences (*N*) and the total number of species (*S*) are known, then the number of species occurring *n* times is given by

$$S_n = \alpha x^n / n$$

Table 9.1   *Occurrence data for modern benthic foraminifera around North America*

	Pacific	Arctic	Atlantic	Gulf of Mexico	Caribbean	Totals
No. of localities	999	368	542	426	338	2673
No. of occurrences	19014	7342	10034	18011	6968	61369
No. of species	965	458	878	849	1188	
α	215	108	232	185	412	
Calculated $n=1$	212	107	226	183	389	
Observed $n=1$	199	139	236	217	391	

The expected number of species occurring once is αx, and because x is a number close to 1 the value of α is close to the expected number of species occurring once. Table 9.1 illustrates the close values of observed $n=1$, calculated (expected) $n=1$ and α in five large geographical regions around North and Central America. Figure 9.3 illustrates how well the observed number of species in $\log_2$ classes fits Fisher's log series curve for the Atlantic continental margin, the Gulf of Mexico and the Caribbean. The nature of this distribution is that most species occur only once, fewer occur twice and so on. On the Atlantic continental margin the most frequently occurring species (>32 occurrences at 542 localities) make up only 8% of the faunal diversity and approximately 25% of 878 species occur at only one locality. Clearly, the majority of species occur rarely. This rareness of occurrence has implications for biogeographical, ecological and palaeobiological studies. On the Atlantic continental margin 236 species occur only once and the probability of finding one of these species in 10 samples is 0.018 (Buzas *et al.*, 1982). Thus, we cannot know whether a rarely occurring species is temporally or spatially widespread, but undetected or restricted to a particular locality where it is found. In addition, rarely occurring species are also (usually) species with low abundances (Buzas *et al.*, 1982). These characteristics led Koch (1987; see also Koch and Morgan, 1988; Koch, 1991) to suggest that rare species be eliminated from consideration in palaeobiological studies due to the inherent sampling problems. If a worker were studying individual rare species, then he/she would be particularly subject to Koch's rareness/sampling pitfall. However, as we shall see later in a discussion of species durations, rarely occurring species can provide reliable information when they are treated as a group and not as individual species.

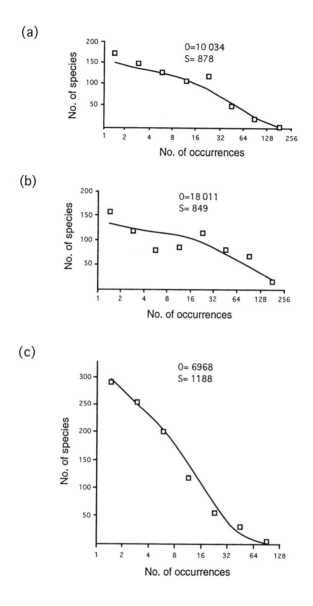

Figure 9.3 *Observed species occurrences (□) and the predicted log series curve for (a) the Atlantic continental margin of North America, (b) the Gulf of Mexico and (c) the Caribbean. O, total number of occurrences; S, total number of species*

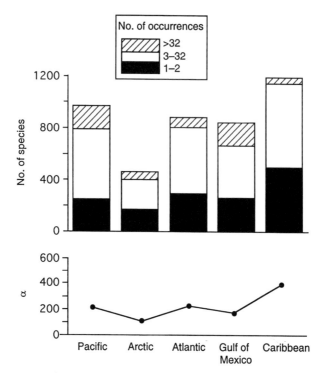

Figure 9.4   *Species diversity (Fisher's α) and number of species in three occurrence classes for five regions around North America. (After Buzas and Culver, 1991)*

Modern Diversity, Ubiquity and Endemism (Data Set 1)

The frequency of species occurrence is intimately related to species diversity. For ease of discussion we arbitrarily define three occurrence classes, 1–2, 3–32 and > 32, and designate these classes as rare, common and abundant, respectively. Figure 9.4 plots α and the number of species in the three occurrence classes for the five geographical regions listed in Table 9.1 (see also Figure 9.2). α is an appropriate measure of diversity because it takes into account the number of occurrences as well as the number of species. The highest value of α is found in the Caribbean followed in descending order by the Atlantic, Pacific, Gulf of Mexico and Arctic. Number of species show a similar pattern (highest in the Caribbean, lowest in the Arctic), although the rank order reverses for the Atlantic and Pacific, probably due to the larger number of localities in the latter (Table 9.1). The high diversity in the Caribbean is due to the large numbers of rare and common species. The Arctic region, which has the lowest value for α, also has the lowest number of rare and common species.

Table 9.2 *Number of endemic and ubiquitous modern benthic foraminiferal species around North and Central America*

	Pacific	Arctic	Atlantic	Gulf of Mexico	Caribbean
Endemic					
Total	420	107	159	182	458
> 32 occurrences	56	10	0	14	0
3–32 occurrences	220	36	56	73	141
1–2 occurrences	144	61	103	95	317
Ubiquitous					
Total	112	112	112	112	112
> 32 occurrences	38	16	35	47	7
3–32 occurrences	53	65	71	42	84
1–2 occurrences	21	31	6	23	21

Therefore, species diversity trends for modern continental margin benthic foraminifera follow the 'classic' pattern of higher diversity in low latitudes and lower diversity in high latitudes. Patterns of endemism and ubiquity are related to those of species diversity. Table 9.2 provides the number of endemic and ubiquitous species around North and Central America in the three occurrence categories; these data are presented graphically in Figures 9.5 and 9.6. Very few species (112 of 2329 = 5%) are ubiquitous (that is, occur in all five regions). However, a large proportion of species (1326 of 2329 = 57%) is endemic (that is, occur in only one of the five regions). Two regions of highest diversity (the Caribbean and Pacific; Figure 9.4) have the greatest number of endemic species (Figure 9.5; Table 9.2). The data for Figure 9.4 are subdivided based on endemicity, or not, in Figure 9.6. Comparison of these two figures indicates that the number of abundantly occurring species proportionally decreases, whereas the number of rarely occurring species proportionally increases. Indeed, there are no abundantly occurring, endemic species in the Atlantic and Caribbean regions (Figure 9.6). The pattern for ubiquitous species is the reverse: abundantly occurring species proportionally increase, whereas the rarely occurring species proportionally decrease (Figure 9.6). In other words, the abundantly occurring species have a wider geographical distribution (by definition) than rarely occurring species and, thus, have a greater chance of dispersal into other areas.

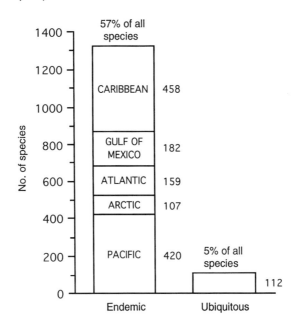

Figure 9.5   *Number of endemic and ubiquitous species of modern benthic foraminifera in five regions around North and Central America*

FOSSIL FORAMINIFERA

Geographical Origin, Dispersal and Duration (Data Set 2)

Our understanding of the nature of modern benthic foraminiferal occurrences (summarized above) provides the framework within which one can interpret patterns in data on fossil foraminifera. The geographical distribution of first occurrences indicates that Cretaceous and Cenozoic benthic foraminifera have originated in all parts of the world oceans without any significant hiatuses (Buzas and Culver, 1989). Although the data do not support the concept of centres of origin, there is a trend of origination increasing in high latitudes during the global cooling phase of the later Cenozoic. This may be due, at least partially, to the distribution of outcrops, but the species duration data lend support to the reality of this trend. Average duration for modern North American Atlantic shelf species south of Cape Hatteras is 20 Myr whereas species restricted to higher latitudes (north of Cape Hatteras) have an average species duration of only 7 Myr, thus indicating higher evolutionary rates in higher latitudes during the late Cenozoic (Buzas and Culver, 1984).

Species durations are of considerable interest because, if species diversity

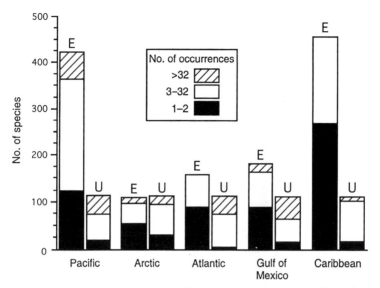

Figure 9.6 *Number of endemic (E) and ubiquitous (U) species of modern benthic foraminifera in three occurrence classes in five regions around North and Central America. See Table 9.2 for counts. (After Buzas and Culver, 1991)*

in a particular environment remains the same through time, then species duration becomes a measure of the rate of evolution. The mean species duration for both abundant and rarely occurring species on the Atlantic continental margin of North America is about 21 Myr (Buzas and Culver, 1989). We noted above the potential sampling problem with rarely occurring species. However, the value of 21 Myr average species duration for rarely occurring species is acceptable because, in this case, we are not discussing the duration of a single rarely occurring species, but the average duration of the entire group of rarely occurring species. Indeed, because of the log series distribution of species occurrence, the rarely occurring group has twice as many species as the abundantly occurring group. Not only are the mean species durations of abundantly and rarely occurring species the same, but also the temporal distribution of origination for these two groups is very similar. Over half of the modern Atlantic shelf species with a fossil record originated in the Miocene, with similar, but lower, proportions originating in other Cenozoic epochs. Thus, evolutionary patterns and rates are apparently the same for rarely and abundantly occurring species groups.

Patterns of species origination, endemism and ubiquity are all dependent on dispersal characteristics. Theoretically, small marine organisms such as benthic foraminifera should be easily dispersed by currents or larger mobile organisms, but how rapid is this dispersal capability? Figure 9.7 presents the number of modern benthic foraminiferal species on the Atlantic continental

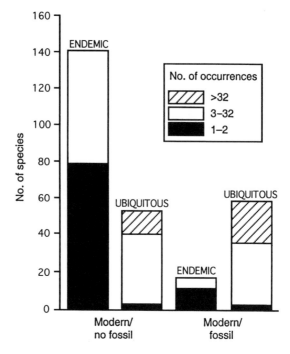

Figure 9.7   *Number of endemic and ubiquitous species of modern benthic foraminifera in three occurrence classes for modern species with no fossil record (modern/no fossil) and modern species with a fossil record (modern/fossil) on the Atlantic continental margin of North America*

margin of North America with and without a fossil record in the ubiquitous and endemic categories discussed earlier. A total of 53 modern species are ubiquitous, but have no fossil record. The most likely explanation is that these species have evolved recently and must, therefore, have dispersed extremely rapidly. The fact that rapid dispersal is necessary to achieve species longevity is further exemplified by the fact that the number of endemic modern species lacking a fossil record is far greater than the number of ubiquitous modern species without a fossil record (Figure 9.7).

Because few species with a fossil record are endemic, the large number of endemic species on the Atlantic continental margin without a fossil record (Figure 9.7) probably evolved recently. Those that disperse are most likely to persist temporally. Thus, species accumulate through time in the commonly and abundantly occurring classes via their success in dispersal. For any particular segment of time, the majority of species that originate in that segment soon become extinct, but those that disperse are more likely to persist. Thus, when we look back from the present day, we see an accumulation of species that originated at different times and that are dispersed

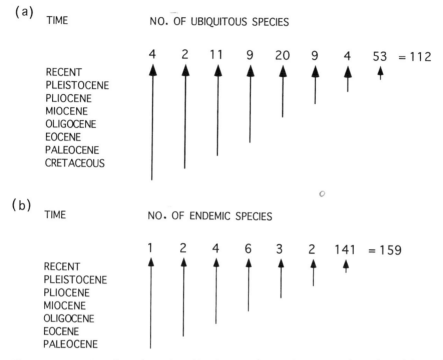

Figure 9.8 *(a) Number of species ubiquitous to five regions around North and Central America, and their epoch-level geological ranges. (b) Number of species endemic to the Atlantic continental margin of North America and their epoch-level geological ranges. (After Buzas and Culver, 1991)*

around North America (Figure 9.8a). The number of endemic species with a fossil record is much smaller, but these species are distributed throughout the fossil record in a pattern similar to that for ubiquitous species (Figure 9.8b).

Community Dynamics (Data Set 3)

For each segment of geological time, some species originate, some become extinct and some carry on from the preceding segment through to the succeeding one. The varying contributions of these arrivals and departures determine the species diversity at any particular moment in time. A ledger consisting of these additions and subtractions of species can be termed an 'evolutionary budget' and can help answer fundamental palaeobiological questions. For example, how do the rates of origination and extinction vary to maintain species diversity? Is species diversity in a particular environment a constant (a species carrying capacity), and how do the rates of origination

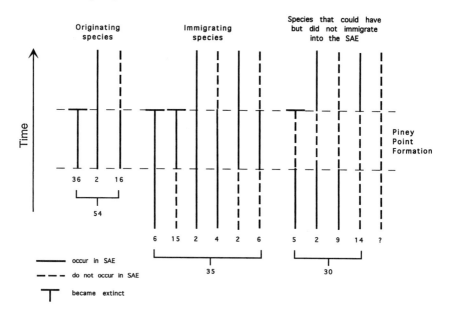

Figure 9.9 *Number of species in all geological range categories (as indicated by vertical lines) for species extant during deposition of the Eocene Piney Point Formation, Salisbury–Albermarle embayment(s) (SAE). Originating species (54) became extinct, or emigrated and returned to SAE, or emigrated elsewhere and did not return. Immigrants (35) originated earlier within the SAE or elsewhere, became extinct, or emigrated and returned, or emigrated elsewhere. Species (30) not recorded in the Piney Point Formation were recorded within the SAE before or after deposition of the Piney Point Formation, or both. An unknown number of species (?) might have immigrated, but never did, into the SAE*

and extinction adjust themselves to maintain it? Are the rates themselves constant or do they, along with species diversity, vary in a predictable way, and so on?

In constructing such a budget it became clear that this data set can inform us about community dynamics through time. Data Set 3, composed of species ranges for six Cenozoic formations from the Atlantic coastal plain of the USA, illustrates the ephemeral nature of neritic foraminiferal communities over the past 55 Myr. Communities inhabiting the Salisbury–Albermarle embayment(s) during each successive transgression (Figure 9.9 gives the data for one of those transgressions, that resulting in deposition of the Eocene Piney Point Formation) are composed of a dynamic mixture of immigrants and newly evolved species (Buzas and Culver, 1994). During regressions, species comprising these communities either emigrated or became extinct. Immigrants proved to be a mixture of former inhabitants of the embayment(s) and species that were formerly living elsewhere, mostly in neritic environments of the Atlantic and Gulf coastal plains. Similarly, emigrants sometimes returned during subsequent transgressions, but many did not (Figure 9.9). Thus, the foraminiferal inhabitants of the Atlantic and Gulf coastal plains/ shelves comprise a species pool from which species are drawn during each transgression. Because dispersal is geologically instantaneous (Buzas and Culver, 1991), any member of the pool could inhabit a new embayment. Indeed, only a few of the 357 species returned to the Salisbury–Albermarle embayment(s) regularly.

Thus, it seems that benthic foraminiferal species respond to changing environmental conditions on an individualistic basis. Although characteristic neritic assemblages are recognizable throughout the Cenozoic, the detailed composition of such communities is a result of the complex interplay of origination, extinction and dispersal characteristics of individual species, changing environmental conditions, and the recruitment (randomly or otherwise) from a species pool, where recruitment is to a considerable degree dependent on frequency of occurrence.

DISCUSSION

Benthic foraminiferal occurrence data show biologically meaningful variations in time and space. The patterns observed in modern data can also be observed in fossil data and both can be interpreted in a meaningful manner. Therefore, it would be inappropriate to suggest that the fossil record of benthic foraminifera is inadequate to address palaeobiological questions. Indeed, even the sampling problem of rarely occurring species can be circumvented if questions are asked of rare species as a group rather than of rare species as individuals. Throughout this paper we have implicitly or explicitly shown the way in which modern data aid interpretation of fossil data, but also how fossil data give us insights into modern data. We can take this a step further and show how modern and fossil data can provide us with a window to the future.

For example, which modern species and habitats are most susceptible to extinction via future environmental perturbations? We have seen that the Caribbean and Pacific regions round North America have the greatest number of endemic species (Figure 9.5), but the Caribbean endemics are much more rarely occurring (Figure 9.6). Natural or anthropogenic habitat disturbance or destruction is thus most likely to cause extinction in the Caribbean (Buzas and Culver, 1991). Indeed, 231 of 1188 species inhabiting the Caribbean have been recorded at only one locality (Figure 9.10). The Arctic has only 46 species occurring at only one locality, even though this region has approximately the same number of localities as the Caribbean (Table 9.1). Thus, localized habitat destruction is more likely to cause species extinction in the low latitudes than in higher latitudes.

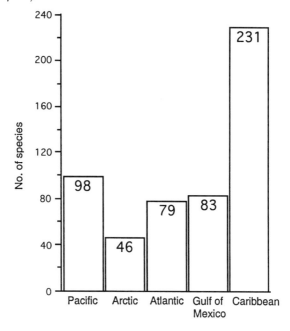

Figure 9.10   *Number of species of modern benthic foraminifera that occur at only one locality around North and Central America. (After Buzas and Culver, 1991)*

We can also use the fossil data to help suggest which foraminiferal communities are most susceptible to disturbance and, hence, change in the future. Recent research (for example, Huntley, 1991; Webb and Bartlein, 1992; Webb *et al.*, 1993) has shown that marine and terrestrial organisms have responded as individual species to post-glacial global warming. Thus, varying migration rates have resulted in changing community composition. Because global warming rates are likely to increase via an anthropogenically enhanced greenhouse effect, and because habitat disturbance and destruction via human activities are likely to be increasingly significant, changes in community composition will occur. The shallow marine (particularly intertidal) communities, which are characteristically persistent through time (for example, Scott *et al.*, 1983; Wightman *et al.*, 1993), are the communities most at risk. This is because the predicted rate of global warming, with all of its associated effects, is much greater than in the past (Peters, 1991; Schneider, 1993a,b) and because the littoral zone is the habitat most affected by human development activities. The neritic species pool will continue as the resource to be drawn upon for recruitment by shelf communities, but the evolutionarily conservative littoral communities, somewhat counter-intuitively, may well be the entities most affected by predicted environmental perturbation in the near future.

QUESTIONS ADDRESSED

The aim of this volume is to investigate whether the fossil record is an adequate resource for addressing palaeobiological questions. We have shown that the fossil record of benthic foraminifera, when considered in the context of modern distributional data, is indeed, more than adequate. To conclude this contribution we list a series of biologically and palaeobiologically significant questions that we have addressed with our data sets, some of which are relevant mainly for benthic foraminifera, but some of which have a much wider relevance. There are further questions that could be addressed with these large databases of fossil foraminifera, but let us be content here with the following fourteen.

1. *Do foraminifera and other organisms have similar patterns?*
   Yes. Increase in species richness with decreasing latitude, distributions of species occurrences and bioprovinces are similar.
2. *Can the traditional qualitative bioprovinces be quantified using species occurrences?*
   Yes. Cluster analyses of occurrence data identify the traditional bioprovinces and new ones at depth, on the North American Atlantic and Pacific margins and in the Gulf of Mexico.
3. *Does the distribution of occurrences adhere to a statistical distribution?*
   Yes. The distribution fits Fisher's log series. Most species occur only once, fewer twice and so on. Fisher's measure of diversity, α, is a good measure of diversity for species occurrences.
4. *Is there a centre of origin for species?*
   No. All areas exhibit evolution all the time. However, rates vary in time and space.
5. *Are observed patterns dependent on dispersal times?*
   No. Dispersal can be geologically instantaneous.
6. *Do species longevities vary with latitude?*
   Yes. Northern species in the North American Atlantic continental margin have shorter species durations.
7. *Do abundantly and rarely occurring species with a fossil record have different species longevities?*
   No. Regardless of the number of occurrences, species with a fossil record may have the same longevity.
8. *Do ubiquitous and endemic species in the modern fauna have different characteristics?*
   Yes. Ubiquitous species are more likely to have a fossil record and occur abundantly. Endemic species are more likely to have no fossil record and occur rarely.
9. *Does the assembly and disassembly of fossil assemblages or communities indi-*

cate that groups of species behave as a unit?
No. Each species has its own distribution in space and time. Neverthe-less, species do not have a haphazard distribution and depth-related assemblages are easily recognized.

10. *Do abundant, common and rare species maintain their abundance with time?*
No. About half change their abundance status with time and show no trend in increasing or decreasing abundance.

11. *Are abundant species more likely to survive extinction events?*
No. Extinction events affect all levels of abundance.

12. *Do fossil communities in the Cenozoic exhibit long-term stasis?*
No. A considerable proportion of origination and extinction occurred in all of the stratigraphic units examined.

13. *Do Cenozoic palaeocommunities exhibit any balance or equilibrium?*
Yes. Although a significant amount of origination and extinction occurs, the numbers of species immigrating and emigrating in some strati-graphic units are nearly equal. However, in the Miocene, origination far exceeds extinction and this interval acts as a diversity pump for the Cenozoic.

14. *Do the foraminiferal data indicate which region in North and Central America might suffer the greatest extinction as a result of a catastrophic environmental perturbation?*
Yes. The Caribbean region has by far the greatest number of species with restricted geographical distributions and is, therefore, most susceptible to extinction.

ACKNOWLEDGEMENTS

We thank F. Rögl, H. Hagn, M. Urlichs, H. Malz, H. Hooyberghs, A. Liebart, H. Gocht, J. Hohenegger, W. Piller, J.W.C. Doppert, M.V. Hounsome, H.B. Whittington, M. Dorling, C.G. Adams, J. Whittaker, R.L. Hodgkinson, L. Collins, J. Jett, G. Knisely, C.F. Koch, L. Koozman, L.W. Ward, S.W. Snyder and J. Swallow for their help. This research was supported by a Smithsonian Scholarly Studies Grant and NSF Grant EAR82-16550, and is a contribution from the NHM/UCL-BkB programme on Global Change and the Biosphere.

REFERENCES

Buzas, M.A., 1978, foraminifera as prey for benthic deposit feeders: results of predator exclusion experiments, *Journal of Marine Research*, **36**: 617–625.
Buzas, M.A. and Culver, S.J., 1984, Species duration and evolution: benthic foraminif-era on the Atlantic continental margin of North America, *Science*, **225**: 829–830.

Buzas, M.A. and Culver, S.J., 1989, Biogeographic and evolutionary patterns of continental margin benthic foraminifera, *Paleobiology*, **15**: 11–19.

Buzas, M.A. and Culver, S.J., 1991, Species diversity and dispersal of foraminifera, *BioScience*, **41**: 483–489.

Buzas, M.A. and Culver, S.J., 1994, Species pool and dynamics of marine paleocommunities, *Science*, **264**: 1439–1441.

Buzas, M.A., Smith, K.K. and Beem, K.A., 1977, Ecology and systematics of foraminifera in two *Thalassia* habitats, Jamaica, West Indies, *Smithsonian Contributions to Paleobiology*, **31**: 139 pp.

Buzas, M.A., Koch, C.F., Culver, S.J. and Sohl, N.F., 1982, On the distribution of species occurrence, *Paleobiology*, **8**: 143–150.

Culver, S.J. and Buzas, M.A., 1980, Distribution of recent benthic foraminifera off the North American Atlantic coast, *Smithsonian Contributions to the Marine Sciences*, **6**: 512 pp.

Culver, S.J. and Buzas, M.A., 1981, Distribution of recent benthic foraminifera in the Gulf of Mexico, *Smithsonian Contributions to the Marine Sciences*, **8**: 898 pp.

Culver, S.J. and Buzas, M.A., 1982, Distribution of recent benthic foraminifera in the Caribbean area, *Smithsonian Contributions to the Marine Sciences*, **14**: 382 pp.

Culver, S.J. and Buzas, M.A., 1985, Distribution of recent benthic foraminifera off the North Pacific coast from Oregan to Alaska, *Smithsonian Contributions to the Marine Sciences*, **26**: 234 pp.

Culver, S.J. and Buzas, M.A., 1986, Distribution of recent benthic foraminifera off the North American Pacific coast from California to Baja, *Smithsonian Contributions to the Marine Sciences*, **28**: 634 pp.

Culver, S.J. and Buzas, M.A., 1987, Distribution of recent benthic foraminifera off the Pacific coast of Mexico and Central America, *Smithsonian Contributions to the Marine Sciences*, **30**: 184 pp.

Culver, S.J., Buzas, M.A. and Collins, L.S., 1987, On the value of taxonomic standardization in evolutionary studies, *Paleobiology*, **13**: 169–176.

Fisher, R.A., Corbett, A.S. and Williams, C.B., 1943, The relation between the number of species and the number of individuals in a random sample of an animal population, *Journal of Animal Ecology*, **12**: 42–58.

Huntley, B., 1991, How plants respond to climate change: migration rates, individualism and the consequences for plant communities, *Annals of Botany*, **67**: 15–22.

Koch, C.F., 1987, Prediction of sample size effects on the measured temporal and geographic distribution patterns of species, *Paleobiology*, **13**: 100–107.

Koch, C.F., 1991, Sampling from the fossil record. *In* N.F. Gilinsky and P.W. Signor (eds), *Analytical Paleobiology, Paleontological Society Short Courses in Paleontology, 4*, University of Tennessee, Knoxville: 4–18.

Koch, C.F. and Morgan, J.P., 1988, On the expected distribution of species ranges, *Paleobiology*, **14**: 126–138.

Loeblich, A.R., Jr and Tappan, H., 1987, *Foraminiferal Genera and their Classification* (2 vols), Van Nostrand Reinhold, New York: 970+212 pp.

Peters, R.L., 1991, Consequences of global warming for biological diversity. *In* R.L. Wyman (ed.), *Global Climate Change and Life on Earth*, Routledge, Chapman and Hall, London: 99–118.

Preston, F.W., 1948, The commonness, and rarity, of species, *Ecology*, **29**: 254–283.

Saidova, K.M., 1967, The biomass and quantitative distribution of live foraminifera in the Kurile–Kamchatka trench area, *Doklady Akademii Nauk SSSR*, **174**: 207–209.

Schneider, S.H., 1993a, Can paleoclimatic and paleoecological analysis validate future global climate and ecological change projections? *In* A. Eddy and A. Oeschger (eds), *Global Changes in the Perspective of the Past*, Wiley, Chichester: 317–340.

Schneider, S.H., 1993b, Degrees of certainty, *National Geographic Research and Exploration*, **9**: 173–190.

Scott, D.B., Gradstein, F., Schafer, C.J., Miller, A. and Williamson, M., 1983, The Recent as a key to the past: does it apply to agglutinated foraminiferal assemblages? *In* J.G. Verdenius, J.E. Van Hinte and A.R. Fortuin (eds), *Proceedings of the First Workshop on Arenaceous Foraminifera, IKV Publication No. 108*, Trondheim: 147–157.

Webb, T., III and Bartlein, P.J., 1992, Global changes during the past million years: climatic controls and biotic responses, *Annual Review of Ecology and Systematics*, **23**: 141–173.

Webb, T., III, Crowley, T.J., Frenzel, B., Gliemeroth, A.-K., Jouzel, J., Labeyrie, L., Prentice, I.C., Rind, O., Ruddiman, W.F., Sarnthein, M. and Zwick, A., 1993, Group report: use of palaeoclimatic data as analogs for future global changes. *In* J.A. Eddy and H. Oeschger (eds), *Global Changes in the Perspective of the Past*, Wiley, Chichester: 51–71.

Wefer, G. and Lutze, G.F., 1976, Benthic foraminifera biomass production in the western Baltic, *Kieler Meeresforschungen*, **3**: 76–81.

Wightman, W.G., Scott, D.B., Medioli, F.S. and Gibling, M.R., 1993, Carboniferous marsh foraminifera from coal-bearing strata at the Sydney basin, Nova Scotia: a new tool for identifying paralic coal-forming environments, *Geology*, **21**: 631–634.

10
The Fossil Record of Cheilostome Bryozoa in the Neogene and Quaternary of Tropical America: Adequacy for Phylogenetic and Evolutionary Studies

Alan H. Cheetham and Jeremy B.C. Jackson

INTRODUCTION

Should the stratigraphic record of species occurrences serve as a source of data for constructing phylogenetic hypotheses, and thus for inferring evolutionary tempo and mode? This question continues to draw attention on both practical and philosophical grounds. One position holds that, because 'true' ancestor–descendant relationships are unknowable, phylogenetic hypotheses need only be consistent with an agreed-upon methodology, such as parsimony, in which it is appropriate to consider 'ancestral diagnoses', but not ancestral taxa (Norell, 1996, and references therein). In contrast, it might be assumed that the ancestors of species can be other species (in the sense of including the organisms from which species descended; Foote, 1996) and that such species could actually occur in the fossil record. Even if this contrasting point of view is adopted, it is still possible to question whether ancestral species occur with sufficient frequency, and, if so, whether their stratigraphic ranges adequately reflect their 'true' (but also unknowable) temporal durations, to provide significant evidence for phylogenetic and evolutionary inference.

The Adequacy of the Fossil Record. Edited by S.K. Donovan and C.R.C. Paul.
© 1998 John Wiley & Sons Ltd.

New methods have been developed in the last few years to address questions of the adequacy of the record by estimating, for example, the probability of preservation of taxa (Foote, 1996; Foote and Raup, 1996) and the completeness of their stratigraphic ranges (Marshall, 1990, 1994, 1995). Using these methods, we here explore the record of a major marine invertebrate group, the cheilostome Bryozoa, in the Neogene and Quaternary of tropical America, with regard to their potential importance in constructing and testing phylogenetic and evolutionary hypotheses. We concentrate on the quality of the record as a whole, leaving aside questions of how stratigraphic information can be incorporated in constructing hypotheses (Alroy, 1994; Fisher, 1994; Huelsenbeck, 1994; Wagner, 1995a, 1996) and questions of how ancestral taxa can be recognized in individual cases (Paul, 1992).

Although the Neogene and Quaternary deposits of tropical America have long been known to contain a rich fossil record of cheilostome bryozoans (Canu and Bassler, 1918, 1919, 1923, 1928), quantitative estimates of its diversity, and spatial and temporal distribution have only recently become possible through detailed sampling of stratigraphic sequences ranging from late Miocene to Pleistocene age in Panama and Costa Rica (approximately 9–1 Ma; Coates *et al.*, 1992; Coates and Obando, 1996) and from early or middle Miocene to early Pliocene age in the Dominican Republic (approximately 17–3 Ma; Saunders *et al.*, 1982, 1986). Previous work on this tightly stratigraphically controlled material has focused on evolutionary patterns in two genera, *Metrarabdotos* and *Stylopoma*, each of which includes more than a dozen species inferred to have originated within the Caribbean Neogene (Cheetham, 1986, 1987; Cheetham and Hayek, 1988; Cheetham *et al.*, 1994; Jackson and Cheetham, 1994; Cheetham and Jackson, 1995, 1996). In each genus, phylogenetic hypotheses based on morphological characters alone were significantly different from those incorporating stratigraphic context to root trees and modify their topology (Cheetham and Hayek, 1988; Jackson and Cheetham, 1994). However, our purpose here is not to examine the effects of stratigraphic information on the construction of particular phylogenetic hypotheses, but rather to estimate its potential significance as a statistical property of the cheilostome fauna as a whole. Therefore, our focus is on data from a recent study of the diversity and distribution of this fauna (Cheetham *et al.*, 1998).

In the following sections, we adopt the terminology of Foote (1996) and Foote and Raup (1996), using *completeness* to refer to the proportion of species that have left a preserved record; *preserved* to refer to species that are both preserved and recovered at least once; and *duration* to refer to the (unknown) time between the origin and extinction of a species, of which its *range* is the portion represented by preserved material. The quality or adequacy of the fossil record of a taxonomic group is thus resolved into estimates of the completeness of taxonomic representation and estimates of durations as extensions of observed ranges.

DIVERSITY OF THE CHEILOSTOME FAUNA AND THE ADEQUACY OF SAMPLING

The cheilostome fauna, as represented in 200 collections from Panama and Costa Rica and in 124 collections from the Dominican Republic, totals 268 species, of which 42% are undescribed (Cheetham *et al.*, 1998). A major consideration in estimates of diversity (species richness) is the degree to which species distinctions based on morphology (especially skeletal morphology, as for most living as well as fossil species of cheilostomes) correspond to genetic differences. For three genera (*Parasmittina*, *Steginoporella* and *Stylopoma*), representing a fairly broad cross-section of the cheilostomes, correspondence between genetic differences based on protein electrophoresis and differences in skeletal morphology is quite high, provided that morphospecies are split to the limits of statistical significance (Jackson and Cheetham, 1990, 1994). To make the estimates of cheilostome diversity in the Neogene and Quaternary of tropical America as consistent as possible with results obtained for these three genera (all of which are richly represented in these deposits), we have used the smallest, consistently observed morphological differences to distinguish the 268 species of this fauna (Cheetham *et al.*, 1998). With this approach, the total number of species is comparable to estimates of the diversity of the living fauna of the tropical western Atlantic (273 species; Schopf, 1973).

To maximize stratigraphic precision, we limit this analysis to species whose first occurrences fall within the stratigraphically well-controlled data set encompassed by the collections from Panama, Costa Rica and the Dominican Republic (Cheetham *et al.*, 1998, table 4). Such species number 246, or 92% of the total, and divide exactly into extant (123) and extinct (123) halves. Last occurrences fall outside the database for only three of the extinct species, in addition to the occurrences in the living fauna for those that are extant. We emphasize first occurrences throughout this analysis because of their critical importance in evaluating the extent of 'ghost lineages' required by phylogenetic hypotheses to connect species to their hypothesized ancestors (Wagner, 1995b).

A rough indication of the completeness of the database is given by cumulative curves of collecting effort (Figure 10.1). The best-collected interval, for both extant and extinct species, is the latest Miocene to mid-Pliocene (6–3 Ma), with each curve levelling off just short of 100 species. This interval is also represented by the greatest number of collections. Older late Miocene collections (9–6 Ma) are slightly fewer in number and lower in diversity,

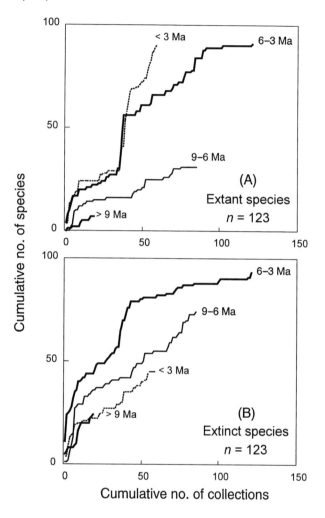

Figure 10.1 *Cumulative curves of collecting effort based on 200 collections from Panama and Costa Rica ranging in age from late Miocene to Pleistocene (9–1 Ma) and on 124 collections from the Dominican Republic ranging in age from early or middle Miocene to early Pliocene (17–3 Ma). For each curve, collections were added by age, starting with the oldest. (A) Species surviving to the present. (B) Extinct species*

especially in extant species, but nevertheless represent a generally good sample. Those from the early or middle Miocene (> 9 Ma) are few and with still lower diversity, especially for extant species, as might be expected. It seems likely that additional collecting in this part of the section would produce an appreciable increase in the number of species. The greatest disparity in collecting curves is between extant and extinct species in collections younger than mid-Pliocene (< 3 Ma). Neither curve levels off, but the rate of climb for extant species, nearly reaching 100, far exceeds that for extinct ones, numbering fewer than 50. Thus, extant species in collections of mid-Pliocene to Pleistocene age are the only component of the cheilostome fauna whose numbers are obviously seriously underestimated, but the completeness of the fauna as a whole might also be less than would justify the use of its stratigraphic distribution as evidence for constructing phylogenetic hypotheses.

OBSERVED RANGES AND ESTIMATES OF TRUNCATION

The median observed stratigraphic range for the 246 cheilostome species, regardless of whether extant or extinct, is 3.55 Myr, less than 25% of the approximately 16–Myr-long early or middle Miocene to Pleistocene interval represented by the database. Observed ranges, grouped by 1.5 Myr intervals in Figure 10.2A, are based on estimated ages (Ma) of oldest and youngest occurrences (Cheetham *et al.*, 1998, table 4). Truncation of true durations due to sampling is likely to be substantial at both ends for extinct species, and for extant species cut-off at the Recent might seem to make the effects of 'artificial' truncation even greater. However, the median range of extant species (4.10 Myr) is 64% *longer* than that of extinct ones (2.50 Myr; Table 10.1). The difference is highly significant (Mann–Whitney $U = 5393.5$, $P = 0.0001$) and is due principally to the numbers of species having the shortest ranges (Figure 10.2A). More than twice as many extinct species (59% of 123) as extant ones (24% of 123) have ranges of 3 Myr or less, but the number having ranges greater than 4.5 Myr is approximately the same (31% of extant species, 32% of extinct ones). This difference has important consequences in the analysis of preservation and completeness (see following section).

To extrapolate from the observed ranges to estimates of species durations, we calculated 95% confidence intervals for the older end-point of each range (that is, the first occurrence), using the relationship $\alpha = (1 - C_1)^{-1/(H-1)} - 1$, where α is the confidence interval expressed as a fraction of the range, C_1 is the confidence level and H is the number of horizons at which the species occurs (Marshall, 1990). Range extensions (α times the stratigraphic range) were calculated for the 91 extant and 91 extinct species known from three or more occurrences (Figure 10.2B), on the assumption that each occurrence represents a separate horizon. This assumption probably results in a tendency to underestimate confidence intervals, but exclusion of species with only one or two occurrences probably lengthens the median value of extended ranges in compensation. (Moreover, the approach employed in the following section, with a different set of assumptions, suggests that range extensions based on 95% confidence intervals may substantially over-

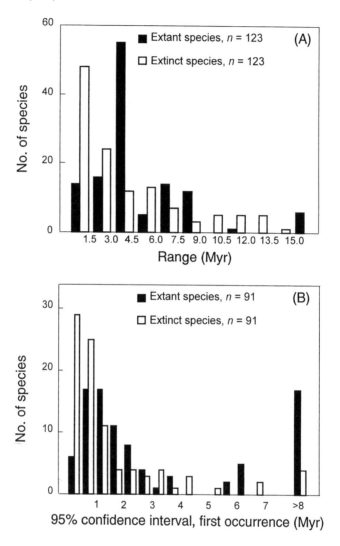

Figure 10.2   *(A) Frequency distribution of observed stratigraphic ranges grouped by 1.5 Myr intervals. (B) Frequency distribution of 95% confidence intervals on first occurrences grouped by 0.5 Myr intervals*

estimate the 'true' median duration.) The median range extension for extant species is more than twice that for extinct ones, further emphasizing the longer ranges of extant species (median 95%-extended range 7.43 Myr compared to 5.68 Myr for extinct taxa). More than half (59%) of the extinct species have ranges extended by 1 Myr or less, compared with only 25% of extant species (Figure 10.2B). (Reducing C_1 to 50% results in median extended

Table 10.1 *Range and first-occurrence statistics for cheilostome species with first occurrences in Neogene and Quaternary collections from Panama, Costa Rica and the Dominican Republic (CI = confidence interval)*

Statistic	Species		
	Extant	Extinct	All
Median range (Myr)	4.10	2.50	3.55
Number of species on which based	123	123	246
Median range extension (Myr), 95% CI	1.78	0.74	1.27
Number of species on which based	91	91	182
Median extended range (Myr)	7.43	5.68	6.26
Median first occurrence (Ma)	4.10	7.10	5.35
Median first occurrence + 95% CI (Ma)	7.43	8.64	8.31

ranges of 4.53, 4.25 and 4.50 Myr, respectively, for extant, extinct and all species; these values not only are closer together than the 95%-extended ranges, but also are much closer to the estimates of median duration based on the assumptions in the following section.)

Extending only the lower end-points of ranges should result, of course, in greater underestimation (or less overestimation) of the ranges of extinct species than of extant ones. However, the median range of extinct species is increased more (2.3 times) by the 95% extensions on first occurrences than that of extant ones (1.8 times). The proportional relationship between extended and observed ranges (Figure 10.3A) is tighter for the extinct species (correlation coefficient $r = 0.83$, $P < 0.001$; Spearman rank-order correlation $r_S = 0.89$, $P < 0.001$) than for extant ones ($r = 0.33$, $P < 0.01$; $r_S = 0.62$, $P < 0.001$); with the two outliers (Figure 10.3A) deleted, the correlation for extant species is increased to the same level of significance as for extinct ones, but is still somewhat lower numerically ($r = 0.63$, $r_S = 0.68$, $P < 0.001$ in both cases). (Reducing C_1 to 50% increases all correlations to more than 0.9.)

First occurrences of extant and extinct species are distributed throughout the approximately 16 Myr-long interval, but those for extant species are especially concentrated at 4.5–3 Ma (Figure 10.3B). Many extinct species also have first occurrences at that time, but approximately as many others first occur at intervals from 6 to >15 Ma. These differences are reflected in the median first occurrences of 4.10 Ma and 7.10 Ma for extant and extinct species, respectively (Table 10.1), a difference of 42%. Because of the longer 95% confidence limits for extant species, median extended first occurrences for extant (7.43 Ma) and extinct (8.64 Ma) species differ by only 16% (Table 10.1).

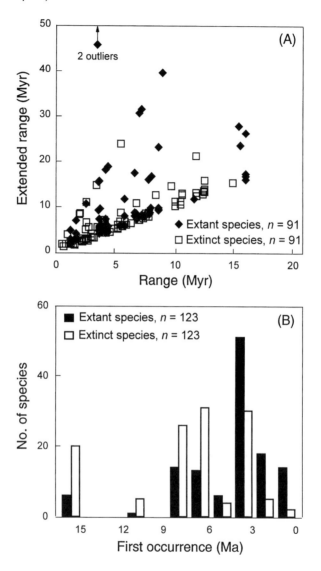

Figure 10.3   *(A) Relation between observed stratigraphic ranges and ranges extended by 95% confidence intervals. Two species with extended ranges of 71 Myr are off the chart. (B) Frequency distribution of observed first occurrences grouped by 1.5 Myr intervals*

COMPLETENESS: HOW WELL PRESERVED IS THE CHEILOSTOME RECORD?

We calculated estimates of preservation probability, completeness and median species duration, following Foote and Raup (1996), from frequency distributions of the observed stratigraphic ranges of extant and extinct species (Figure 10.4). (It should be noted that these ranges are compiled from occurrences in discrete time intervals, as in Foote and Raup (1996, fig. 1), rather than from the ages of first and last occurrence used for calculations in the preceding section; thus, the frequencies in Figures 10.2 and 10.4 are not closely comparable.) Preservation estimates are based on the assumption that true durations are distributed exponentially with a stochastically constant extinction rate (q). This approach considers only the intervals containing the end-points of ranges, not the occurrences in between, and thus yields estimates of median duration that are independent of those calculated in the preceding section. Calculations were made for extant and extinct species separately, and for both combined (Table 10.2). Despite a large apparent difference in preservation probability per stratigraphic interval between extant and extinct species (40% versus 85%), high estimates of completeness, 76% and 95%, respectively, were obtained for both groups (Table 10.2). Estimates of median duration, 4.31 Myr for extant species and 3.23 Myr for extinct ones, are 42% and 43% shorter, respectively, than those obtained by extending first occurrences with 95% confidence intervals (Table 10.1). However, all of these values are likely to be affected by sampling biases (M. Foote, personal communication, 1996), considered below.

Bin size, that is, the length of the shortest stratigraphic interval used to form the frequency distribution of ranges, affects the precision with which preservation and duration are estimated (Foote and Raup, 1996). For the cheilostome data, intervals of 1.5 Myr are the shortest that yield no empty bins for the 246 species, that is, $f(t) > 0$ for all values of t, with t = observed range measured in the chosen bin units. We used the same units for extant and extinct species (Figure 10.4). This bin size is less than half the median duration, 35% for extant species and 46% for extinct ones (Table 10.1), and thus should yield estimates of preservation and duration that are relatively stable (Foote and Raup, 1996). However, coarsening of resolution by binning increases median stratigraphic ranges (Table 10.2) by 10% for extant species (4.5 Myr) and 20% for extinct ones (3 Myr) over their 'raw' values (Table 10.1).

For estimates of preservation probability, R (Foote and Raup, 1996), that is, the probability that a species is preserved in a 1.5 Myr interval (Table 10.2), we used the range–frequency ratio, $FreqRat = f(2)^2/(f(1)f(3))$ (Foote and Raup, 1996). $FreqRat$ for the extinct species (0.85) is double that for extant ones (0.40), largely because the latter include so many ranging through three adjacent 1.5 Myr intervals ($f(3)$); $f(3)$ exceeds not only $f(2)$, but also $f(1)$,

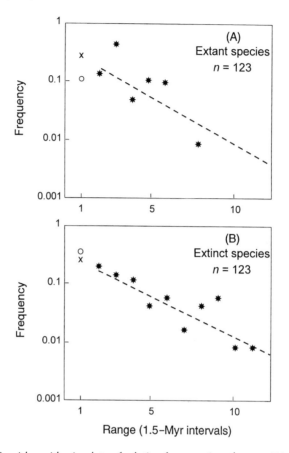

Figure 10.4   *Semi-logarithmic plots of relative frequencies of extant (A) and extinct (B) cheilostome species ranges measured in 1.5 Myr intervals. O, observed frequency of species with ranges of one 1.5 Myr interval (single hits); X, expected frequency of single hits; dashed lines are least-squares regressions through points representing ranges of two or more intervals (*). See Table 10.2 for statistics based on these plots*

contrary to expectation under the assumption of constant q (Figure 10.4A). However, despite the difference in *FreqRat*, the proportion of unfilled 1.5 Myr intervals within the observed stratigraphic ranges is nearly the same in extant and extinct species (43% and 46%, respectively; Table 10.2), implying that the actual preservation probability R may be less different between the two groups. The apparent deficiency in extant species ranging through one or two 1.5 Myr intervals, relative to those ranging through three (Figure 10.4A), is consistent with the probable underestimation of diversity of extant species in collections younger than 3 Ma (Figure 10.1A). However, there is no reason to suspect as important a deficit in the number of single hits, $f(1)$, for

Table 10.2 *Range, extinction and preservation statistics for cheilostome species with first occurrences in Neogene and Quaternary collections from Panama, Costa Rica and the Dominican Republic*

Statistic	Species		
	Extant	Extinct	All
Median range (1.5 Myr intervals)	3	2	3
Equivalent in Myr	4.50	3.00	4.50
Median duration, ln2/q (Myr)	4.31	3.23	3.03
Extinction rate, q (1.5 Myr intervals)	0.2411	0.3221	0.3433
Equivalent in Myr	0.1607	0.2148	0.2289
Goodness of fit (squared correlation)	0.3704	0.7609	0.6260
Preservation probability per interval, *FreqRat*	0.4042	0.8471	0.4405
Completeness, P_P	0.7600	0.9526	0.7034
Range intervals unfilled (%)	43.0	45.6	44.7

the extinct species because of the good fit between expected and observed values (Figure 10.4B).

Estimates of extinction rate were obtained by least-squares regression on the log-transformed frequencies (Figure 10.4) of species with ranges of two or more 1.5 Myr intervals, that is, excluding single hits (ln$f(t)$, $t > 1$; Foote and Raup, 1996). The regression for extinct species yields a higher squared correlation (goodness of fit, Table 10.2) than those for the extant species or both combined (although the z-transformed correlations are not significantly different, Student's $t = 0.66$, $P > 0.3$). Moreover, the frequency of observed single hits agrees more closely with expectation based on the regression for the extinct species than for the extant ones (Figure 10.4). Despite the differences in how well the two sets of species fit the model of constant extinction rate, the extinction rates themselves (q) are not far apart (0.16 per Myr for extant species, 0.21 per Myr for extinct ones; Table 10.2).

To complete the estimation of duration and preservation statistics, we used the relationship $S_T = S_0 \exp(-qT)$, where S_T is the number of species with true duration T (Raup, 1975), and the assumption that preservation probability and extinction rate are stochastically constant (Foote and Raup, 1996). For $S_T = 0.5 S_0$, $T = \ln 2/q$, the median duration, in either 1.5 Myr intervals or Myr, depending on the units in which q is expressed (Table 10.2). The calculation of P_P, the proportion of species preserved (completeness), is more complicated, involving the variables $P_1(T) = 1 - (1 - R)^T$, the probability that a species is preserved at least once, and $P_D(T) = \exp(-q(T-1)) - \exp(-qT)$, the probability that species durations equal T (Foote and Raup, 1996). Using *FreqRat* as the estimate of R and 1.5 Myr intervals for q and T, we calculated P_P by summing the product of these two variables to a very large value of T (approximately 8000) by copying formulae to the bottom of a spreadsheet.

Table 10.3 *Range, extinction and preservation statistics for cheilostome species grouped by mode of life*

Statistic	Species		
	Encrusting	Erect	Free-living
Number of species	169	54	23
Median occurrences per species	5	10.5	28
Median abundance per occurrence	2.76	3.25	15.14
Median range (Myr)	3.55	4.1	5.15
Median range extension (Myr), 95% CI	1.78	0.94	0.4
Number of species on which based	115	45	22
Median extended range (Myr)	6.16	6.26	6.3
Median range (1.5 Myr intervals)	3	3	4
Equivalent in Myr	4.5	4.5	6
Median duration, $\ln 2/q$ (Myr)	3.6	7.52	11.59
Extinction rate, q (1.5 Myr intervals)	0.289	0.1384	0.0598
Equivalent in Myr	0.1927	0.0922	0.0399
Goodness of fit (squared correlation)	0.6228	0.4658	0.2058
Preservation probability per interval, *FreqRat*	0.3846	0.6667	1.125
Completeness, P_P	0.7135	0.9393	1.0065
Range intervals unfilled (%)	47	45.2	40.9

With increasing departure from a homogeneous probability of preservation R, extinction rate (and thus estimates of median duration) are little affected, but completeness can be substantially overestimated, especially as the total number of taxa goes down (Foote and Raup, 1996; M. Foote, personal communication, 1996). Preservation potential is probably quite heterogeneous among cheilostome species. Degrees of robustness of cheilostome skeletons (and to a lesser extent differences in carbonate composition) are likely to contribute to differences in preservability, but the chances of being preserved (and recovered) may be even more closely related to modes of life. Encrusting species in the tropical American Neogene and Quaternary are far more numerous than erect and free-living species, but have fewer occurrences per species and much less abundance per occurrence (Table 10.3). Collecting curves in Cheetham *et al.* (1998, figs 7–9) show that encrusting species are the least well-sampled component of the cheilostome fauna, especially in stratigraphically younger collections.

To explore the possible effect of differences in frequency of preservation among the encrusting, erect and free-living species, we calculated range, extinction and preservation statistics for each of the three groups (Table 10.3). Completeness estimates for encrusting species ($P_P = 0.71$), 63% of which are extant, and for erect species ($P_P P = 0.94$), 72% of which are extinct, are similar to those for extant species and extinct species as a whole (0.76 and 0.95, respectively; Table 10.2); P_P for free-living species (96% of which are extinct) is unrealistically high, probably because of the extremely small number of species (Table 10.3). Small numbers also make the estimates of q and median duration suspect for free-living and erect species (Table 10.3), but the other range statistics are also quite similar to those for the extant and extinct species as a whole. These results suggest that differences in preservation potential may not produce substantial overestimates of completeness.

Foote (personal communication, 1996) has found that estimates of extinction rate (q), probability of preservation (R) and, therefore, completeness (P_P) are biased by finite sample size and by the fact that taxa cannot be followed out to infinite time. With a maximum likelihood approach and using the frequencies for the extinct cheilostome species (Figure 10.4B), he calculated $q = 0.17$ per Myr (with implied median duration $= 3.94$ Myr), $FreqRat = 0.73$ and $P_P = 0.91$ (M. Foote, personal communication, 1996), values that are 22%, 14% and 4% lower, respectively, than those obtained by unmodified calculations for extinct species (Table 10.2). If the modified estimates are typical, corresponding values for extant species would be median duration $=$ 5.26 Myr and completeness $= 0.73$. Completeness thus remains quite high. Although estimates of median duration are revised substantially upward from approximately 3–4 Myr to about 4–5 Myr, they are still 41–44% *lower* than the median ranges extended by 95% confidence intervals (Table 10.1). To approximate the 4–5 Myr durations, extended ranges would have to be based on confidence levels (C_1) of approximately 50%, rather than 95% as calculated in the preceding section.

IMPLICATIONS FOR PHYLOGENETIC AND EVOLUTIONARY STUDIES

If the completeness of the cheilostome record in the Neogene and Quaternary of tropical America does exceed 70% as these results suggest, the probability that some species recovered from the record could be the ancestors of others also recovered is excellent. (It must be emphasized that completeness, as calculated here, refers to the region encompassed by our collections, not global completeness, which is probably substantially lower because, for example, of loss of fossiliferous rock; Foote and Raup, 1996.) Foote (1996) derived a relationship between the probability of finding ancestor–descendant pairs (P_A) and the proportion of species preserved at least once (P_P), for each of three models of origination and extinction. If the patterns identified in *Metrarabdotos* and *Stylopoma* are indicative, cheilostomes appear to conform to his budding model, in which species may persist after producing descendants (Cheetham, 1986; Jackson and Cheetham, 1994). However, regardless of which model is assumed (budding, bifurcation, or phyletic transformation; Foote, 1996), P_A for the cheilostome data can be estimated to fall only slightly below the maximum value of 0.5 (Foote, 1996, fig. 3). (The models are based on an assumption that origination and extinction rates are equal and stochastically constant, so that half of all species become extinct without issue; thus, $P_A = 0.5$ when $P_P = 1$, that is, when all species are preserved; Foote, 1996.) These estimates of completeness then must be accepted as strong evidence that the cheilostome record is an appropriate venue, on practical if not on philosophical grounds, for seeking information on the morphological identity of species ancestors.

The extent to which the record offers stratigraphic evidence for ancestor–descendant relationships may be somewhat more variable. On average, preserved stratigraphic ranges could be as much as 76–86% of true durations if estimates based on an assumption of stochastically constant extinction rates (and adjusted by maximum likelihood methods) are accepted. Alternatively, if caution suggests preference for durations based on 95% confidence intervals, ranges may average as little as 44–55% of true durations. However, even in this 'worst' case, it is encouraging to note that many of the species shown in Figure 10.3A are concentrated near the lower bound of the distribution, with 58% having stratigraphic ranges within one 1.5 Myr interval of their durations as estimated by 95% confidence limits. Moreover, the majority of these species (62%) are the extinct ones that should, by probability, more often be potential ancestors. This relationship is consistent with results we obtained in the two genera studied in detail: 95% confidence intervals in *Metrarabdotos*, in which most species are extinct, average less than 0.5 Myr compared with more than 1 Myr in *Stylopoma*, in which almost all species are extant (Cheetham and Jackson, 1995), even though both genera fall among those cheilostome taxa with generally good agreement between ranges and durations.

Thus, some clades within the Neogene and Quaternary cheilostomes of tropical America will be better candidates than others for the use of fossil species and their stratigraphic occurrence as evidence in constructing phylogenetic hypotheses and inferring evolutionary patterns. The same can probably be said, however, for any other major taxonomic group with an excellent fossil record.

ACKNOWLEDGEMENTS

We thank Mike Foote for much help with calculation and interpretation, including comments and calculations based on his work in preparation; Lee-Ann Hayek for reviewing the manuscript; and Steve Donovan and Chris Paul for inviting us to contribute to this book and providing helpful comments on the manuscript. This work was supported by the Marie Bohrn Abbott Fund of the National Museum of Natural History and by STRI; other sources of support for the data on which this paper is based are acknowledged in Cheetham *et al.* (1998).

REFERENCES

Alroy, J., 1994, Continuous track analysis: a new phylogenetic and biogeographic method, *Systematic Biology*, **44**: 152–178.

Canu, F. and Bassler, R.S., 1918, Bryozoa of the Canal Zone and related areas, *Bulletin of the United States National Museum*, **103**: 117–122.

Canu, F. and Bassler, R.S., 1919, Fossil Bryozoa from the West Indies, *Publication of the Carnegie Institution of Washington*, **291**: 73–102.

Canu, F. and Bassler, R.S., 1923, North American later Tertiary and Quaternary Bryozoa, *Bulletin of the United States National Museum*, **125**: 302 pp.

Canu, F. and Bassler, R.S., 1928, Fossil and Recent Bryozoa of the Gulf of Mexico region, *Proceedings of the United States National Museum*, **72**: 199 pp.

Cheetham, A.H., 1986, Tempo of evolution in a Neogene bryozoan: rates of morphologic change within and across species boundaries, *Paleobiology*, **12**: 190–202.

Cheetham, A.H., 1987, Tempo of evolution in a Neogene bryozoan: are trends in single morphologic characters misleading? *Paleobiology*, **13**: 286–296.

Cheetham, A.H. and Hayek, L.C., 1988, Phylogeny reconstruction in the Neogene bryozoan *Metrarabdotos*: a paleontologic evaluation of methodology, *Historical Biology*, **1**: 65–83.

Cheetham, A.H. and Jackson, J.B.C., 1995, Process from pattern: tests for selection versus random change in punctuated bryozoan speciation. *In* D.H. Erwin and R.L. Anstey (eds), *New Approaches to Speciation in the Fossil Record*, Columbia University Press, New York: 184–207.

Cheetham, A.H. and Jackson, J.B.C., 1996, Speciation, extinction, and the decline of arborescent growth in Neogene and Quaternary cheilostome Bryozoa of tropical America. *In* J.B.C. Jackson, A.F. Budd and A.G. Coates (eds), *Evolution and Environment in Tropical America*, University of Chicago Press, Chicago: 205–233.

Cheetham, A.H., Jackson, J.B.C. and Hayek, L.C., 1994, Quantitative genetics of bryozoan phenotypic evolution: II. Analysis of selection and random change in fossil species using reconstructed genetic parameters, *Evolution*, **48**: 360–375.

Cheetham, A.H., Jackson, J.B.C., Sanner, J. and Ventocilla, Y., 1998, Neogene and Quaternary cheilostome Bryozoa of tropical America: contrast between the Central American isthmus (Panama, Costa Rica) and the north-central Caribbean (Dominican Republic), *Bulletins of American Paleontology*.

Coates, A.G. and Obando, J., 1996, The geologic evolution of the Central American isthmus. *In* J.B.C. Jackson, A.F. Budd and A.G. Coates (eds), *Evolution and Environment in Tropical America*, University of Chicago Press, Chicago: 21–56.

Coates, A.G., Jackson, J.B.C., Collins, L.S., Cronin, T.M., Dowsett, H.J., Bybell, L.M., Jung, P. and Obando, J.A., 1992, Closure of the Isthmus of Panama: the near-shore marine record of Costa Rica and western Panama, *Geological Society of America Bulletin*, **104**: 814–828.

Fisher, D.C., 1994, Stratocladistics: morphological and temporal patterns and their relation to phylogenetic processes. *In* L. Grande and O. Rieppel (eds), *Interpreting the Hierarchy of Nature – From Systematic Patterns to Evolutionary Theories*, Academic Press, Orlando: 133–171.

Foote, M., 1996, On the probability of ancestors in the fossil record, *Paleobiology*, **22**: 141–151.

Foote, M. and Raup, D.M., 1996, Fossil preservation and the stratigraphic ranges of taxa, *Paleobiology*, **22**: 121–140.

Huelsenbeck, J.P., 1994, Comparing the stratigraphic record to estimates of phylogeny, *Paleobiology*, **20**: 470–483.

Jackson, J.B.C. and Cheetham, A.H., 1990, Evolutionary significance of morpho-species: a test with cheilostome Bryozoa, *Science*, **248**: 579–583.

Jackson, J.B.C. and Cheetham, A.H., 1994, Phylogeny reconstruction and the tempo of speciation in cheilostome Bryozoa, *Paleobiology*, **20**: 407–423.

Marshall, C.R., 1990, Confidence intervals on stratigraphic ranges, *Paleobiology*, **16**: 1–10.

Marshall, C.R., 1994, Confidence intervals on stratigraphic ranges: partial relaxation of the assumption of randomly distributed fossil horizons, *Paleobiology*, **20**: 459–469.

Marshall, C.R., 1995, Stratigraphy, the true order of species' originations and extinctions, and testing ancestor–descendant hypotheses among Caribbean bryozoans. *In* D.H. Erwin and R.L. Anstey (eds), *New Approaches to Speciation in the Fossil Record*, Columbia University Press, New York: 208–236.

Norell, M.A., 1996, Ghost taxa, ancestors, and assumptions: a comment on Wagner, *Paleobiology*, **22**: 453–455.

Paul, C.R.C., 1992, The recognition of ancestors, *Historical Biology*, **6**: 239–250.

Raup, D.M., 1975, Taxonomic survivorship curves and Van Valen's Law, *Paleobiology*, **1**: 82–96.

Saunders, J.B., Jung, P., Geister, J. and Biju-Duval, B., 1982, The Neogene of the south flank of the Cibao Valley, Dominican Republic: a stratigraphic study. *In* W. Snow, N. Gil, R. Llinas, R. Rodriguez-Torres, M. Seaward and I. Tavares (eds), *Transactions of the Ninth Caribbean Geological Conference, Santo Domingo, Dominican Republic, 16–20 August 1980*, **1**: 151–160.

Saunders, J.B., Jung, P. and Biju-Duval, B., 1986, Neogene paleontology in the northern Dominican Republic: 1. Field surveys, lithology, environment, and age, *Bulletins of American Paleontology*, **89**: 1–79.

Schopf, T.J.M., 1973, Ergonomics of polymorphism: its relation to the colony as the unit of natural selection in species of the phylum Ectoprocta. *In* R.S. Boardman, A.H. Cheetham and W.A. Oliver, Jr (eds), *Animal Colonies: Development and Function Through Time*, Dowden, Hutchison and Ross, Stroudsburg, Pennsylvania: 247–294.

Wagner, P.J., 1995a, Stratigraphic tests of cladistic hypotheses, *Paleobiology*, **21**: 153–178.

Wagner, P.J., 1995b, Diversity patterns among early gastropods: contrasting taxonomic and phylogenetic patterns, *Paleobiology*, **21**: 410–439.

Wagner, P.J., 1996, Ghost taxa, ancestors, assumptions, and expectations: a reply to Norell, *Paleobiology*, **22**: 456–460.

11
The Fossil Record of Bivalve Molluscs

Elizabeth M. Harper

INTRODUCTION

Members of the class Bivalvia are often some of the first taxa introduced to the budding palaeontologist. There are at least two reasons for this. On the one hand, many Recent members of the class are relatively familiar from excursions to the beach or as culinary ingredients; on the other, they have a long and relatively good fossil record, with many post-Cambrian shallow marine sedimentary rocks yielding bivalve fossils. Even the most modest knowledge of the life habits and soft-part anatomy of living bivalves enables the student to make inferences about the life habits of fossil shells and at least partially reclothe their empty shells with tissue.

It is generally, but not universally (Yochelson, 1978, 1981), accepted that the earliest bivalves are found in rocks of early Cambrian age (Jell, 1980; Runnegar, 1996). These bivalves are small, only a few millimetres in length and probably lived as shallow burrowers within fine-grained shelf sediments. However, from these modest beginnings, the bivalves underwent dramatic change; most post-Cambrian bivalves are significantly larger, with the Mesozoic inoceramids and rudists, and the Recent giant clam, *Tridacna*, topping the league and reaching valve heights of extraordinary proportions (up to 2 m). Bivalves have also undergone an almost relentless increase in diversity, checked by the major mass extinction events (Figure 11.1). They have invaded a large variety of niches, both marine and non-marine, and exploited a number of varied life habits. A swift glance at Recent bivalves shows that many taxa have abandoned the burrowing habit to attach to

The Adequacy of the Fossil Record. Edited by S.K. Donovan and C.R.C. Paul.
© 1998 John Wiley & Sons Ltd.

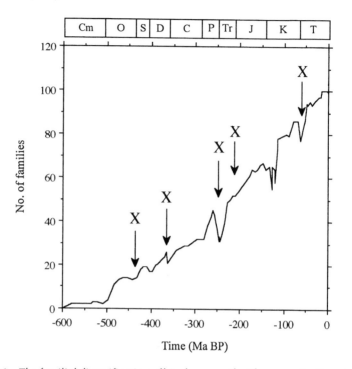

Figure 11.1 *The familial diversification of bivalves over the Phanerozoic. X, major mass extinction events. Cm, Cambrian; O, Ordovician; S, Silurian; D, Devonian; C, Carboniferous; P, Permian; Tr, Triassic; J, Jurassic; K, Cretaceous; T, Tertiary. (Data after Skelton and Benton, 1993)*

surfaces by flexible byssal threads or rigid calcareous cement, or to bore into rock and wood, or to lie free on the sea floor. Some even have the capacity to swim. In the deep oceans septibranchs have become predatory, whilst a number of other taxa have invaded freshwater systems where dreissenids have become a particular menace to man: their introduction to North America is costing US\$4 billion per year in efforts to stop fouling damage (Britton, 1991). Perhaps one of the most startling innovations, yet little remarked upon, is that of the south-east Asian anomiid bivalve *Enigmonia*, which is able to live subaerially on mangroves and, when immersed at high tide, to crawl, limpet-like (Yonge, 1977). Many new habits have been acquired at times of increased familial diversification, most notably in the Ordovician, early Mesozoic and early Cenozoic, and thus the Bivalvia are often lauded as an excellent example of adaptive radiation (Skelton, 1993).

Many of the new habits exploited by the bivalves have been acquired polyphyletically. For example, the ability to attach by cementation has evolved in over 20 clades of bivalves (Harper, 1991), whilst the ability to bore into hard substrata has been acquired at least seven times (Vermeij, 1987). The constraints of the bivalve bauplan, such as the necessity to grow by marginal accretion, mean that there are really very few ways of achieving change, thus leading to solutions that are strikingly similar and favouring marked morphological convergence. Morphological convergence may mask polyphyletic acquisition of traits, particularly where a trait has arisen independently in a very small group of taxa. For example, within the bivalve family Pectinidae, cementation has arisen from the primitive byssate habit several times (Harper and Palmer, 1993) and their morphological similarity has made identification of different clades difficult. Although convergence and polyphyly pose many problems in the study of bivalve evolution, it is precisely these phenomena which make the study of these organisms so fascinating and provide us with a means of understanding the general evolutionary processes.

Tradition has it that the Mollusca in general, and the bivalves in particular, have an excellent fossil record (Raup, 1979; Pease, 1985), but just how good is it? What are the biases that affect it and can we use the fossil record to gain any useful insight into bivalve evolution? Given that bivalves appear to have a greater diversity in Recent seas than they have ever had in the geological past (Figure 11.1), is the study of their fossil record necessary at all, or are the combined neontological forces of classical comparative anatomy and the battery of relatively modern techniques, such as molecular phylogenetics, sufficient to elucidate the relationships between bivalve taxa? In order to address the adequacy of the bivalve record, I will first consider the completeness of that record and then consider various areas where it could be used.

COMPLETENESS OF THE BIVALVE FOSSIL RECORD

As argued by Paul (1982, 1985, 1990), completeness and adequacy are two entirely different concepts. Of course, the bivalve fossil record, as that of all other taxonomic groups, is incomplete, although it is arguably rather more complete than many others. Although the soft parts of bivalves are seldom preserved, their generally robust, calcareous valves have a reasonably good preservation potential. This is not the place to undertake a full review of taphonomic processes as they affect the bivalves; indeed, an excellent review and discussion of this topic is provided by Kidwell and Bosence (1991). However, some appreciation of these are critical if we are to understand the biases that afflict the fossil record and hence affect its adequacy. Specifically, there are a number of questions that can usefully be asked: (1) Since some bivalves are composed wholly of aragonite, whilst others are at least partially made-up of calcite (the more stable polymorph), do the latter have a more complete fossil record?; (2) Do individuals that exploit certain life habits, such as deep burrowers, have better fossil records than those which are more exposed, as has been proposed for echinoids (Kier, 1977)? (3) Are bivalves that inhabit environments such as the deep sea or fresh water less likely to be fossilized than those in shallow marine settings, where the sedimentation rate is more rapid and the waters are less corrosive?

There are many ways of assessing the completeness of the fossil record. Allmon (1989) used eight different methods, most of them based on Paul (1982), to assess the completeness of the record of Paleogene turritellid gastropods from the Atlantic and Gulf coastal plains of the USA. Not all of these methods are applicable when considering the entire fossil record of a higher taxon, and are instead better suited to a consideration of shorter periods of time in rather more restricted geographical regions. For example, comparison of the geographical range of living members with that of their fossil counterparts will be inappropriate when a group has undergone significant spatial radiation, which presumably was the case in the early history of each taxon, or suffered dramatic spatial restriction. An example of the latter would be the trigoniid bivalves, which during the Mesozoic were abundant and diverse in shallow waters globally, but today comprise a single genus restricted to Australasian waters. Paul (1982) also suggested a number of more rigorous methods of assessing completeness based on measured sections, but these again are not suitable for considering a higher taxon, world-wide.

In this survey I have sought to assess the completeness of the bivalve fossil record in a number of different ways, concentrating where appropriate on data at class, family and genus level culled from a number of sources. It is first necessary to consider available databases and to make some comment on their adequacy.

The three-part *Treatise on Invertebrate Paleontology* (Moore *et al.*, 1969–1971), which provides stratigraphic data for genera and families, is now clearly dated, as are many of its concepts of phylogenetic relationships. Since its publication, however, a number of significant databases (both published and unpublished) have been compiled.

Family-level Data

Currently there are available two recently published databases that provide information on the stratigraphic ranges of bivalve families (Sepkoski, 1992; Skelton and Benton, 1993). Both aim to supply the date of first and last appearances of a particular taxon in the fossil record, mostly resolved at stage level. Sepkoski (1992) dealt only with marine taxa that have a fossil record, whilst Skelton and Benton (1993) also tackled those taxa which inhabit fresh water and those which are not known as fossils. There are, of course, problems when compiling this type of database, much of it in debate over the constituent membership of each family, and it would be difficult indeed to produce a family-level classification with which all could concur, which necessarily causes problems when trying to identify first and last occurrences. It may simply be too ambitious for a single author, or small team, to be sufficiently conversant with the familial details of a class as large as the bivalves to be accurate in all respects; an editor's note in Skelton and Benton (1993) warns prospective users that its quality might be 'patchy'.

Despite the fact that the two family databases were published at similar times, there are a number of significant differences between them; of the 161 families recognized by both works, only 53% have identical data entries. The ranges of the remaining families differ, mostly by more than a stage, often quite considerably. These discrepancies probably stem from the manner in which the databases were compiled. Sepkoski's data are derived from a careful review of the published literature, whilst Skelton and Benton adopted a less exhaustive, but a more critical, approach, particularly in regard to the movement of taxa between families and to unpublished findings (29 of their records are based on personal communication). In the following analyses I have used a modified and revised form of Skelton and Benton (1993) in which many of the discrepant entries have been resolved and excluding the Tuarangiidae, which are now believed not to be bivalves (Runnegar and Pojeta, 1992).

Generic-level Data

If it is difficult to compile a meaningful family database for the bivalves, the problem is an order of magnitude worse for the acquisition of generic data. Although Vokes (1980) produced a list of generic names for the class, he included no stratigraphic information and as a result the most recently published compilation of data remains the *Treatise on Invertebrate Paleontology* (Moore *et al.*, 1969–1971). However, in order to undertake their studies into changing diversity in marine organisms (including bivalves) during the Phanerozoic, Sepkoski and his co-authors have used an unpublished database (Raup and Sepkoski, 1986; Miller and Sepkoski, 1988; Sepkoski, 1989, and unpublished). These data were procured using the *Treatise on Invertebrate Paleontology* as a base, and revising it by including data from more recently published papers and monographs. In order to produce a reliable database it is necessary not only to scour the literature for the description of new genera and for discoveries that prolong the stratigraphic ranges of known genera (but may not be reported as such), but also to take account of, and critically assess, claims of synonymy. The problems in compiling a generic database for a relatively large class of organisms, such as the Bivalvia, are manifest. Sepkoski's 1995 version of the database contained over 2500 entries and excluded genera with no fossil record. In the following analyses I have used a combination of the Sepkoski database, Vokes (1980) and the list of new generic names supplied by the *Zoological Record* (post-1980); even so, it is inevitably imperfect.

The clades used in the following analyses are those adopted by Skelton *et al.* (1990): protobranchs, arcoids, mytiloids, pteriomorphs, the heteroconchs (*sensu* Morris, 1978), lucinoids (*sensu* Pojeta, 1978, plus leptonoideans and cyamioideans) and anomalodesmatans. Life habits are as recorded in the literature or inferred from morphological criteria following Stanley (1970), whilst shell mineralogy data are based on Taylor *et al.* (1969, 1973) and Carter (1990).

Estimates Based on Knowledge of Recent Taxa

One simple measure of the completeness of the fossil record is provided by an analysis of the number of Recent taxa that have a known fossil record. Of the 222 bivalve families listed by Skelton and Benton (1993), 109 are extant, of which only 7, that is 6%, have no recorded fossil representatives. Each of these is wholly aragonitic and for most it is possible to suggest reasons for poor preservation. Three families, the Pristoglomidae, Siliculidae and Lametelidae, are deep-sea protobranch bivalves found at bathyal and abyssal depths (Allen and Sanders, 1973; Sanders and Allen, 1973). The Chlamydoconchidae, Bernardinidae and Anatinellidae are small and/or possess thin, fragile shells, unlikely to be preserved. The chlamydoconchiids are among the rarest of bivalves and their unusual condition of possession of an internal shell and the presence of dwarf males (Morton, 1981) make fossilization even more unlikely. In any case it should be noted that the status of the Chlamydoconchidae as a family is uncertain: Morton (1981) preferred, on the basis of reproductive characters and reduction of the shell, to place the single known genus *Chlamydoconcha* into the family Galeommatidae, which is represented in the fossil record. Perhaps the only surprising family absence from the fossil record is that of the Glauconomidae. The single genus *Glauconome* has a robust shell, and is widespread and abundant throughout the Indo-Pacific and south-east Asia (Ong Che and Morton, 1994).

I estimate, on the basis of data derived from Sepkoski (unpublished), Vokes (1980) and the *Zoological Record*, that there are 773 extant bivalve genera. Of these, 128 (17%) have no published fossil record (Table 11.1). If these 'missing' genera are split into their respective clades, it is striking that very few are epifaunal pteriomorphs, although more are epifaunal mytiloids. Both the heteroconchs and lucinoids have high number of genera with no fossil record, but, given the relative numbers of extant families in these

Table 11.1 Numbers of extant genera with no known fossil record in each of the seven major bivalve clades

Clade	No. of extant genera with no fossil record
Protobranchs	11
Arcoids	2
Mytiloids	11
Pteriomorphs	2
Lucinoids	46
Heteroconchs	47
Anomalodesmatans	9

clades, 47 families and 14 families, respectively, it is clear that the record of the latter is inferior. Most of the lucinoid genera with no fossil record belong to cryptobyssate families such as the Leptonidae, Neoleptonidae and Cyamiidae. Further analysis of the generic data reveals that of 103 living genera with calcite in their shells only 6 (6%) have no fossil record, whilst of the wholly aragonitic examples 122 (18%) are not known as fossils. Interestingly, Raup (1979) undertook a similar analysis and discovered a rather higher level of completeness of the family record, because his study pre-dated the recognition of the extant deep-sea protobranch families, and a rather lower value at generic level (which is less easy to explain).

Another way of using living taxa to assess the completeness of the fossil record is to consider their preservation potential. This approach was used by Schopf (1978) on the intertidal fauna, including bivalves, of the Friday Harbor region of Washington State, USA. He based his estimates on three criteria: possession of skeletal elements likely to be fossilizable, occurrence of genera in the fossil record and the occurrence of recognizable fragments in modern coastal sediments. Unsurprisingly, all the bivalve genera studied, regardless of whether they came from muddy, sandy or rocky environments, had an excellent fossilization potential matched only by other molluscs. However, it should be noted that intertidal organisms tend to be robust and often calcitic, and may be more likely to endure taphonomic processes than more fragile subtidal taxa. Valentine (1989) performed a similar analysis based on a comparison of the Recent and Pleistocene malacofauna of the Californian biotic province, including intertidal and subtidal taxa from a wide range of water depths. He discovered that 90.6% of bivalve families, 84.7% of genera and 80.1% of species living today were also known as fossils from the same region. Each of these figures is slightly higher than those he obtained for the gastropods. Many of the fossil taxa recorded from the province, but not represented in the Recent, are extinct, whilst others are found alive outside the Californian province, presumably due to a change in regional climatic or other environmental conditions since the Pleistocene. The living taxa that have no fossil representatives fall into three categories: (1) those which live in deep water, (2) those which are small and fragile, and (3) those which appear genuinely rare today.

Gap Analysis

Paul (1982) used this simple method to assess the completeness of the Paleozoic record of cystoid echinoderms by examining the entire recorded range of each family and determining the proportion of intervening series for which representatives of each are recognized. There are obviously problems with using this approach for the entire Phanerozoic. For example, it places equal weight on stage duration, but the effect of this may be alleviated by using radiometric dates to calculate the percentage of 'absolute' time that is unrepresented. Also, complete family ranges are taken to represent generic range data that are consecutive or overlapping, but may themselves not be continuous. Additionally, gaps may be underestimated as the finding of a particular taxon in a stage is taken to indicate its presence at both the beginning and the end of that stage.

Here, I have undertaken a gap analysis using data for the entire class of Bivalva and for three important clades within it, namely the pteriomorphs, the heteroconchs and the lucinoids. These particular clades were chosen for the following reasons. The pteriomorphs are exclusively epifaunal, containing many of the key exponents of this habit (with the notable exception of the Mytiloidea), and have robust shells containing a high proportion of calcite. The heteroconchs are largely infaunal (except the Petricolidae, Carditidae and Chamidae, which are epifaunal) and contain many of the most successful of the shallow burrowers (Morton, 1996). With the exception of two chamid species, they are composed exclusively of aragonitic shell microstructures. Although the lucinoids are also wholly aragonitic and often shallow burrowers, they were selected as some of the families are cryptobyssate forms with small and fragile shells (Yonge and Thompson, 1976), and others have acquired specialist adaptations that have enabled them to exploit sulphide-rich sediments, which may not be the most promising places for fossilization. The analysis was undertaken by examining the fossil record of each family within the class and chosen clades using Sepkoski's (unpublished) database of generic longevity. As mentioned above, this use of a generic database does not preclude the fact that individual genera may have incomplete fossil records. Nevertheless, in most of the families investigated there was considerable overlap in generic ranges.

Data for the entire class reveal that no verified fossil bivalves have been recovered from rocks of the Middle and Upper Cambrian. Using the early Paleozoic chronology of Tucker and McKerrow (1995), this stratigraphic gap represents at least 23 Myr, that is, over 4% of the total history of the class. This startling gap is well known and its undoubted importance in our understanding of bivalve evolution will be discussed later.

Of the 48 families of pteriomorphs, 39 appear to have complete ranges. Of the nine with incomplete ranges (Inoceramidae, Bakevellidae, Pergamiidae, Deltopectinidae, Terquemiidae, Propeamussidae, Buchiidae, Neitheidae and Anomiidae), many gaps occur in the early Mesozoic: for example, Crame (1995) described an early Cretaceous form that probably belongs to the pergamiid genus *Manticula*, which extends the range of both the genus and the family from Upper Triassic to Lower Cretaceous, but leaves a significant gap (over 64% of the range) for which there are no known records. Crame also pointed out that the Buchiidae also appear to have a significant stratigraphic gap between the Upper Triassic and Upper Jurassic. Many of these problems result from uncertainties in the classification of the early forms. Uncertainties in 'inoceramid' classification suggest a gap between Permian *Atomodesma*, and the Jurassic *Retroceramus* and Cretaceous *Inoceramus* (J.A. Crame, personal communication). Another interesting example of incompleteness within the pteriomorph record is that of the Anomiidae. The earliest known true anomiid is the Bathonian *Eonomia timida* (Fürsich and Palmer, 1982), but the only other certain Jurassic record is *Juranomia calcibyssata* from the Kimmeridgian (Fürsich and Werner, 1989). Although there have been other claims for Jurassic anomiids, most can easily be shown to be other taxa that happen to have a hole bearing a superficial resemblance to the anomiid byssal notch. In true anomiids this notch is a large circular embayment in the thin right valve that significantly weakens it, and it is to be anticipated that the right valves, at least, would have a poor fossil record; indeed, Fürsich and Werner (1989) noted that, apart from a few articulated specimens, the right valve of *J. calcibyssata* is known from only five specimens as opposed to over 300 left valves. It is perhaps conceivable that the left valves, which frequently bear a xenomorphic sculpture, could be mistaken for those of oysters, although they do have a highly distinctive muscle scar pattern (Yonge, 1977). The problem of missing anomiids may be yet more acute, as the superfamily Anomioidea also contains the monospecific family Permanomiidae, known only from the Upper Permian (Carter, 1990), which leaves a substantial stratigraphic gap between the two families.

Of the 48 families of heteroconch that do have a fossil record, most (43, or 90%) have complete stratigraphic records. Gaps appear in the records of the Carditidae, Crassatellidae, Icanotidae, Trapeziidae and Dicerocardiidae, although the last is acknowledged as probably polyphyletic (Skelton and Benton, 1993).

The lucinoids have the most incomplete fossil record. Substantial gaps occur in three of the 14 families (21%) namely the Lucinidae, Mactromyidae and Fimbriidae. As noted earlier, many of the lucinoids have life habits that may not promote fossilization. It is also interesting that many of the lucinoid families first appeared in the Tertiary, and one can speculate whether this is the true timing of their radiation or whether many of the families have significantly earlier origins, but are unknown.

Monotypy

Paul (1982) argued that a very incomplete fossil record would result in most fossils being so distinct that they would be placed into relatively high taxa, hence yielding a record rich in monotypic families. Thus, it is possible to make a crude measurement of completeness of the bivalve record by calculating the proportion of families that are monogeneric for their entire stratigraphic range.

Of the 222 families listed by Skelton and Benton (1993), only 31 (that is, 14%), are monogeneric and, of these, 11 are based on a single species (Table 11.2). These include epifaunal, infaunal and freshwater families. The validity of some of these is, perhaps, rather dubious and therefore completeness may be underestimated by this method; for example, the family Dattidae is based on a single internal mould (Skelton and Benton, 1993). Paul (1982) pointed out that a further source of underestimation comes from the fact that some taxa were likely to have been genuinely monotypic. This point is underlined with reference to the Bivalvia by the fact that of the monogeneric families five are still extant. The soft parts of each of these have been subject to study by neontologists, who have been unable either to assign them to other families or to split them into more than one genus.

Paul (1982) suggested that incompleteness might also be assessed in a related way by counting taxa known from material restricted to either a single stratigraphic interval or a single locality. Again, there is a problem from underestimating completeness by considering taxa that were genuinely spatially and temporally restricted. Excluding those families which are known from living taxa, only 11 (5% of the total bivalve families) are restricted to a single stratigraphic stage. These not surprisingly, but not inevitably, include those families known only from a single species. Two of these families, the Pitchleridae and Lithiotidae, are restricted to a single stratigraphic stage, but are known to have two genera. The lithiotid genera *Lithiotis* and *Cochlearites*, bizarre oyster-like forms, are restricted to Pliensbachian carbonates in southern Tethys (Reis, 1903; Chinzei, 1982), but are reasonably robust, with at least some calcite in the shells, and were capable of forming bioherms (Nauss and Smith, 1988). Indeed, they were in many respects rather similar to the rudists that did have a good fossil record.

Table 11.2 *Monotypic bivalve families*

Family	Monogeneric	Monospecific
Thoraliidae	•	•
Butovicellidae	•	•
Isoarcidae	•	
Eurydesmidae	•	
Monopteridae	•	
Umburridae	•	•
Dattidae	•	•
Myodakryotidae	•	•
Plicatostylidae	•	•
Permanomiidae	•	•
Babinkiidae	•	
Chlamydoconchidae*†	•	?
Redoniidae	•	
Carydiidae	•	
Eoschizodiidae	•	•
Schizodiidae	•	
Scaphellidae	•	
Margaritiferidae*(f)	•	
Trigonioidae(f)	•	
Anatinellidae*†	•	
Uniocardiopsidae	•	
Euloxiidae	•	•
Glauconomidae*†	•	
Ichthyosarcolitidae	•	
Sinodoridae	•	
Hippopodiidae	•	
Myopholadidae	•	
Pleuromyidae	•	
Cleidothaeridae*	•	
Pleurodesmatidae	•	
Ferganoconchidae(f)	•	•

*Extant family; † taxon with no fossil record; (f), freshwater taxa.

Survivorship

Paul (1982) also suggested that survivorship might be used to gauge completeness of the record: the more incomplete the record, the shorter survivorship of taxa. Survivorship of the Bivalvia has been explored by a number of authors (Simpson, 1944; Van Valen, 1973; Gilinsky, 1988). All have found that survivorship for extant genera and families is rather greater than for those of extinct taxa. Although it is possible to argue that this might result from the fact that longer-lived taxa are more likely than shorter-lived ones to survive into the Recent (Raup, 1975), Gilinsky (1988) has argued that this, and related objections, are not sufficient to explain the disparity and that the explanation lies in an as yet unidentified biological cause. One thing that is clear from the curves produced by Gilinsky (1988) for both extinct and living taxa is that they are not very short-lived: 80% of all extinct and living genera have longevities of 40 Myr and 80 Myr, respectively, whilst comparable values for families are 170 Myr and 320 Myr.

Freshwater Bivalves

Freshwater systems, with their high acidity, low net sedimentation rate, and spread over geographically restricted and often ephemeral areas, are un-promising preservation traps (Gray, 1988). Cummins (1994) has reviewed the taphonomy of freshwater molluscs. Corrosive waters may cause extreme dissolution damage even ante-mortem (Tevesz and Carter, 1980), and Taylor (1988) noted that the freshwater fossil record is inevitably biased against small, thin individuals and taxa. Gray (1988) and Taylor (1988) documented the known freshwater bivalve record: 18 families are, or were, truly fresh-water (as opposed to euryhaline), 13 of these are extant and of these five lack a fossil record (in each of these there is only one genus out of the entire family that is known to live in fresh water; for example, the Arcidae and Mytilidae are both diverse families of mainly marine genera, but also contain the freshwater genera *Scaphula* and *Sinomytilus*, respectively). All of the principal extant families of freshwater bivalves have fossil records: the Unionidae, Margaritiferidae, Mutelidae, Corbiculidae and Sphaeriidae first appear in either Triassic or Jurassic strata, whilst the dreissenids appear first in the Oligocene and the Etheriidae (possibly a polyphyletic assemblage of three genera, *Etheria*, *Acostaea* and *Pseudomulleria*) date back to the Pliocene. A further five families of freshwater bivalves have no extant representatives, including the early Devonian Cycloconchidae, which are the oldest known freshwater bivalves (Morris, 1985).

Summary

From each of the above analyses, summarized in Table 11.3, it appears that the fossil record of the bivalves is very good. There is a certain consistency in the taxa that are found to be under-represented. In terms of the questions posed at the beginning of the section:

1. The fossil record of calcitic taxa is better than that of taxa which are wholly aragonitic.
2. In terms of mere representation, there is no evidence that the fossil record of infaunal taxa is superior to that of epifaunal bivalves; indeed, all the

non-lucinoid members of the living families not represented in the fossil record are shallow burrowers. The fact that many epifaunal taxa have robust shells as an adaptation to their exposure and have calcitic shell layers, whilst burrowers are universally wholly aragonitic, has probably enhanced their preservation potential. The most fragile members of the class are those which have cryptobyssate or boring life habits, but of these the latter are favoured by being trapped and preserved in their hard substrata. Many cryptobyssate taxa live as commensals with a variety of soft-bodied organisms, which affords them protection during life, but not after death; for example, the leptonids *Montacutona compacta* and *M. olivacea* live in association with coelenterates and phoronids (Morton, 1980).

3. There is good evidence that freshwater and deep-sea bivalves are less well represented in the fossil record. With the increase in accessibility, and thus interest, in deep-sea vent faunas, it is likely that significant discoveries are yet to be made as regards the fossil and extant bivalves associated with them.

ADEQUACY OF THE BIVALVE FOSSIL RECORD

It is clear that the adequacy of the bivalve fossil record will vary depending upon the problem being addressed; data that are adequate for one problem may not be sufficient for another. I shall consider separately the adequacy of the bivalve fossil record for our understanding of a number of different evolutionary problems: the origin of the bivalves, their early evolution, use in phylogenetic reconstruction, microevolutionary studies and whether the general lack of preserved soft parts is a critical stumbling block in our understanding of bivalve palaeobiology.

Origins and Early Evolution

There are a number of different interpretations for the origins of the molluscs, summarized by Runnegar (1996). Among these there are two main models that shed light on the origin of the bivalves: the view expounded by Pojeta and Runnegar (Runnegar and Pojeta, 1974; Pojeta and Runnegar, 1985) that there are two subphyla of the Mollusca of which one, the Diasoma, contains the extinct rostroconchs, and two classes derived from these, the Scaphopoda and the Bivalvia; and that of Peel (1991), which places the bivalves and rostroconchs in entirely separate groups deriving them from exogastric tergomyans and endogastric helcionellids, respectively. Both models are based on the study of the same taxa, of which there are a reasonable number

Table 11.3 *Summary of the completeness values for the bivalve fossil record as calculated by a number of different methods explained in the text*

Method	Percentage completeness
Extant families with a fossil record	94
Calcite-bearing	100
Wholly aragonite	93
Freshwater	62
Extant genera with a fossil record	84
Calcite	94
Aragonite	82
Entire class gap analysis	96
Pteriomorphs with no range gaps	81
Heteroconchs with no range gaps	90
Lucinoids with no range gaps	79
Monotypy (that is, percentage families not monogeneric)	86
Families based on specimens collected from more than one site	95

of specimens, and differ in the interpretation of the function of apertural
structures and, hence, the direction of flow of inhalant and exhalant currents.
The debate has become unnecessarily bitter (Runnegar, 1996) and one suspects that its resolution requires relevant soft-part preservation.

The earliest known bivalve, *Fordilla troyensis*, has been known for over a
hundred years since its description by Barrande (1881) from the Tommotian
of North America. Although Barrande recognized it as a bivalve, later
workers, including Walcott (1887), regarded it as a bivalved crustacean, until
its detailed description and reinstatement by Pojeta *et al.* (1973). Another,
apparently coeval bivalve, but geographically isolated, *Pojetaia runnegari*,
was described from Australia by Jell (1980).

Much is now known about these early Cambrian bivalves. Neither taxon is
known from isolated specimens of dubious preservation. Pojeta (1975, 1978)
studied 350 specimens of *Fordilla troyensis* from New York State (USA) and
Conception Bay (Newfoundland) alone, and the species is now known from
a variety of localities in North America, western Europe and Greenland. Jell
(1980) based his description of *Pojetaia runnegari* on over 70 specimens and,
although further material has been studied (Runnegar and Bentley, 1983), the
taxon is known only from Australia. Information gleaned from these specimens has furnished us with a wealth of detail concerning the musculature,
dentition, shell microstructure and likely infaunal mode of life of these
earliest bivalves (but see Tevesz and McCall, 1976). Based on this information

many workers (for example, Runnegar and Pojeta, 1992) unite *Pojetaia* and *Fordilla* into a monophyletic order, although some earlier papers (such as Runnegar and Bentley, 1983) had suggested that the two belonged to the different subclasses, Isofillibranchia and Palaeotaxodonta. Assuming the former view were to be correct, it would be painfully clear that a substantial part of the bivalve record was missing. As it is, the appearance of these two genera differs in many respects from the primitive bivalve envisaged by neontologists (Morton and Yonge, 1964; Morton, 1996), not least in their predictions of an isomyarian *Nucula*-like bivalve with an epifaunal crawling mode of life.

Unfortunately, as noted earlier, there is a distinct gap in the bivalve fossil record covering the middle and late Cambrian. This gap at such an important time in the radiation of the group poses severe problems, for there are important differences between the very small *Pojetaia* and *Fordilla* (only a few millimetres long) and the bivalves from the Ordovician, which are, in general, much larger (a centimetre or more in length). There is still much to be learnt about early Ordovician bivalves, but certain facts have become clear. Between the early Cambrian and late Ordovician, taxonomic diversity of bivalves exploded, such that, by the close of the Ordovician, all extant subclasses are recognizable (Pojeta, 1978). The range of life habits exploited also greatly increased: by the close of the period only the cryptobyssate and cementing habits had not been utilized (Skelton *et al.*, 1990). The cause of the gap is intriguing. Yochelson (principally 1978, 1981) has used it to promote the view that the Bivalvia are in fact a post-Cambrian group and that Cambrian records belong to extinct clades, whilst Runnegar (1983) suggested various reasons for the gap including the prevalence of discontinuity in the stratigraphic record and the fact that small bivalves could have been easily overlooked. Cope (1996) has suggested that the gap may be at least partly due to facies control, pointing out that the near-shore silty muds and sands that were home to the early Ordovician bivalves are seldom preserved in the rock record. However, it is worth noting that there is no shortage of brachiopod or rostroconch material of middle and late Cambrian age; indeed, there have been several examples of these taxa that have been considered as contenders for middle and late Cambrian bivalves, but have subsequently been rejected (see Runnegar and Pojeta, 1992).

Phylogenetic Reconstruction

Pure cladists discount the fossil record in their pursuit of phylogenies but, as Fortey and Jefferies (1982) have argued, a good fossil record ($> 70\%$ complete) may play an important part in the recognition of ancestors and primitive character states, and in the provision of a time frame for synapomorphies. Indeed, when attempting to construct the complete tree for groups that include extinct taxa, the character states of the latter must necessarily be derived from fossil material. The fossil record also provides a means of discriminating between the multitude of possible trees thrown up by cladistic analysis. Since the bivalve record appears good, with more than 70% completeness, it would seem surprising that, despite the widespread use of cladistics in the study of other groups, bivalve workers, both neontologists and palaeontologists, have been extremely slow to embrace its methods. Schneider (1992, 1995) is the only author to have published rigorously described, computer-driven cladistic studies on any bivalve group, whilst Waller (1978) is the only author to have attempted to use cladistics to examine higher-order phylogeny within the class. At least part of this reluctance is explained by the notion expounded by Seilacher (1984) that morphological convergence and parallelism were so rife within the class as to make cladistic analysis impossible. However, as Schneider (1995) pointed out, far from masking phylogenetic relationships by failing to distinguish polyphyletic traits, rigorous use of cladistics may identify hitherto unrecognized instances of convergence. Skelton *et al.* (1990) and Skelton and Benton (1993) further suggested that bivalves lack sufficient characters to use in cladistic analysis, but it is hard to accept such an argument, for even in the hard parts alone there are a wealth of characters based on microstructural arrangement, dentition, ligamenture, adductor and pedal musculature, and sculpture that are readily defined. Indeed, the range of states for most of these characters seems rather more promising than for brachiopods and gastropods. Schneider's study of the Cardiidae used 16 characters (53 character states), of which nine were based on hard parts. The lack of soft-part characters are not hazards restricted to the study of fossils; Paul (1985) has remarked that many Recent gastropods are known only as hard parts. The same is true of many bivalves and, in fact, with the exception of taxa that have been studied extensively due to their maricultural importance, few Recent bivalve taxa are anything other than morphospecies.

Microevolutionary Trends

Over the last couple of decades, microevolutionary patterns and processes have received much attention and, where it is possible to collect abundant body fossils bed by bed over a continuous stratigraphic section, the fossil record can provide useful information on these (Sheldon, 1987). Bivalves have been used in such analyses and one family, the gryphaeid oysters in the Jurassic, has been studied extensively. Indeed, *Gryphaea* is often heralded as the palaeontologist's *Drosophila* (Gould, 1972). There are two main reasons for this. First, gryphaeids were abundant in shallow-shelf seas during the

Mesozoic and had an excellent preservation potential imparted by their thick, robust shells composed almost entirely of calcite. Thus, they are relatively easy to collect in the large numbers necessary for this type of study. Secondly, they underwent a series of morphological changes over the course of the Jurassic, from flat forms with large attachment scars to more tightly coiled free-recliners. Numerous papers have been published on these evolutionary changes (see Johnson and Lennon, 1990, and references therein), but there is much disagreement as to the extent of polyphyly and the identification of gradualistic versus punctuated equilibrium styles of evolution. However, it is clear that much of this stems more from lack of agreement over biometric manipulation and the selection of characters to be examined (made difficult by the immense ecophenotypic variation displayed by most oysters) than inadequacy of the gryphaeid fossil record.

Other groups of bivalves have also lent themselves to microevolutionary studies. Johnson (1985) found evidence of sudden, trans-specific changes in Jurassic pectinids. Stanley and Yang (1987), in a mammoth study of the morphometrics of 19 lineages of Neogene bivalves over a 4 Myr period, based on 24 variables and utilizing over 43 000 specimens, argued for stasis. In a study of Miocene bivalves, Kelley (1989) described gradual morphological changes in traits, such as valve thickness, which she linked to the concomitant increase in predation pressure from naticid gastropods.

Clearly, carefully chosen bivalve taxa can be utilized for microevolutionary studies. However, the problem may not be so much the adequacy of the fossil record of a particular taxon, but whether it possesses characters that are both easily defined and easily measured.

Significance of the General Lack of Soft Parts to Palaeobiology

Many of the significant strides in our understanding of the phylogenetic relationships between bivalves have been made by biologists working on various aspects of bivalve anatomy, for example the ctenidia (Atkins, 1936–1938) and stomach (Purchon, 1987), and yet here, surely, the fossil record must be virtually mute. In 'normal' conditions we might anticipate a bivalve to be either preyed upon or scavenged and any soft parts left to be degraded rapidly by bacterial decay. None of the celebrated *Lagerstätten* are famed for their preservation of bivalves, but soft-part preservation is known from rather less glamorous sedimentary settings. Gill preservation in trigoniid bivalves has been reported from Portlandian limestones (see Wilby and Whyte, 1995), phosphatized adductor muscle fibres from an Upper Jurassic gryphaeid oyster (Harper and Todd, 1995), Liassic internal moulds of nuculid gut (Cox, 1960), and ligament of rudists (Skelton, 1979), trigoniids and pectinids (Harper *et al.*, 1996). Wilby and Whyte (1995) and Harper and

Todd (1995) commented that soft-part preservation is most likely in bivalves that remain tightly shut after death, promoted by phosphate ions liberated by partial decay (Allison, 1988). The types of tissue preserved are also important. As Wilby and Whyte (1995) pointed out, trigoniids have phosphatic gill supports, and adductor muscles, with their rich supply of ATP, presumably promote phosphatization, whilst gut preservation may be facilitated in deposit-feeders. One suspects that soft-part preservation may be more common in bivalves than is generally believed, and although it is frowned upon, and rather impractical, to cut open too many tightly closed articulated specimens, there are probably wonders to be found in such material.

Many of the major adaptive radiative events of the bivalves were associated with key innovations in their soft-part anatomy. Arguably the most important of these were the evolution of a feeding gill, mantle fusion to produce siphons and the retention of the juvenile byssus in some adult forms. Skelton (1991) argued that adaptive radiation requires intrinsic and extrinsic factors, one of which will be the effective cue to the radiation event, the other, by virtue of its pre-existence, the enabling circumstance. Therefore, there is scope for the determination of whether soft-part innovations were pre-adaptive and, thus, the enabling circumstance in any radiative event or whether they were the direct trigger.

The evolution of a gill system that was used in feeding as well as respiration freed the foot from its role in scooping up organic detritus, and allowed its use in burrowing and/or byssal attachment (Morton, 1996). Although some preserved gill tissue is known, none is of the critical age that could supply useful information here. Nevertheless, Cope (1996) has suggested an ingenious way of identifying the functional switch in the fossil record, proposing that the evolution of the feeding gill would have caused significant problems to the bivalve, as the attendant deceleration of the inhalant current over the gills would have resulted in the deposition of suspended material within the mantle cavity. This fouling material would tend to be expelled as pseudofaeces by adduction movements and Cope suggested that this frequent adduction favoured the acquisition of large, strongly differentiated hinge teeth that prevented the valves from shearing even when the valves opened widely. On the basis of this, Cope proposed that acquisition of the lamellibranch feeding gill appeared and radiated in the early to middle Ordovician.

Mantle fusion, which allowed the formation of siphons and assisted foot extrusion, allowed bivalves to burrow deep within the sediment, and as a result infaunal bivalves underwent significant radiation at the generic level during the early Mesozoic (Stanley, 1968; Skelton *et al.*, 1990). Although it is possible to infer the extensive mantle fusion necessary for siphon formation from the possession of the pallial sinus on the interior surface of the shell – the earliest known example is the Ordovician *Lyrodesma* (Pojeta, 1978) – we have very little information regarding siphonal evolution. Indeed, since taxa with very short siphons lack a pallial sinus, it is likely that any estimates of the timing of the key innovation that allowed siphon formation will postdate, perhaps quite considerably, the actual event. Moreover, it is clear that mantle fusion in extant bivalves is achieved in a number of different ways, involving different mantle folds, which are not distinguishable from hard-part evidence alone, suggesting strongly that siphon formation was polyphyletic (Yonge, 1982). Similarly, byssal threads used in the anchorage of most bivalve larvae were polyphyletically and neotenously retained into maturity by a wide range of epifaunal and shallow infaunal bivalves (Yonge, 1962). In this case, however, the evolution of the habit may be tracked in the fossil record from byssal notches and pedal musculature.

CONCLUSION

The generally robust shells of bivalves have ensured a good fossil record. It is certainly not complete: soft parts are rarely preserved and the hard-part record is demonstrably impoverished in some cases. However, it is clear that the record is good enough to be of major assistance in our understanding of bivalve evolution. It is also clear that many of the apparent imperfections are due to the inadequacy of our knowledge of the fossil record rather than the record itself. Koch (1978) has shown how concerted efforts to collect and identify bivalve and gastropod fossils from the well-known Western Interior of North America resulted in a fivefold increase in the numbers of these taxa identified, and thus in a reorganization of the relative importance of taxa in the overall fauna. Even in some of the most well-known successions there are new discoveries to be made. Middle and Upper Jurassic sections of France and England, the malacofaunas of which were so well studied by Arkell, Morris and Lycett to name but a few, have relatively recently yielded *Eonomia timida*, which considerably extended the range of anomiid bivalves (Fürsich and Palmer, 1982), and cemented right valves of *Eopecten*, a Jurassic scallop known previously almost exclusively from left valves and of previously unknown life habit (Harper and Palmer, 1993). Neither taxon is particularly rare at the localities in which they have been found and both are quite distinctive. There is thus considerable scope for the 'improvement' of the bivalve fossil record by new discoveries that fill existing gaps in our knowledge, in particular in such critical areas as the early Paleozoic.

The bivalve record has more to offer than merely allowing the evolution of this class to be traced, albeit an important class whose extant members are economically and environmentally important. Their shells, in particular in calcitic taxa with their excellent preservation of original shell material, also provide a record of seawater chemistry and palaeoclimatic data (see, for example, Jones and Quitmyer, 1996). The record is good enough to be used in more general studies of microevolution and adaptive radiation. The fact that the class is still extant and highly diverse is a great boon to the palaeomalacologist, allowing comparison of extinct forms with living taxa and providing the possibility of testing evolutionary hypotheses on modern faunas. For example, the hypothesis that there was a causal link between the 'sudden' polyphyletic evolution of the cemented habit in a number of byssate clades and the increased predation pressure of the Mesozoic Marine Revolution (Vermeij, 1987) was tested and accepted by Harper (1991) by offering modern predators the choice between byssate and cemented bivalve prey.

Our understanding of bivalve evolution to date is a tribute to the endeavours of both zoologists and palaeontologists. Zoologists, such as C.M. Yonge and his 'school', have dwelt upon anatomy for their phylogenetic analyses, whereas palaeontologists have concentrated, by necessity, on hard-part characters. It is the integration of these two approaches that has led to our current depth of understanding (see Morton, 1996), which could not have been achieved by either set of workers on their own and in which neither group has been dominant. There is much promise for further understanding in which the fossil record has a great part to play. Clearly, there is much that molecular phylogeneticists will add to understanding of phylogenetic relationships within the class, but the fossil record will be important in providing a means of discriminating between competing hypotheses and providing a time scale for the branching events. Other studies that will seek to understand the 'anatomy' of the various bivalve adaptive radiations (that is, the nature of the pre-adaptations and cues concerned) and to understand the parts played in them by now extinct groups can only proceed with recourse to the fossil record. Carefully chosen, many aspects of the bivalve fossil record are adequate enough to support such endeavours.

ACKNOWLEDGEMENTS

I am grateful to the Royal Society for my University Research Fellowship, to J.J. Sepkoski for the kind access to his unpublished database and to J.A. Crame for discussion. This is Cambridge Earth Sciences Publication 4896.

REFERENCES

Allen, J.A. and Sanders, H.L., 1973, Studies on deep-sea Protobranchia (Bivalvia); the families Siliculidae and Lametilidae, *Bulletin of the Museum of Comparative Zoology, Harvard*, **145**: 263–310.
Allison, P.A., 1988, Konservat-Lagerstätten: cause and classification, *Paleobiology*, **14**:

331–344.

Allmon, W.D., 1989, Paleontological completeness of the record of lower Tertiary mollusks, US Gulf and Atlantic coastal plains: implications for phylogenetic studies, *Historical Biology*, **3**: 141–158.

Atkins, D., 1936–1938, On the ciliary mechanisms and interrelationships of lamellibranchs, *Quarterly Journal of Microscopical Science*, **79**: 181–308, 339–445; **80**: 321–436.

Barrande, J., 1881, *Système Silurien du centre de la Bohême, 6, Ancéphalés*, Paris: 342 pp.

Britton, J.C., 1991, Pathways and consequences of the introduction of non-indigenous freshwater, terrestrial and estuarine mollusks in the United States, *A Report to the Office of Technical Assessment, Congress of the United States*: 66 pp.

Carter, J.G., 1990, *Skeletal Biomineralization: Patterns, Processes and Evolutionary Trends*, Van Nostrand and Reinhold, New York: 832 pp.

Chinzei, K., 1982, Morphological and structural adaptations to soft substrates in the early Jurassic monomyarians *Lithiotis* and *Cochlearites*, *Lethaia*, **15**: 179–197.

Cope, J.C.W., 1996, The early evolution of the Bivalvia. *In* J.D. Taylor (ed.), *Origin and Evolutionary Radiation of the Mollusca*, Oxford University Press, Oxford: 361–370.

Cox, L.R., 1960, The preservation of moulds of the intestine in fossil *Nuculana* (Lamellibranchia) from the Lias of England, *Palaeontology*, **2**: 262–269.

Crame, J.A., 1995, Occurrence of the bivalve genus *Manticula* in the early Cretaceous of Antarctica, *Palaeontology*, **38**: 299–312.

Cummins, R.H., 1994, Taphonomic processes in modern freshwater death assemblages: implications for the freshwater fossil record, *Palaeogeography, Palaeoclimatolology, Palaeoecology*, **108**: 55–73.

Fortey, R.A. and Jefferies, R.P.S., 1982, Fossils and phylogeny: a compromise approach. *In* K.A. Joysey and A.E. Friday (eds), *Problems of Phylogenetic Reconstruction, Systematics Association Special Volume 21*, Academic Press, London: 197–234.

Fürsich, F.T. and Palmer, T.J., 1982, The first true anomiid bivalve? *Palaeontology*, **25**: 897–903.

Fürsich, F.T. and Werner, W., 1989, Taxonomy and ecology of *Juranomia calcibyssata* gen. et sp. nov.: a widespread anomiid bivalve from the Upper Jurassic of Portugal, *Geobios*, **22**: 325–338.

Gilinsky, N.L., 1988, Survivorship in the Bivalvia: comparing living and extinct genera and families, *Paleobiology*, **14**: 370–386.

Gould, S.J., 1972, Allometrical fallacies and the evolution of *Gryphaea*: a new interpretation based on White's criterion of geometric similarity, *Evolutionary Biology*, **6**: 91–118.

Gray, J., 1988, Evolution of the freshwater ecosystems: the fossil record, *Palaeogeography, Palaeoclimatolology, Palaeoecology*, **62**: 1–214.

Harper, E.M., 1991, The role of predation in the evolution of the cemented habit in bivalves, *Palaeontology*, **34**: 455–460.

Harper, E.M. and Palmer, T.J., 1993, Middle Jurassic cemented pectinids and the missing right valves of *Eopecten*, *Journal of Molluscan Studies*, **59**: 63–72.

Harper, E.M. and Todd, J.A., 1995, Preservation of the adductor muscle of an Upper Jurassic oyster, *Paläontologische Zeitschrift*, **69**: 55–59.

Harper, E.M., Radley, J.D. and Palmer, T.J., 1996, Early Cretaceous cementing pectinid bivalves, *Cretaceous Research*, **17**: 135–150.

Jell, P.A., 1980, Earliest known pelecypod on Earth: a new early Cambrian genus from South Australia, *Alcheringa*, **4**: 233–239.

Johnson, A.L.A., 1985, The rate of evolutionary change in European Jurassic scallops, *Special Papers in Palaeontology*, **33**: 91–102.

Johnson, A.L.A. and Lennon, C.D., 1990, Evolution of gryphaeate oysters in the mid-Jurassic of Western Europe, *Palaeontology*, **33**: 453–485.

Jones, D.S. and Quitmyer, I.R., 1996, Marking time with bivalve shells: oxygen isotopes and season of annual increment formation, *Palaios*, **11**: 340–346.

Kelley, P.H., 1989, Evolutionary trends within bivalve prey of Chesapeake Group naticid gastropods, *Historical Biology*, **2**: 139–156.

Kidwell, S.M. and Bosence, W.J., 1991, Taphonomy and time-averaging of marine shelly faunas. *In* P.A. Allison and D.E.G. Briggs (eds), *Taphonomy: Releasing Data Locked in the Fossil Record*, Plenum Press, New York: 115–209.

Kier, P., 1977, The poor fossil record of the regular echinoid, *Paleobiology*, **3**: 168–174.

Koch, C.F., 1978, Bias in the published fossil record, *Paleobiology*, **4**: 367–372.

Miller, A.I. and Sepkoski, J.J., Jr, 1988, Modeling bivalve diversification: the effect of interaction on a macroevolutionary system, *Paleobiology*, **14**: 364–369.

Moore, R.C. (ed.), 1969–1971, *Treatise on Invertebrate Paleontology, Part N (1–3)*, Geological Society of America and University of Kansas Press: 1224 pp.

Morris, N.J., 1978, The infaunal descendants of the Cycloconchidae: an outline of the evolutionary history and taxonomy of the Heteroconchia, superfamily Cycloconchacea to Chamacea, *Philosophical Transactions of the Royal Society of London*, **B284**: 259–275.

Morris, N.J., 1985, Other non-marine invertebrates, *Philosophical Transactions of the Royal Society of London*, **B309**: 239–240.

Morton, B., 1980, Some aspects of the biology and functional morphology (including the presence of a ligamental lithodesma) of *Montacutona compacta* and *M. olivacea* (Bivalvia: Leptonacea) associated with coelenterates in Hong Kong, *Journal of the Zoological Society, London*, **192**: 431–455.

Morton, B., 1981, The biology and functional morphology of *Chlamydoconcha orcutti* with a discussion on the taxonomic status of the Chlamydoconchacea (Molluscsa Bivalvia), *Journal of the Zoological Society, London*, **195**: 81–121.

Morton, B., 1996, The evolutionary history of the Bivalvia. *In* J.D. Taylor (ed.), *Origin and Evolutionary Radiation of the Mollusca*, Oxford University Press, Oxford: 337–356.

Morton, J.E. and Yonge, C.M., 1964, Classification and structure of the Mollusca. *In* K.M. Wilbur and C.M. Yonge (eds), *Physiology of Mollusca, 1*, Academic Press, New York: 1–58.

Nauss, A.L. and Smith, P.L., 1988, *Lithiotis* (Bivalvia) bioherms in the Lower Jurassic , east central Oregon, USA, *Palaeogeography, Palaeoclimatology, Palaeoecology*, **65**: 255–268.

Ong Che, R. and Morton, B., 1994, Spatial and seasonal variations in the intertidal sandflat community of Tai Tam Bay, Hong Kong, *Asian Marine Biology*, **11**: 89–101.

Paul, C.R.C., 1982, The adequacy of the fossil record. *In* K.A. Joysey and A.E. Friday (eds), *Problems of Phylogenetic Reconstruction, Systematics Association Special Volume 21*, Academic Press, London: 75–117.

Paul, C.R.C., 1985, The adequacy of the fossil record reconsidered. *In* J.C.W. Cope and P.W. Skelton (eds), *Evolutionary Case Histories from the Fossil Record, Special Papers in Palaeontology*, **33**: 7–15.

Paul, C.R.C., 1990, Completeness of the fossil record. *In* D.E.G. Briggs and P.R. Crowther (eds), *Palaeobiology: A Synthesis*, Blackwell, Oxford: 298–303.

Pease, C.M., 1985, Biases in the durations and diversities of fossil taxa, *Paleobiology*, **11**: 272–292.

Peel, J.S., 1991, Functional morphology of the class Helcionelloida and the early evolution of the Mollusca. *In* A. Simonetta and S. Conway Morris (eds), *The Early Evolution of Metazoa and Significance of Problematic Taxa*, Cambridge University

Press, Cambridge: 157–177.

Pojeta, J., 1975, *Fordilla troyensis* Barrande and early pelecypod phylogeny, *Bulletin of American Paleontology*, **67**: 363–379.

Pojeta, J., 1978, The origin and taxonomic diversification of pelecypods, *Philosophical Transactions of the Royal Society, London*, **B284**: 225–243.

Pojeta, J. and Runnegar, B., 1985, The early evolution of diasome molluscs. *In* E.R. Trueman and M.R. Clarke (eds), *The Mollusca, 10, Evolution*, Academic Press, New York: 295–336.

Pojeta, J., Runnegar, B. and Kriz, J., 1973, *Fordilla troyensis* Barrande: the oldest known pelecypod, *Science*, **180**: 866–868.

Purchon, R.D., 1987, The stomach in the Bivalvia, *Philosophical Transactions of the Royal Society of London*, **B316**: 183–276.

Raup, D.M., 1975, Cohort analysis of generic survivorship and Van Valen's law, *Paleobiology*, **1**: 82–96.

Raup, D.M., 1979, Biases in the fossil record of species and genera, *Bulletin of the Carnegie Museum of Natural History*, **13**: 85–91.

Raup, D.M. and Sepkoski, J.J., Jr, 1986, Periodic extinction of families and genera, *Science*, **231**: 833–836.

Reis, O.M., 1903, Über Lithioden, *Abhandlungen der Kaiserlich-Königlichen, Geologischen Reichsanstalt*, **17**: 1–44.

Runnegar, B., 1983, Molluscan phylogeny revisited, *Memoirs of the Association of Australasian Palaeontologists*, **1**: 121–144.

Runnegar, B., 1996, Early evolution of the Mollusca: the fossil record. *In* J.D. Taylor (ed.), *Origin and Evolutionary Radiation of the Mollusca*, Oxford University Press, Oxford: 77–87.

Runnegar, B. and Bentley, C., 1983, Anatomy, ecology and affinities of the Australian early Cambrian bivalve *Pojetaia runnegari* Jell, *Journal of Paleontology*, **57**: 73–92.

Runnegar, B. and Pojeta, J., 1974, Molluscan phylogeny: the paleontological viewpoint, *Science*, **186**: 311–317.

Runnegar, B. and Pojeta, J., 1992, The earliest bivalves and their Ordovician descendants, *Bulletin of American Malacolologists*, **9**: 117–122.

Sanders, H.L. and Allen, J.A., 1973, Studies on deep-sea Protobranchia (Bivalvia); prologue and the Pristoglomidae, *Bulletin of the Museum of Comparative Zoology, Harvard*, **145**: 237–262.

Schneider, J.A., 1992, Preliminary cladistic analysis of the bivalve family Cardiidae, *Bulletin of American Malacologists*, **9**: 145–155.

Schneider, J.A., 1995, Phylogeny of the Cardiidae (Mollusca, Bivalvia): Protocardiinae, Laevicardiinae, Lahilliinae, Tulogocardiinae subfam. n. and Pleuriocardiinae subfam. n., *Scripta Zoologica*, **24**: 321–346.

Schopf, T.J.M., 1978, Fossilization potential of an intertidal fauna: Friday Harbor, Washington, *Paleobiology*, **4**: 261–270.

Seilacher, A., 1984, Constructional morphology of bivalves: evolutionary pathways in primary versus secondary soft-bottom dwellers, *Palaeontology*, **27**: 207–237.

Sepkoski, J.J. Jr, 1989, Periodicity in extinction and the problem of catastrophism in the history of life, *Journal of Geological Society, London*, **146**: 7–19.

Sepkoski, J.J. Jr, 1992, A compendium of fossil marine animal families, *Milwaukee Public Museum, Contributions in Biology and Geology*, **83**: 156pp.

Sepkoski, J.J. Jr, unpublished, Generic longevity database for the Bivalvia.

Sheldon, P.R., 1987, Parallel gradualistic evolution of Ordovician trilobites, *Nature*, **330**: 561–563.

Simpson, G.G., 1944, *Tempo and Mode in Evolution*, Columbia University Press, New

York: 237 pp.

Skelton, P.W., 1979, Preserved ligament in a radiolitid rudist bivalve and its implication of mantle marginal feeding in the group, *Paleobiology*, **5**: 90–106.

Skelton, P.W., 1991, Morphogenetic versus environmental cues for adaptive radiations. *In* N. Schmidt-Kittler and K. Vogel (eds), *Constructional Morphology and Evolution*, Springer-Verlag, Berlin: 375–387.

Skelton, P.W., 1993, Adaptive radiation: definition and diagnostic tests. *In* D.R. Lees and D. Edwards (eds), *Evolutionary Patterns and Processes, Linnean Society Symposium Series, No. 14*, Academic Press, London: 45–58.

Skelton, P.W. and Benton, M.J., 1993, Mollusca: Rostroconchia, Scaphopoda and Bivalvia. *In* M.J. Benton (ed.), *The Fossil Record*, 2, Chapman and Hall, London: 237–263.

Skelton, P.W., Crame, J.A., Morris, N.J. and Harper, E.M. 1990, Adaptive divergence and taxonomic radiation in post-Palaeozoic bivalves. *In* P.D. Taylor and G.P. Larwood (eds), *Major Evolutionary Radiations, Systematics Association Special Volume 42*, Clarendon Press, Oxford: 91–117.

Stanley, S.M., 1968, Post-Paleozoic adaptive radiation of infaunal bivalve molluscs: a consequence of mantle fusion and siphon formation, *Journal of Paleontology*, **42**: 214–229.

Stanley, S.M., 1970, Relation of shell form to life habits of the Bivalvia, *Geological Society of America Memoir*, **125**: 296 pp.

Stanley, S.M. and Yang, X., 1987, Approximate evolutionary stasis for bivalve morphology over millions of years: a multivariate, multilineage study, *Paleobiology*, **13**: 113–139.

Taylor, D.W., 1988, Aspects of freshwater mollusc ecological biogeography, *Palaeogeography, Palaeoclimatology, Palaeoecology*, **62**: 511–576.

Taylor, J.D., Kennedy, W.J. and Hall, A., 1969, The shell structure and mineralogy of the Bivalvia. Introduction, Nuculacea–Trigonacea, *Bulletin of the British Museum (Natural History), Zoology, Suppl.*, **3**: 125 pp.

Taylor, J.D., Kennedy, W.J. and Hall, A., 1973, The shell structure and mineralogy of the Bivalvia: II. Lucinacea–Clavegellacea, Conclusion, *Bulletin of the British Museum (Natural History), Zoology*, **22**: 253–284.

Tevesz, M.J. and Carter, J.G., 1980, Environmental relationships of shell form and structure of unionacean bivalves. *In* D.C. Rhoads and R.A Lutz (eds), *Skeletal Growth of Aquatic Organisms*, Plenum Press, New York: 295–322.

Tevesz, M.J. and McCall, P.L., 1976, Primitive life habits and adaptive significance of the pelecypod form, *Paleobiology*, **2**: 183–190.

Tucker, R.D. and McKerrow, W.S., 1995, Early Paleozoic chronology: a review in light of new U–Pb zircon ages from Newfoundland and Britain, *Canadian Journal of Earth Sciences*, **32**: 368–379.

Valentine, J.W., 1989, How good was the fossil record? Clues from the Californian Pleistocene, *Paleobiology*, **15**: 83–94.

Van Valen, L., 1973, A new evolutionary law, *Evolutionary Theory*, **1**: 1–30.

Vermeij, G.J., 1987, *Evolution and Escalation: An Ecological History of Life*, Princeton University Press, Princeton, New Jersey: 523 pp.

Vokes, H.E., 1980, *Genera of the Bivalvia: A Systematic and Bibliographic Catalogue*, Paleontological Research Institution, New York: 292 pp.

Walcott, C.D., 1887, Fauna of the 'Upper Taconic' of Emmans, in Washington County, New York, *American Journal of Science*, **38**: 29–42.

Waller, T.R., 1978, Morphology, morphoclines and a new classification of the Pteriomorphia (Mollusca: Bivalvia), *Philosophical Transactions of the Royal Society,*

London, **B284**: 345–365.

Wilby, P.R. and Whyte, M.A., 1995, Phosphatized soft tissues in bivalves from the Portland Roach of Dorset (Upper Jurassic), *Geological Magazine*, **132**: 117–120.

Yochelson, E.L., 1978, An alternative approach to the interpretation of ancient mollusks, *Malacologia*, **17**: 165–191.

Yochelson, E.L., 1981, *Fordilla troyensis* Barrande: 'the oldest known pelecypod' may not be a pelecypod, *Journal of Paleontology*, **55**: 113–125.

Yonge, C.M., 1962, On the primitive significance of the byssus in the Bivalvia and its effects in evolution, *Journal of the Marine Biological Association, UK*, **42**: 112–125.

Yonge, C.M., 1977, Form and evolution in the Anomiacea (Mollusca: Bivalvia): *Pododesmus, Anomia, Enigmonia* (Anomiidae) and *Placunomia, Placuna* (Placunidae), *Philosophical Transactions of the Royal Society, London*, **B276**: 453–523.

Yonge, C.M., 1982, Mantle margins with a revision of siphonal types in the Bivalvia, *Journal of Molluscan Studies*, **48**: 102–103.

Yonge, C.M. and Thompson, T.E., 1976, *Living Marine Molluscs*, Collins, London: 288 pp.

12
The Quality of the Fossil Record of the Vertebrates

Michael J. Benton

INTRODUCTION

Fossil vertebrates include a great diversity of animals of all sizes and shapes, ranging in age back to the Cambrian. The history of the vertebrates has been recounted many times (for example, Romer, 1966; Carroll, 1987; Benton, 1990a, 1997a) and the outlines of the story are well known. These broad outlines were worked out during the nineteenth century, and the sequence includes the armoured ostracoderms and placoderms of the Devonian, Carboniferous amphibians, Permian mammal-like reptiles, Mesozoic dinosaurs, ichthyosaurs, plesiosaurs, pterosaurs and birds, Tertiary mammals and Plio-Pleistocene hominids. This succession is usually recalled as a one-way progression from essentially toothed worm-like creatures of the early Palaeozoic to humans, even though such a vision is merely a didactic device, and does not properly depict the branching bushy pattern of vertebrate evolution.

How good is this record of the diversification of vertebrates? There are two intuitive answers. One is to suggest that the record is good because nothing much has changed in our understanding of the timing of events in the past 100 years of research. The other is to say that the record is terrible because many vertebrates, particularly tetrapods, live on land, and they are much less likely to be preserved than marine-shelf invertebrates. This criticism cannot, of course, apply to fishes.

The purpose of this paper is to summarize current knowledge of the history of backboned animals, but only in the simplest of outline forms, and to present some recent work on the quality of the fossil record of vertebrates.

The Adequacy of the Fossil Record. Edited by S.K. Donovan and C.R.C. Paul.
© 1998 John Wiley & Sons Ltd.

A GOOD FOSSIL RECORD?

It is likely that the numbers of fossils, localities and vertebrate palaeontologists have multiplied by several orders of magnitude during the 20th century, and yet there have been no surprises in the accepted broad-scale pattern of vertebrate evolution. Of course, the impressive efforts of collectors have pushed the origins of various groups backwards in time, but these stratigraphic range extensions have all been predictable. This century, the origin of agnathan fishes has been pushed back from the Silurian to the Cambrian, the origin of amphibians from the early Carboniferous to the latest Devonian, the origin of reptiles from the early Permian to the mid-Carboniferous, and the origin of mammals back from the mid-Jurassic to the late Triassic. Arguably, the origin of birds has remained unchanged in the latest Jurassic, since the work of Owen and Huxley in the 1860s and 1870s. However, if *Protoavis* is a bird (Chatterjee, 1995), then the point of origin of the group moves back to the late Triassic, and that would distort many parts of the phylogeny, not only of birds, but also of Dinosauria in general.

A critic of the quality of the vertebrate fossil record might have expected more surprises. Recall that Charles Lyell, a supporter of the idea that time proceeded in cycles, and an opponent of the idea of progression, or unidirectional change through time, quite expected in the 1830s that human fossils might be found in the Silurian. He campaigned hard in the 1850s to convince colleagues that new discoveries of Silurian arthropod tracks from North America had actually been made by land vertebrates. He also lent his strong support, in the early 1850s, to the view that the aeolian yellow sandstones round Elgin in north-east Scotland, which had just yielded supposed turtle tracks and the skeleton of an apparently lizard-like animal, were actually Devonian in age, rather than Triassic (Benton, 1983). Had he been right, then the generally accepted pattern of vertebrate evolution would have looked very different. So far, in 150 years of searching, palaeontologists have not found human remains in the Silurian, nor have they found modern-style reptiles in the Silurian or Devonian. It can be asserted that, the longer such out-of-place fossils do not turn up, the greater the likelihood that our knowledge of vertebrate evolution approximates the truth.

The notion of a good fossil record of vertebrates was confirmed in a quantitative analysis by Maxwell and Benton (1990). These authors compared several stages in the development of knowledge about the history of tetrapods over the past 100 years. They used a number of publications, dated 1900, 1933, 1945, 1966 and 1987, as snapshots of then-current knowledge of the former diversities and distributions in time of families of fossil tetrapods. There had certainly been huge changes in palaeontological knowledge from 1900 to 1987, not least a doubling of the known diversities of all groups, presumably as a result largely of intensive collecting efforts. In addition, the snapshots of palaeontological understanding included revisions of stratigraphy and taxonomy. However, the results of all these changes appeared to be randomly distributed with respect to time. Global diversities essentially doubled throughout the whole fossil record of tetrapods, from the late Devonian to the present-day, but without any biases becoming evident. The overall pattern of diversification, and the timing and magnitudes of major extinction events, were unchanged. Maxwell and Benton (1990) concluded that all the changes in understanding of the tetrapod fossil record in the past 100 years had not altered the broad-scale macroevolutionary patterns derived from it. These findings have been confirmed for marine animals in an analogous study by Sepkoski (1993).

A POOR FOSSIL RECORD?

Perhaps a commoner intuitive view of the vertebrate fossil record is that it is poor or very poor, especially when compared with the fossil records of marine-shelf skeletized invertebrates (for example, Valentine, 1969; Raup, 1979; Benton, 1985; Cowen, 1990; Flessa, 1990; Jablonski, 1991). This assumption has been made by scaling up from field observations. Typically, limestones and clastic rocks laid down on the shallow continental shelf yield abundant fossils of skeletized invertebrates, such as brachiopods, molluscs, corals, arthropods, bryozoans and echinoderms (Kidwell, 1986; Fürsich, 1990). Continental sedimentary rocks, on the other hand, generally yield much less abundant fossil faunas of freshwater fishes and molluscs, and of terrestrial insects and vertebrates (Behrensmeyer and Hill, 1980; Retallack, 1984).

This differentiation may be largely an effect of the nature of the sedimentary rocks: sedimentation in river systems and lakes is highly episodic compared with the more continuous deposition on marine shelves, and particularly in abyssal areas of oceans (Sadler, 1981). In addition, there may be biological factors. Many groups of skeletized marine-shelf invertebrates include forms with relatively short life spans, forms that live in huge abundances, and some that moult and hence produce several potential body fossils during a lifetime. Many vertebrates, and tetrapods in particular, often have life spans lasting several years, and populations are often not counted in the thousands or millions.

Many of these observations are qualitative, but recent taphonomic and palaeoecological studies (for example, papers in Briggs and Crowther, 1990; Allison and Briggs, 1991; Donovan, 1991) show huge differences in the abundance and closeness of spacing between fossil species of marine invertebrates and planktic forms on the one hand, and continental vertebrates on the other.

In conclusion, there are two intuitive views of the quality of the fossil record of vertebrates, and each is supported by observational evidence and by quantitative studies. How can they be reconciled?

THE PATTERN OF THE EVOLUTION OF VERTEBRATES

The evolution of any group can be represented diagrammatically in various ways. One useful kind of graphic presentation is a 'spindle diagram', in which the evolution of a group is represented by symmetrical spindle shapes that indicate the waxing and waning of a group through time. The y-axis is proportional to time and the x-axis to species numbers, or some similar measure. Each group originates as a narrow point, and then typically expands into a wider spindle as it radiates. Any individual spindle may remain narrow (low diversity) through time, or it may expand, or vary in width through time, depending on the relative fortunes of the group.

For vertebrates, the latest compilation of data (chapters in Benton, 1993; also http://palaeo.gly.bris.ac.uk/palaeo/frwhole/fr2.html) gives a diagram (Figure 12.1) that has not changed much since earlier comparable attempts (for example, Romer, 1966; Carroll, 1987). The relative widths of different spindles have changed somewhat, as have also the currently oldest indications of some of the groups. Certain mass extinctions, particularly those in the late Devonian, late Permian and at the end of the Cretaceous, are highlighted by relatively rapid contractions in the widths of several of the spindles. This implies high rates of extinction of several groups at the same time.

These data on diversity may be rendered as more precise plots of actual counts of families through time (Figure 12.2). In this case, the same database (Benton, 1993) was trawled for data on the diversities of families of the major groups. The data are plotted separately for fishes (Figure 12.2a) and tetrapods (Figure 12.2b), and, in each case, the presentation is cumulative, with each labelled curve adding on top of the curves below. The upper curve represents the total sum of diversity for all fishes or all tetrapods at any time in the past 500 Myr.

In detail, the pattern of diversification of the fishes (Figure 12.2a) showed a rapid rise in the early Palaeozoic, to a broad peak, at about 50–70 families, in the Devonian. This initial rise in diversity may be more apparent than real; Sansom *et al.* (1996) have argued that the pre-Devonian record of fishes is unnaturally impoverished and that true diversity was much higher than is currently believed. Diversity declined at the end of the Devonian and diminished further through the late Permian, before slowly climbing during the Mesozoic. The diversity of fishes reached Devonian levels of 50–70 families again in the Cretaceous and then diversification was apparently explosive

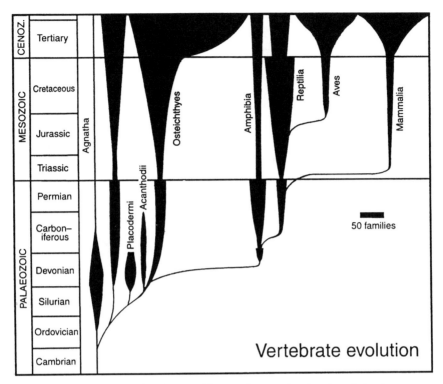

Figure 12.1 *The pattern of evolution of the vertebrates, showing the relative importance of the major groups through time. This is a 'spindle diagram', in which the vertical axis represents geological time and the horizontal axis represents the diversity of each group. In this case, the horizontal dimension is proportional to the number of families in each group. The groups include some clades (that is, monophyletic groups), such as Chondrichthyes, Placodermi, Acanthodii, Aves and Mammalia, but the others are paraphyletic groups (that is, a group that includes the ancestor, but not all of the descendants of that ancestor). All groups are treated in their traditional sense. Mass extinctions show up in the late Devonian, late Permian, and end-Cretaceous, indicated by relatively rapid contractions in the diversities of several clades. (From various authors in Benton, 1993)*

through the Tertiary, although apparently slowing down in the last 40 Myr.

The high levels of fish diversity in the Devonian are dominated by ostracoderm agnathans and placoderms, both of which groups were hard hit by the late Devonian mass extinction. However, bear in mind that 'fishes' are a paraphyletic group, and the drop in post-Devonian diversity was matched by the diversification of Palaeozoic tetrapods, just another branch of fish diversity that happened to move partly or fully on to land, through the late Devonian and Carboniferous. The subsequent diversification of fishes was dominated by the chondrichthyans (sharks and rays) and actinopterygians (bony fishes), and the dramatic early Tertiary burst in diversification was

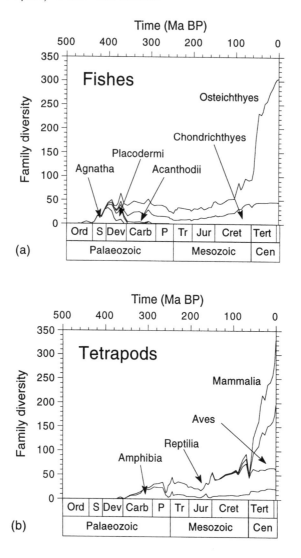

Figure 12.2   *Patterns of the diversification of (a) fishes and (b) tetrapods through time, based on counts of numbers of families present during each geological stage. Major groups, some monophyletic, some paraphyletic, are shown, and the upper curve in each case is the sum of the family diversities of these groups. The effects of mass extinctions may be detected in the late Devonian (390 Ma), late Permian (250 Ma), late Triassic (225, 205 Ma) and end-Cretaceous (65 Ma). Ord, Ordovician; S, Silurian; Dev, Devonian; Carb, Carboniferous; P, Permian; Tr, Triassic; Jur, Jurassic; Cret, Cretaceous; Tert, Tertiary; Cen, Cenozoic. (Data from Benton, 1993)*

driven largely by the radiation of two clades, the Neoselachii (modern sharks) and the Teleostei (the majority of modern bony fishes).

The diversification of tetrapods (Figure 12.2b) does not show such a rapid early rate of increase and there was a steady rise to a total of about 40 families by the end of the Palaeozoic. Diversity levels remained roughly constant at that level throughout most of the Triassic and Jurassic, but there was a steady rise from 50 to 70 families during the early Cretaceous, followed by a rapid increase during the late Cretaceous to 100 families and a further accelerating rate of increase during the Cenozoic.

The Palaeozoic diversity record is dominated by 'amphibians' and these basal tetrapods declined in diversity dramatically during the Mesozoic, finally disappearing in the Cretaceous. The post-Palaeozoic record of amphibians represents the clade Lissamphibia and their diversity has risen slowly from Mesozoic levels of about 10 families to about twice that at the end of the Cenozoic. Note that the diversification pattern of 'Amphibia' focuses on a paraphyletic group in the Palaeozoic (that is, all tetrapods except amniotes) and a descendant clade, the Lissamphibia, after the Palaeozoic, so there is no need to seek special reasons for the double-peaked pattern of diversification of the 'Amphibia', as Carroll (1977) attempted. The pattern is an artefact of the conventional classification of the group and need not represent an unusual gap in the fossil record.

The diversification pattern for reptiles is similarly unrealistic, but the groups are represented in this way to indicate the common understanding of these terms. Through most of the Mesozoic at least, the patterns are meaningful, although the descendant clades, the birds and mammals, rise to prominence especially in the late Cretaceous. Reptile diversity increased marginally through the Mesozoic from 20–30 families in the Triassic to 50–60 in the late Cretaceous, mainly dinosaurs. Birds and mammals existed through much of the Mesozoic, but at low diversities, and both groups showed dramatically accelerating rates of diversification through the Cenozoic.

Changes in diversity may be tracked also by documenting origination and extinction rates. Here, the percentage rates are presented (Figures 12.3 and 12.4); in other words, these are the numbers of families arising or becoming extinct in a geological stage as a proportion of the numbers extant at the time. This measure of origination and extinction gives a measure of risk, but it is not normalized to geological time. This has not been done since the stratigraphic stages are largely of comparable length, typically 5–10 Myr, although the precise durations of many of these stages are not known with confidence.

Origination rates for fishes (Figure 12.3) were particularly high in the late Cambrian (radiation of agnathans, including conodonts), during the Silurian, Devonian and Carboniferous, in the early Triassic (after the late Permian extinction), and in the mid-Cretaceous and Eocene (after the Cretaceous/ Tertiary (K/T) extinction event). Tetrapods also show high rates of origina-

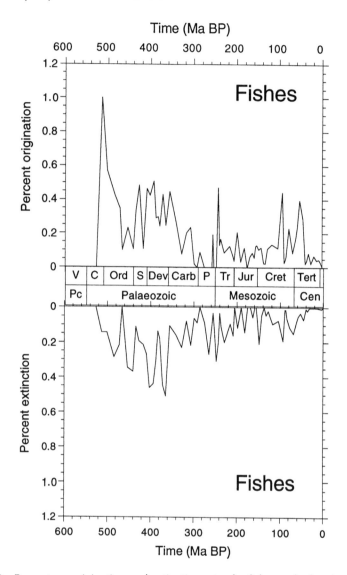

Figure 12.3 *Percentage origination and extinction rates for fishes, calculated as numbers of originations/extinctions per stage in proportion to total diversity at the time. V, Vendian; C, Cambrian; Ord, Ordovician; S, Silurian; Dev, Devonian; Carb, Carboniferous; P, Permian; Tr, Triassic; Jur, Jurassic; Cret, Cretaceous; Tert, Tertiary; Pc, Precambrian; Cen, Cenozoic*

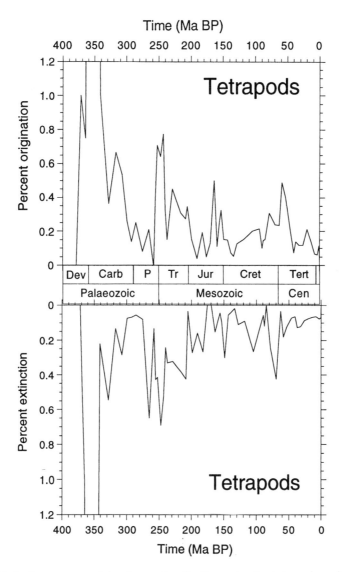

Figure 12.4 *Percentage origination and extinction rates for tetrapods, calculated as numbers of originations/extinctions per stage in proportion to total diversity at the time. Dev, Devonian; Carb, Carboniferous; P, Permian; Tr, Triassic; Jur, Jurassic; Cret, Cretaceous; Tert, Tertiary; Cen, Cenozoic*

tion in the late Devonian and early Carboniferous, just after the late Permian extinction and just after the K/T event. The late Jurassic peaks may be real or may reflect some sites of exceptionally good preservation.

The fossil records of both fishes and tetrapods show the influence of mass extinctions (Figures 12.2–12.4). The late Devonian mass extinction affected fishes severely, with ostracoderm agnathans and placoderms virtually wiped out. The diversity of fishes, essentially chondrichthyans and sarcopterygians, fell roughly to half during the whole of the Permian, and the end-Permian mass extinction did not particularly affect them. The effects of the end-Permian event were much greater on tetrapods, with a reduction in amphibian diversity from about 20 families to five, and in reptile diversity from 10 families to about five. Fish diversity was apparently little affected by Mesozoic events, showing some small reductions, but nothing profound, and seemingly very little at the K/T boundary, 65 Myr ago. Tetrapods, on the other hand, were affected by a number of the Mesozoic extinction events, notably the two late Triassic events, and possibly an event at the end of the Jurassic, 150 Myr ago (although that drop is preceded by a sudden rise, possibly reflecting the exceptional preservation of the Solnhofen beds of southern Germany). The K/T event was profound, marked by a drop in the diversity of families of tetrapods from over 90 to 60.

There is some evidence for a coupling of patterns of origination and extinction (Figures 12.3 and 12.4), as noted before for vertebrates (Benton, 1989), hence suggesting a possible 'Lagerstätten effect'. (If a fossil record is affected excessively by the influence of specific localities of exceptional preservation, the patterns of origination and extinction are often tightly coupled.) For fishes, apparent high rates of turnover (high origination and high extinction rates) occur in the mid-Devonian and early late Cretaceous (Figure 12.3). However, the evidence for coupling is not strong, and more often high extinction rates are followed by high origination rates. The same appears to be largely true for tetrapods (Figure 12.4), although there were coupled high rates of origination and extinction in the early Triassic and late Jurassic. The coupling seems to be less than it was in earlier analyses (Benton, 1985, 1989); perhaps our knowledge of the background fossil record is improving to the extent that Lagerstätten are no longer distorting the patterns unduly.

The overall diversification curves both appear to follow exponential patterns of increase, from relatively low numbers in the Palaeozoic, through slightly increasing diversities in the Jurassic and Cretaceous, to dramatically accelerating diversifications from the late Cretaceous onwards. It would be hard to interpret these curves as logistic, as Sepkoski (1984) has done for the diversification of families of marine animals. The diversity curves for both fishes and tetrapods show a good fit to an exponential curve (Figure 12.5). The fit is markedly better than one to a straight line (linear model), and the

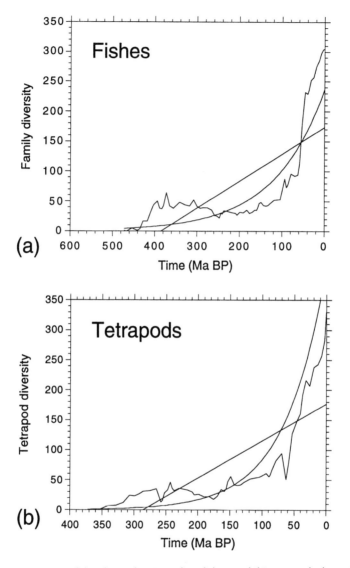

Figure 12.5 *Patterns of the diversification of (a) fishes and (b) tetrapods through time, with best-fitting linear and exponential models. In both cases the exponential model fits better: for fishes, the linear model ($y = 173.7 - 0.45x$, $r = 0.719$) is a poorer match to the data than the exponential model ($y = 236.9^{-0.0085x}$, $r = 0.925$); and for tetrapods, the linear model ($y = 177.0 - 0.62x$, $r = 0.779$) is also a poorer match to the data than the exponential model ($y = 409.5^{-0.0159x}$, $r = 0.979$)*

distribution of points prevents the calculation of any kind of meaningful logistic fit. Benton (1995a) found the same result for all continental organisms, including essentially tetrapods, insects and land plants.

EXPONENTIAL OR EQUILIBRIAL SYSTEMS

The implications of exponential, rather than logistic, curves of increase in diversity are profound (Benton, 1997b, 1998; Figure 12.6). An exponential curve, allowing for the temporary reverses caused by mass extinction events, implies that there is no limit to the global diversity that can be achieved, or at least that vertebrates (or life on land, or all life) have yet to approach that maximum global carrying capacity for life. A logistic curve, or series of logistic curves, implies that there are global-scale caps to diversity, that the world reaches a stage where all ecospace is full, and this limiting level, or steady state, can be breached only by some major revolution (a mass extinction, the origin of a substantial new adaptive complex, a dramatic environmental shift). Which is the true interpretation of the history of life, or of vertebrates at least?

Logistic Models

The logistic model for the expansion of the diversity of life has developed from an influential body of ecological theory. MacArthur and Wilson (1967) presented their 'theory of island biogeography' as a simple means of estimating the rate of filling of an island or other defined patch of living space. They showed how the rate of arrival of organisms is initially high, but that, as the space fills up, the rate of successful colonization diminishes and the rate of local extinction increases. At some point, the two rates, colonization and extinction, stabilize and this dynamic equilibrium represents the carrying capacity, or ideal diversity, of life on the island. The carrying capacity is not fixed, but can be modified by major changes: for example, movements in the relative position of the island with respect to the nearest sources of species, changes in topography, climatic shifts, and the like.

Rosenzweig (1975, 1995) and others have extended this island-scale model to regional and global scales, allowing essentially for intraclade competition. Thus, MacArthur and Wilson's island becomes the world, and their local rates of colonization and extinction become global rates of origination and extinction of species, or higher taxa. Time scales move from tens or hundreds of years to millions.

Sepkoski (1978, 1979, 1984, 1996) has developed an equivalent system of modelling, but one in which he has focused on more complex intraspecific

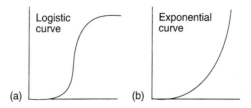

Figure 12.6 *The basic shape of (a) a logistic and (b) an exponential curve*

competition rather than intraclade competition. However, the end-result is similar and he has found evidence for equilibrium levels in the Cambrian and in the post-Cambrian Palaeozoic, and these are interpreted as representing real steady states. The post-Palaeozoic record of diversification of marine animals, according to his analysis, shows no further equilibrium level, and hence the past 250 Ma is interpreted as the rising part of a logistic curve. Courtillot and Gaudemer (1996), on the other hand, in a re-analysis of Benton's (1995a) data, confirmed a logistic model for the diversification of life, but they found evidence for only one equilibrium level in the Palaeozoic, and believed they had identified the beginning of a slow-down in diversification rates in the late Cenozoic, and hence a hint that life is approaching a new equilibrium level.

Discussion in this area has been dominated by the logistic model and the implications of global steady states or equilibria. Sepkoski's interpretation has been reproduced in many textbooks (for example, Allen and Briggs, 1989; Clarkson 1993; Skelton, 1993; Benton and Harper, 1997). However, some of the key assumptions behind such a view have been criticized (see, for example, Hoffman, 1985).

The critical issue with assumptions of equilibrium levels, or even less-stable steady states, is the notion of a carrying capacity, or ideal number of species that can be accommodated on the Earth. Rieppel (1984), for example, showed that the idea of a global steady state for species is equivalent to the old 'principle of plenitude'. This was a pre-Darwinian idea, founded on the key assumption of natural theology, that God had created all organisms perfect and that he had fitted some plant or animal to each available task (we would say niche) in the economy of nature. There was no space left.

Darwin borrowed the principle of plenitude and made it explicitly evolutionary in his analogy of a barrel of apples. In his unpublished *Natural Selection* (see Stauffer, 1975, p. 208), Darwin compared the present-day diversity of species to a number of apples floating on the surface of a barrel filled with water. The surface is covered by exactly the right number of apples, and it is impossible to add a new apple without displacing one that is already there. Similarly, he argued, each species had been honed by evolution to fit its niche and, if a new species arises, it has to displace a pre-existing species before it can become established. Walker and Valentine (1984) questioned Darwin's assumption, that all niches are full. This has also typically been an assumption of regional- and global-scale equilibrium models, but it is unnecessary. These authors estimated that the mean proportion of empty niches ranged from 12% to 54% for eight marine invertebrate groups, and that species turnover could occur readily and rapidly, but without the need to assume that there is constant evolutionary pressure from competitively superior species.

Should we believe mathematical models? There is no question that a logistic model is a much better fit to the pattern of diversification of families of Palaeozoic marine animals than a straight-line, exponential or power-law curve (Sepkoski, 1979, 1984, 1996; Courtillot and Gaudemer, 1996). However, if the basic ecological and evolutionary assumptions behind such a logistic model are contradicted, then perhaps a fresh consideration is justified.

Exponential Models

The question of mathematics versus basic observations does not arise for vertebrates, since a logistic curve, or series of logistic curves, would be hard to fit to the known patterns of diversification of fishes or tetrapods (Figures 12.2 and 12.5). The patterns as a whole suggest expansion, with especially rapid expansion over the past 50–100 Myr, the part of the fossil record that is probably better known and better dated than earlier segments.

Intuitive observations of the fossil record of vertebrates confirm the idea of continuing expansion based on evolutionary innovation. For example, the bursts of radiation among fishes (Figure 12.2a) may be related to specific new adaptations, such as extensive armour in ostracoderm agnathans and placoderms in the late Silurian and Devonian, jaws in placoderms and acanthodians in the Devonian, increased swimming speeds and efficiency of jaws in Jurassic bony fishes (and especially the teleosts), expansion of trophic levels in the sea (especially the modern sharks, Neoselachii), adaptations to feeding on specific foods (such as plankton, corals, buried bivalves and echinoderms), and parasitic modes.

A similar sequence of dramatic expansions of ecospace characterizes the evolution of tetrapods: terrestrial adaptations among early tetrapods in the late Devonian and Carboniferous, insect-eating in Carboniferous tetrapods, the cleidoic (enclosed) egg in late Carboniferous amniotes, herbivory in certain early Permian amniotes, large size in some late Permian herbivorous mammal-like reptiles, fully upright posture in several amniote groups in the late Triassic, true flapping flight in late Triassic pterosaurs and late Jurassic birds, endothermy in Mesozoic mammals and birds, very large size in Jurassic and Cretaceous dinosaurs, new marine top-predator roles in the

Mesozoic (ichthyosaurs, plesiosaurs, mosasaurs), burrowing and tree-climbing among some Mesozoic and Cenozoic amniote groups, further expansion of niches to include polar regions and nocturnal habits among Cenozoic mammals, and so on.

These additions of dietary modes and habitats have been a key feature of the evolution of tetrapods (Benton, 1990b) and, presumably, also of fishes, and indeed many other expanding clades. The tetrapods shifted from being essentially 100% fish-eaters and 100% inhabitants of fresh water and adjacent land areas in the late Devonian and early Carboniferous, to a much wider array of niches. Indeed, fish-eating and amphibious freshwater niches fell to a steady 10% of all families from the late Mesozoic onwards (Benton, 1990b). With a fine magnifying glass, perhaps one could argue that there were 'steady state' conditions world-wide between each of these adaptive bursts of radiation (Rosenzweig, 1995). However, the sum total of new adaptations that can be identified is larger than the lists just given (for example, colonial nesting in mole rats, feeding on garbage by various urbanizing mammals, ant-eating, etc.). How small do the 'steady states' have to become before they evaporate altogether into a picture of opportunistic expansion in diversification?

Can the expansion of diversity go on forever? Of course, there is an ultimate limit to the numbers of families, or other taxa, that can inhabit the Earth at any time; such a limit would be caused not least by the amount of standing room on the Ark. Presumably, though, if a limit of living space were approached, ever-smaller organisms might perhaps be favoured by evolution. Equally, as has happened so many times during evolution, organisms would take unexpected measures to survive by, for example, occupying the air, burrowing into sediments and, in the case of some bacteria, living deep within the Earth's crust. With size reduction, the ultimate limit to the diversification of life might then become the availability of the chemical components of life, principally carbon. Of course, other partial escape mechanisms from such a limiting factor are to speed up the rate of cycling of carbon through biogeochemical cycles, to retain such chemicals in the organic realm for longer and to reduce the amount of time they remain buried.

Clade Replacements

An expectation of the logistic model for diversification is that there is a phase of rapid increase in diversity, followed by a levelling-off phase as the gradient of the curve diminishes (Figure 12.6a). This occurs as diversity approaches the equilibrium level and the rate of increase diminishes progressively the closer the diversity approaches that level. Just as the rate of diversification declines, so the phase of rapid expansion (filling of ecospace) switches to a prolonged phase of dynamic equilibrium; new taxa may arise, but they will tend to displace pre-existing taxa (Darwin's barrel of apples analogy). In a real case of the evolution of a major clade, one might expect to find a change in the nature of taxon originations, from expansionist to equilibrial, from taxa moving into unoccupied ecospace, to more and more competitive displacement.

Is there any evidence for such a switch from expansionist to competitive originations among tetrapods, say in the later Palaeozoic, or perhaps in the late Cenozoic, if Courtillot and Gaudemer (1996) were correct in suggesting that the diversification of 'all life' is entering the levelling-off phase of a logistic curve? Benton (1996a,b) has carried out a comprehensive census of all 840 families of tetrapods that have a fossil record and which include more than a single species. He found that 13% of familial origins could be (but need not be) explained by competitive interaction with a pre-existing family on a reasonable estimate. Even using a maximal estimate of possible competitive interactions, where it was assumed that the fossil record was extraordinarily incomplete, the proportion of familial originations that were candidate competitive replacements (CCRs) rose to 26%. In a plot of CCRs (Figure 12.7), the distribution of peaks appears to be largely random with respect to time. There is no evidence for a rise in CCRs in the late Palaeozoic or the late Cenozoic and, hence, no suggestion that tetrapod niches were filling up at either of those times. This is evidence against a logistic model of diversity increase.

QUALITY OF THE FOSSIL RECORD OF VERTEBRATES

The quality of the fossil record, or of some segment of it, may be assessed in a qualitative way and assertions may be made based upon field experience, or upon overall surveys of a group. These intuitive approaches have been discussed above, especially with regard to the fossil record of vertebrates. However, these observations are often hard to justify, especially to non-palaeontologists, and they may give misleading evidence; cases were set out above that the fossil record of vertebrates is either good (stability in our understanding; no real surprises) or poor (incompleteness of sedimentary record; rarity of fossils). Of course, ultimately a decisive test on the quality of any fossil record cannot be applied, since no mortal can know what really happened in the past. Without a yardstick of the truth, it is clearly impossible to test assertions about the fossil record once and for all. However, three quantitative approaches may shed some light on completeness.

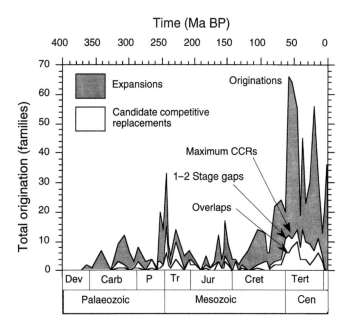

Figure 12.7 *Distribution in time of originations of tetrapods of all habitats, showing also the occurrences of overlaps and of maximum candidate competitive replacements (CCRs). CCRs were identified by comparison of pairs of families. First, stratigraphic range charts were plotted for each combination of body size, diet and habitat, and maximum geographical ranges of each family were noted. Then, the point of origin of each of the 840 families was scrutinized to determine whether it was a CCR or an expansion. CCR cases were subdivided into overlaps, where the stratigraphic range of the family over-lapped another family, and situations where the family apparently originated at the precise time of extinction of another (gap 0), or after a gap of one or two (gap 1, 2) stratigraphic stages. Overlaps give evidence that the two families could have encoun-tered each other, while the gap 0, 1 and 2 cases allow for possible incompleteness of the fossil record. CCRs are plotted as overlaps and maximum CCRs (the sum of all overlaps and gap 0, 1 and 2 cases). Dev, Devonian; Carb, Carboniferous; P, Permian; Tr, Triassic; Jur, Jurassic; Cret, Cretaceous; Tert, Tertiary; Cen, Cenozoic. (Data from Benton, 1996a,b)*

Taphonomic Tests of the Completeness of the Fossil Record

The first quantitative approach to assessing the completeness of the fossil record is to compare like with like through time. For example, it is a fair assumption that deposits containing exceptionally preserved fossils (*Lager-stätten*) from the Cambrian are not necessarily worse than equivalent sedi-mentary settings of Jurassic or Eocene age. *Lagerstätten* of specific types may then be treated as comparable snapshots of the true diversity of life at particular times and in specific environments/regions, since they include soft parts and entirely soft-bodied organisms that are otherwise missed. They may then act as standards against which other, more typical, deposits (those that do not preserve soft parts and soft-bodied organisms) may be compared. This may be a fruitful approach (Allison and Briggs, 1993; and see Chapter 3) and it can provide a semi-quantitative assessment of how the quality of 'normal' fossil deposits has varied through time.

Comparing like with like need not stop at *Lagerstätten*. Certain other kinds of fossil accumulations may be treated as equivalent and unaffected by time-related destructive phenomena (see also Chapters 1, 5 and 11). For example, coquinas, or winnowed accumulations of fossil shells, may survive in equally unmetamorphosed and uncrushed condition from the early Palaeozoic and the late Cenozoic. Kidwell and Brenchley (1996) found no diminution of the quality of preservation backwards in time when they compared large samples of Ordovician–Silurian, Jurassic and Neogene coquinas. They did find other time-related trends, some of them actually making the Mesozoic fossil record poorer than that of the Palaeozoic; for example, a dramatic increase in the diversity and effectiveness of predatory organisms that crush shells, and increases in the diversity of organisms that burrow and bore through sea-bed sediments. Further, in line with the evolution of new and ever-more fiendish groups of shell-crushers and shell-borers, most shells of the potential prey became thicker and, hence, the younger coquinas are themselves thicker (more and thicker shells are introduced into the shell beds), but they are also probably more time-averaged (that is, representing a longer time period of accumulation).

Historical Tests of the Completeness of the Fossil Record

Historical approaches may also be helpful. The 'collector curve' (Figure 12.8) is a useful approach that is often used in ecology and may be used in various palaeontological studies. This technique was devised as a time-saving approach to help field ecologists decide when to stop collecting specimens. A primary requirement of an ecological survey is to produce a list of all the species in an area. At first, the collecting goes well and most specimens that are picked up represent additions to the list. In time, the 'hit' rate declines, and it becomes harder and harder to find a new species that has not already been identified. The collector curve simply quantifies the collecting effort (assessed as either number of specimens picked up or time) against the identifications of new species. When the rate declines markedly, it is assumed that the collector is approaching the true total diversity and he/she then decides to stop collecting at the 90% or 95% level; the effort required to find the very last species in an area might well equal the effort expended in finding all the others.

Palaeontologists can readily use such an approach when collecting fossils

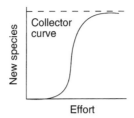

Figure 12.8 *The basic collector curve is a way of estimating when to stop collecting. At first, the 'effort' devoted to collecting is richly rewarded with discoveries of new species, but as the collector approaches the maximum possible total (dashed line), the effort expended to find a new species increases dramatically*

from a new locality. This also provides a technique for standardizing collecting effort when a palaeontologist is attempting to draw up range charts and assessments of the relative abundance of particular groups through a section. The collector curve may also provide an approach to assessing the quality of the fossil record.

Maxwell and Benton (1990) suggested that, given enough time, all the fossils that are out there in the rocks will be collected. This sample of fossils in the rocks does not, of course, represent all of the life of the past, since multitudes of species must have come and gone, and yet never have been fossilized. None the less, those organisms that were fossilized are potentially knowable. Of course, like the ecologist setting out to compile a faunal list on a new tropical island, the palaeontologist has to exert ever more effort to find those last rare specimens (see Chapter 8).

Large-scale palaeontological collector curves may be compiled in various ways. One simple approach is to use years of study as the measure of effort (since it would be hard to count accurately the numbers of specimens palaeontologists have inspected and perhaps discarded) and then to count the rate of accretion of new species. One caveat is that the rate of accretion of new species can be distorted horribly by an enthusiastic splitter (a taxonomist who names species based on small differences).

One example is a preliminary study of the rate of discovery of new dinosaurs (Figure 12.9). Here, the effort scale runs from 1824, the date of publication of the first formal dinosaur name, *Megalosaurus*, by William Buckland, up to 1990. The 'new species' axis represents 'net new species', that is, actual new species named, minus those new species that were later synonymized. If the literal number of new species had been plotted, the final total diversity would have been much higher, but any interpretations would then have been based partly on fantasy. Admittedly, there is no guarantee that later synonymies were always justified, nor that all currently accepted dinosaur names are actually valid. In addition, the synonymy rate falls off for newer dinosaur names since these have not all gone through the normal

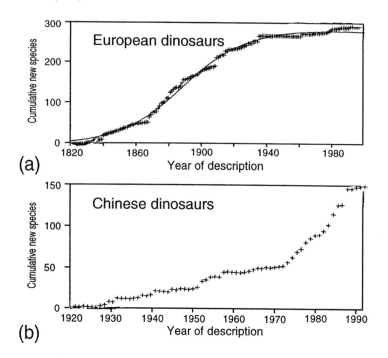

Figure 12.9   *Collector curves and discovery of dinosaurs in (a) Europe and (b) China. In Europe, the rate of recovery of new dinosaurs has been fairly static since the 1920s, and perhaps all the species that are in the rocks to be found have been found. In China, on the other hand, collecting did not begin until the 1920s; the rate of determination of new species increased in the 1970s, with an equivocal flattening in the 1990s*

processes of reassessment by other workers.

Dinosaurs from two broad regions are shown, those from Europe (Figure 12.9a) and from China (Figure 12.9b), approximately equivalent areas of the Earth's surface and, thus, potentially roughly equally likely to yield dinosaurs of different kinds, and potentially likely to yield comparable total numbers of dinosaur species. Intuitively, palaeontologists would expect Europe to be a 'mature' region in terms of the discovery of new species of dinosaurs, since the first finds were made there, and there is a record of nearly 200 years of collection in generally populous areas that are actively combed by large numbers of collectors and academics. China, on the other hand, is a mere juvenile, since the first dinosaurs were identified there only in the 1920s, much of the area is remote, and there have been far fewer collectors and palaeontologists in China than in Europe. Indeed, the intuitive expectation is borne out: the European collector curve shows a fully developed logistic pattern and the rate of determination of new species of dinosaurs has been levelling off since the 1920s. On the other hand, in China the long slow phase of accretion, from the 1920s to the 1960s, has been followed by the accelerating phase since the 1970s and there is only an equivocal hint of a levelling off in the 1990s. These figures are also rather more immature than the European totals, since more revision and synonymy of the Chinese material may occur in the future. Up to 1990, about 220 species of dinosaurs had been identified from Europe and 160 species from China. Can Chinese palaeontologists expect to push their totals to 220 or even higher? Until the logistic curve definitely begins to bend over, no one can tell!

The historical approach was used, in a slightly different form, in Maxwell and Benton's (1990) assessment of the fossil record of tetrapods and in Sepkoski's (1993) assessment of the fossil record of marine animals. In the former case, the historical time span under investigation ran from 1900 to 1987, and especially from 1966 to 1987. In the latter case, the time span was from 1982 to 1992. The premise in both cases was the same, that many changes had occurred in our knowledge of the fossil record (new specimens, revised stratigraphy, revised taxonomy), but had these had a significant effect on palaeontological knowledge as a whole? In both studies, it was shown that the changes were randomly distributed with respect to time and major clades; in other words, many changes, but the overall pattern stays the same. Specific changes were that overall diversity increased and that extinction events became sharper (as new fossils were found that filled gaps). Thus, the fossil record is good enough to read empirically, as a valid indicator of the true history of life, but how much change has there been?

Maxwell and Benton (1990) hoped that palaeontological knowledge was improving. They, and Sepkoski (1993), were only able to show, however, that palaeontological knowledge was changing. Perhaps all the new fossils, new stratigraphies and new taxonomies were actually making things worse. How could it be shown that the changes in knowledge were tending in the right direction, towards a full knowledge of the fossil record? This required a different approach, the use of some external yardstick and, luckily, such a yardstick exists.

Phylogenetic Tests of the Completeness of the Fossil Record

The fossil record may be tested against cladistic and molecular phylogenies, since phylogenies are independent of stratigraphy. Cladograms are constructed by the search for patterns in the distribution of characters among fossil or living organisms (Platnick, 1979), and there is no test of geological age in the assessment of a character or a taxon. Likewise, molecular phylogenies are based on comparisons of sequence or general similarity data, and stratigraphy is not involved. This means that the order and distribution of fossils in

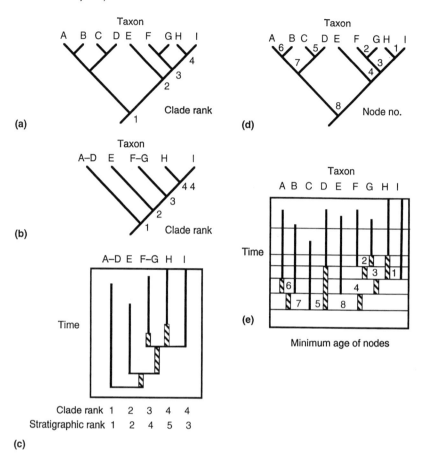

Figure 12.10   *Techniques for assessing the quality of the fossil record. Comparisons are made between branching order in cladograms and stratigraphic data (a–e), and between the relative amount of gap and the known record (e). The example is a cladogram with nine terminal branches (A–I). For comparisons of clade order and age order, cladistic rank is determined by counting the sequence of primary nodes in a cladogram (a): nodes are numbered from one (basal node) upwards to the ultimate node. In cases of non-pectinate cladograms (a), the cladogram is reduced to pectinate form (b), and groups of taxa that meet the main axis at the same point are combined and treated as a single unit. The stratigraphic sequence of clade appearance is assessed from the earliest known fossil representative of sister groups, and clade rank and stratigraphic rank may then be compared (c). Matching of clade rank and stratigraphic rank may be tested by Spearman rank correlation (SRC). SRC coefficients may range from 1.0 (perfect correlation) through 0 (no correlation) to −1.0 (perfect negative correlation). For assessing the proportion of ghost range, or minimum implied gap (MIG), and known stratigraphic range (SRL), the whole cladogram is used (e). The MIG (striped bar) is the difference between the age of the first representative of a lineage and that of its sister, as oldest known fossils of sister groups are rarely of the same age. The proportion of MIG to known range is assessed*

the rocks may be compared with morphological cladograms or molecular phylogenies, and these independent sources of data on the true phylogeny of life may be assessed for congruence (Benton, 1994, 1995b; Benton and Hitchin, 1996, 1997).

The assessment metrics

There are a variety of metrics for comparing phylogenies and fossil records (Figure 12.10), such as Spearman rank correlation (SRC), the Relative Completeness Index (RCI), and the Stratigraphic Consistency Index (SCI). The SRC is an established non-parametric statistical test and it has been used in comparing the order of fossils in the rocks with the implied order of appearance of groups based on the sequence of nodes (branching points) in a cladogram. The first applications of the SRC test for this purpose were by Gauthier *et al.* (1988) and Norell and Novacek (1992a,b).

The RCI was proposed (Benton, 1994; Benton and Storrs, 1994) as an additional metric that took account of the actual time spans between branching points and of implied gaps before the oldest-known fossils of lineages. Sister groups, by definition, originated from an immediate common ancestor and diverged from that ancestor. Thus, both sister groups should have fossil records that start at essentially the same time. In reality, usually the oldest fossil of one lineage will be older than the oldest fossil of its sister lineage. The time gap between these two oldest fossils is the 'ghost range' or minimal cladistically implied gap. The RCI assesses the ratio of ghost range to known range, and high values imply that ghost ranges are short and, hence, that the fossil record is good.

(It has been suggested (Paul, 1992; Wagner, 1995) that ghost ranges may be an artefact of the cladistic technique, which assumes that sister taxa generally originate at the same time (by dichotomous branching). However, if one sister includes ancestors of the other and branching occurs after the origin of one of the sister taxa, the ghost range might disappear and, hence, the RCI

---

using the relative completeness index (RCI), according to the formula:

$$RCI = \left\{ 1 - \frac{\Sigma(MIG)}{\Sigma(SRL)} \right\} \times 100\%$$

RCI values may range from 100% (no ghost range) through 0 (ghost range = known range) to high negative values (ghost range ≫ *known range). Stratigraphic consistency is assessed (d, e) as a comparison of the ratio of nodes that are younger than, or of equal age to, the node immediately below (consistent), compared to those that are apparently older (inconsistent). The stratigraphic consistency index (SCI) is assessed on the full cladogram (d, e). SCI values range from 1.0 (all nodes stratigraphically consistent) to 0 (no nodes stratigraphically consistent). (Data from Benton and Hitchin, 1997)*

technique would be invalid. It is not clear whether the ancestor model applies to a majority of cases or not. Certainly, the criticism of ghost ranges is not valid for groups analysed at low taxonomic levels where the fossil record is patchy, since the taxa available for analysis are necessarily only a sample of all those that ever existed and the chances of hitting on a true ancestor are small. This is probably the case for the vertebrates, echinoderms, arthropods and other groups for which cladograms are abundant, and on which we have based our tests. On the other hand, many fossil molluscs and foraminifera may be ancestors of other known forms and the RCI technique would not perhaps work for them. If most sister taxa split at a single point of origin, as asserted by cladists, then the technique works.)

The SCI was proposed by Huelsenbeck (1994) to test how well the nodes in cladograms corresponded to the known fossil record. Nodes are dated by the oldest known fossils of either sister group subtended from the node. Each node is compared with the node immediately below it. If the upper node is younger than, or equal in age to, the node below, the node is said to be stratigraphically consistent. If the node below is younger, the upper node is stratigraphically inconsistent. The SCI for a cladogram compares the ratio of the sums of stratigraphically consistent to inconsistent nodes. SCI values can indicate cladograms whose nodes are all in line with stratigraphic expectations through to cladograms that imply a sequence of events that is entirely opposite to the known fossil record.

These metrics may be applied to individual phylogenetic problems – that is, which of these 10 cladograms of marsupial relationships provides the best fit to current stratigraphic evidence? – or they may be used to assess large samples of cladograms. In the cases to be presented here, the latter approach is used. The assumption here is that variations in cladistic or molecular techniques, variations in taxic level and stratigraphic variations are subsumed in the overall variation within a large sample. So far, the sample of cladograms assessed in these ways amounts to 384, composed of 174 cladograms of tetrapods, 147 cladograms of fishes and 63 cladograms of echinoderms (Benton and Hitchin, 1996, 1997; see also http://palaeo.gly.bris.ac.uk/palaeo/cladestrat.html).

Testing fossil record quality: branching order

The key question to be answered by this regime of testing is whether the fossil record is good enough. First results were encouraging: Norell and Novacek (1992a) found that 18 out of 24 test cases (75%) of family-level and generic-level cladograms of vertebrates gave statistically significant ($P <$ 0.05) correlations of clade and age data, using the SRC test. In larger samples, Norell and Novacek (1992b) found significant correlation in 24 of 33 test cases (73%), while Benton and Storrs (1994) found significant correlation in 41 of 74 test cases (55%). In other words, for tetrapods, there was apparently good agreement between stratigraphic and cladistic evidence in most cases.

However, subsequent assessments, based on larger samples of cladograms, provided more disappointing results. For echinoderms, Benton and Hitchin (1996) found that only 24 out of 63 cladograms (38%) showed statistically significant ($P < 0.05$) matching of clade rank and age rank data. For fishes, the figure was 37 out of 147 cladograms (25%) and for tetrapods it was 87 out of 174 cladograms (50%). The results for all cladograms in the test sample was that 148 out of 384 showed significant SRC values (38%). Does this mean that only a minority, something from one-quarter to one-half, of cladograms are congruent with fossil records, that the fossil record is equally likely to give the wrong order of fossils as it is to give the correct order?

Two points may be made. The comparisons by Benton and Hitchin (1996) were made on large samples of cladograms and these included some categories of cladogram that performed badly in the SRC test (Benton and Hitchin, 1997):

1. *Cladograms based on species or genera* Taxonomically low-level cladograms generally gave very poor matches to stratigraphy, largely because the fossil records were not adequate. Often only one or two out of 10 or 15 genera had any fossil representatives.
2. *Rejected cladograms* Our sample of cladograms was comprehensive and we included a number of cladograms that had been published as possible, but not preferred, solutions. Many of these performed badly.
3. *Cladograms with small numbers of terminal taxa* When there are only four or five terminal taxa, the SCI, in particular, has only two or three nodes to compare with the basal node. The sequence has to be perfect in order to achieve a good result and one mismatch gives a very low value. For the SRC test also, cladograms with fewer than five or six terminal taxa must have a perfect match of the sequences of clade and age order before they produce a statistically significant coefficient. For both the SCI and SRC metrics, acceptable scores may be achieved by larger cladograms even if there are some mismatches.

Testing fossil record quality: RCI and SCI metrics

Results of the tests of the quality of the fossil records of echinoderms, fishes and tetrapods using the SRC statistic were disappointing. However, much better results were obtained with the RCI and the SCI metrics (Benton and Hitchin, 1996, 1997). It may seem unusual that many cladograms that apparently failed the SRC test of matching between age and clade data should pass with another metric. The reason is that the RCI and the SCI measure different aspects of cladogram and fossil record quality, and perhaps the SRC test is too indiscriminate for many purposes. The SRC simply compares raw orders of fossils and branching points. It takes no account of the overall amounts of time involved, nor especially of the seriousness of a mismatch. We found that many cladograms failed the SRC test because they had many nodes packed within a narrow time band. Some of the nodes were out of sequence by only 1–2 Myr or less, which is an insignificant amount of time in most analyses.

For all three groups assessed, more cladograms have RCI values ≥ 0.5, than values < 0.5 (Figure 12.11). The pass rates are 49 out of 63 cladograms (78%) for echinoderms, 124 out of 147 cladograms (84%) for fishes, and 128 out of 174 cladograms (74%) for tetrapods (Figure 12.12). The pass rate for all cladograms is 78%. In other words, 301 of the 384 cladograms tested have more than twice as much of their ranges represented by fossils than represented by ghost range. The differences in mean values of RCI for echinoderms (mean, 62.3%) and fishes (mean, 69.4%) are modest, but continental tetrapods have a much lower mean value (mean, 49.8%).

The pass rates are similarly favourable for the SCI measure (Figure 12.11). In these cases, all three sets of cladograms have significantly more than half their nodes showing stratigraphic consistency than inconsistency. The pass rates are 60 out of 63 cladograms (95%) for echinoderms, 102 out of 147 cladograms (69%) for fishes, and 152 out of 174 cladograms (87%) for tetrapods (Figure 12.12). The pass rate for all cladograms is 82%, based on 314 of the 384 cladograms (the SCI metric could not be calculated for 70 small cladograms in the full sample). The reasons for significantly higher mean values of SCI for echinoderms (mean, 0.78) than for fishes (mean, 0.55) and tetrapods (mean, 0.66) are not immediately evident.

The results of the RCI and SCI metrics show that fossil records are on the whole good for echinoderms, fishes and tetrapods. Comparisons among the three groups show that none of them consistently has a better fossil record, or better cladistic resolution, than the others. Each of the animal groups under study could be said to have the best fossil record since each is supported by one of the three tests: tetrapods by the SRC, fishes by the RCI and echinoderms by the SCI (Figure 12.12). Only one group comes out worst on two of the tests: fishes have the poorest showing according to the SRC and SCI metrics. Tetrapods have the worst fossil records according to the RCI metric, while echinoderms are not worst of the three animal groups according to any of the metrics.

Comparing continental and marine habitats

Benton and Simms (1995) obtained some counter-intuitive results when they showed that continental tetrapods have a fossil record that is as good as, or better than, that of echinoderms, based on comparisons of results obtained with the SRC and RCI metrics. This surprising result could not have been

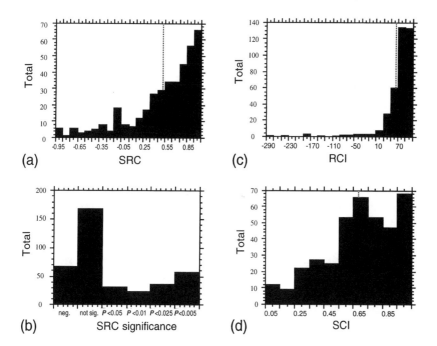

Figure 12.11 *Assessments of congruence between stratigraphic and cladistic data show highly skewed distributions. Values for three metrics calculated on a sample of 384 cladograms of echinoderms, fishes and tetrapods: (a) Spearman rank correlation (SRC) coefficients; (b) measures of the significance of those SRC coefficients, which take account of cladogram size; (c) Relative Completeness Index (RCI) values; and (d) Stratigraphic Consistency Index (SCI) values. Mean values for each sample are indicated by dotted lines. (Data from Benton and Hitchin, 1997)*

predicted from observations of the field occurrence of both groups: tetrapods are found in sporadic and unpredictable sedimentary settings, while echinoderm remains are hugely abundant in many marine-shelf deposits.

A more detailed comparison of SRC, RCI and SCI metrics for all marine cladograms and all continental cladograms yielded mixed results (Benton and Hitchin, 1996; Figure 12.13). The SRC test showed that 87 out of 174 continental cladograms (50%) had significant matching of age and clade order, while the value for marine groups was only 61 out of 210 cladograms (29%), much worse than the results for echinoderms alone reported by Benton and Simms (1995). The pass rate for RCI values was much more comparable, with 173 of the 210 marine cladograms (82%) yielding values higher than 50%, compared to 128 of 174 continental cladograms (74%). Mean values confirmed that marine cladograms (mean RCI, 67.3%) show a lower proportion of implied gaps than do continental cladograms (mean

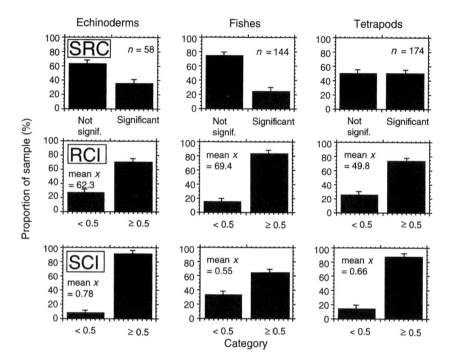

Figure 12.12   *Summary of the metrics for comparison of cladogram data and strati-graphic age data. Metrics indicated are Spearman rank correlation (SRC) of age and clade data; the Relative Completeness Index (RCI), based on comparisons of known and implied stratigraphic ranges; and the Stratigraphic Consistency Index (SCI) of nodes in cladograms. The metrics have been applied to large samples of cladograms (n = number of cladograms in sample) for echinoderms, fishes and tetrapods. Comparisons are be-tween significant and non-significant SRC coefficients, and between frequencies of values of the RCI above and below 50%, and frequencies of values of the SCI above and below 0.5. The differences in values among the three groups are significant, based on comparison of the binomial error bars. (Data from Benton and Hitchin, 1996)*

RCI, 49.8%). The pass rate for SCI values, on the other hand, favoured the continental cladograms, where 152 of 174 cladograms (87%) had values ≥ 0.5, compared with 162 of the 210 marine cladograms (77%). Mean values suggested that continental cladograms (mean SCI, 0.66) perform slightly better than marine cladograms (mean SCI, 0.62) in the SCI test.

The finding that continental vertebrates have a fossil record of similar quality to marine echinoderms and fishes suggests two observations. (1) The relative abundance of specimens at individual fossil localities is no indicator of the completeness of their fossil record on a large scale; this depends on the number of stratigraphic horizons that have yielded fossils and on the packing of those horizons in time. (2) The fossil record of continental tetrapods has probably been more intensively studied than has that of echinoderms and fishes. Hence, current knowledge of the tetrapod fossil record is now higher on the collector curve (numbers of taxa vs. effort), and may be assumed to approach closer to the level of complete sampling and full knowledge of all fossil taxa that exist in the rocks.

COMPARING CHANGES IN KNOWLEDGE OF THE FOSSIL RECORD

The SRC and RCI metrics have been used to compare historical aspects of the understanding of the fossil record. It might be expected that the addition of new fossil finds and re-analysis of older ones would improve the fit of age data to a fixed sample of cladograms, by the filling of gaps, and corrections of former taxonomic assignments. However, in a comparison of a 1967 data set (Harland *et al.*, 1967) with one from 1993 (Benton, 1993), Benton and Storrs (1994, 1996) found no change at all in the proportions of cladograms that showed statistically significant ($P < 0.05$ and $P < 0.01$) matching of clade and age order, although there had been a change in the status of 28 of the 71 cladograms compared (39%; Figure 12.14). In other words, as a result of 26 years of work, new discoveries and reassignments had improved the fit in 20% of cases, but had caused mismatches of clade and age data in a further 20% of cases. Sometimes, a new fossil does not fill a gap, but creates additional gaps on other branches of a cladogram.

This discovery of a lack of improvement in the congruence of clade versus age rank order is important, since it highlights the fact that mismatches may arise from subtle changes in knowledge. Non-correlation may result from minor variations in fossil dating, and may not imply wildly different evidence about the history of life from cladograms and from fossil occurrences.

However, the RCI metric detected a significant improvement in knowledge of the tetrapod fossil record from 1967 to 1993. In their study of cladograms of vertebrates, Benton and Storrs (1994, 1996) found that the mean RCI value shifted from 67.9% to 72.3%, a statistically significant difference, according to a Wilcoxon signed ranks test ($P = 0.026$). In other words, comparisons of the relative completeness of cladograms shows a significant improvement, by about 5%, in knowledge of the fossil record over the past 26 years of research. Hence, new fossil discoveries, and reassignments of older ones, do positively affect the amount of ghost range, although such changes in knowledge did not apparently affect the match of clade and age rank order, as assessed by the SRC test.

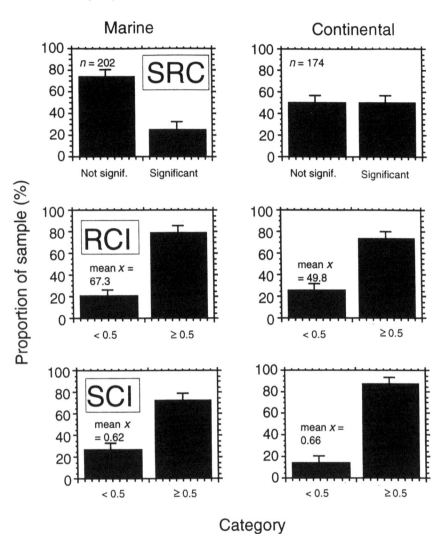

Figure 12.13 *Summary of the metrics for comparison of cladogram data and strati-graphic age data. Comparisons of cladograms of marine (echinoderm + fish) and conti-nental (tetrapod) cladograms according to the Spearman rank correlation (SRC), Relative Completeness Index (RCI) and Stratigraphic Consistency Index (SCI) metrics. The dif-ferences in values among the three groups are significant, based on comparison of the binomial error bars. (Data from Benton and Hitchin, 1996)*

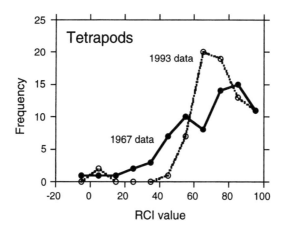

Figure 12.14 *Relative improvement in fossil record quality from 1967 (Harland et al., 1967) to 1993 (Benton, 1993). During these 26 years, gaps in the record were filled and there is a clear shift in the distribution of Relative Completeness Index (RCI) values to the right from 1967 to 1993, indicating improvement in palaeontological knowledge (significant shift at P < 0.05; t-test and non-parametric signs and Wilcoxon signed ranks tests). (Data from Benton and Storrs, 1994)*

CONCLUSIONS

The adequacy of the fossil record is hard to assess, as the various contributions in this book indicate. In a strict sense, it will never be possible to assess the adequacy of any segment of the fossil record, since the true picture will be forever unknown. Perhaps there have been whole phyla, or even kingdoms, of extraordinary organisms that lived at different times in the past, but which have left no fossil indications. One could imagine whole tribes of giant purple worms with bodies 100 m long, squirming around on Carboniferous forest floors, or an entirely unknown kingdom of photosynthesizing organisms that lived in Cambrian seas and moved by means of floppy wheels made from protoplasm. Such organisms are not impossible, but they are unlikely. The unlikelihood increases day by day as ever more palaeontological effort fails to turn up any hint of such unknown major groups of macroscopic organisms.

There are a variety of powerful new techniques for assessing the quality of the fossil record. One approach is to adopt a uniformitarian approach to specific kinds of fossil-preservation sites and to compare like with like across vast spans of geological time. The kind of intuitive argument presented in the previous paragraph represents a second approach, where the pattern of discovery over research time is investigated. It is statistically valid to quantify effort against discovery rate and to assert that the longer some unknown organism remains undiscovered, the less likelihood there is of its former existence. (This statement assumes that there is a premium attached to finding such an unknown organism and this is certainly very much the case.)

The third new approach, comparing phylogenetic and stratigraphic evidence to assess the adequacy of the fossil record, has been applied with considerable success to vertebrates. This approach is based on the observation that phylogenies (cladograms founded on morphological characters and molecular phylogenies) are constructed independently of geological evidence. The order and timing of splitting events in phylogenies may then be cross-compared with stratigraphic evidence on the order and timing of the appearance of groups in the fossil record in order to assess the degree of congruence. Good matching of the data sets implies that both the fossil record and the phylogeny are probably good, while a mismatch implies either a misleading fossil record or an inaccurate phylogenetic hypothesis.

The phylogenetic congruence assessments have indicated that there is no evidence that vertebrates have a fossil record that is either any better, or indeed any worse, than that of any other major group of animals. In addition, there is no evidence that the record of continental (that is, terrestrial and freshwater) tetrapods is worse than that of marine echinoderms or fishes. These kinds of assessments have proved highly fruitful and they provide a sound quantified answer to the old cry of 'the fossil record is pretty incomplete and uninformative'.

ACKNOWLEDGEMENTS

I thank Gilles Cuny, Steve Donovan, Becky Hitchin, Chris Paul and David Unwin for comments on the manuscript of this chapter, and the Leverhulme Trust (Grant F182/AK) for funding.

REFERENCES

Allen, K.C. and Briggs, D.E.G., 1989, *Evolution and the Fossil Record*, Belhaven Press, London: 265 pp.

Allison, P.A. and Briggs, D.E.G., 1991, *Taphonomy: Releasing the Data Locked in the Fossil Record*, Plenum, New York: 560 pp.

Allison, P.A. and Briggs, D.E.G., 1993, Exceptional fossil record: distribution of soft-tissue preservation through the Phanerozoic, *Geology*, **21**: 527–530.

Behrensmeyer, A.K. and Hill, A.P., 1980, *Fossils in the Making: Vertebrate Taphonomy and Paleoecology*, University of Chicago Press, Chicago: 338 pp.

Benton, M.J., 1983, Progressionism in the 1850s: Lyell, Owen, Mantell and the Elgin fossil reptile *Leptopleuron* (*Telerpeton*), *Archives of Natural History*, **11**: 123–136.

Benton, M.J., 1985, Mass extinction among non-marine tetrapods, *Nature*, **316**: 811–814.

Benton, M.J., 1989, Patterns of evolution and extinction in vertebrates. *In* K.C. Allen and D.E.G. Briggs (eds), *Evolution and the Fossil Record*, Belhaven, London: 218–241.

Benton, M.J., 1990a, The causes of the diversification of life. *In* P.D. Taylor and G.P. Larwood (ed.), *Major Evolutionary Radiations, Systematics Association Special Volume 42*, Clarendon Press, Oxford: 409–430.

Benton, M.J., 1990b, *Vertebrate Palaeontology*, Unwin Hyman, London: 377 pp.

Benton, M.J., (ed,), 1993, *The Fossil Record, 2*, Chapman and Hall, London: 839 pp.

Benton, M.J., 1994, Palaeontological data, and identifying mass extinctions, *Trends in Ecology and Evolution*, **9**: 181–185.

Benton, M.J., 1995a, Diversification and extinction in the history of life, *Science*, **268**: 52–58.

Benton, M.J., 1995b, Testing the time axis of phylogenies, *Philosophical Transactions of the Royal Society, London*, **B348**: 5–10.

Benton, M.J., 1996a, On the nonprevalence of competitive replacement in the evolution of tetrapods. *In* D. Jablonski, D.H. Erwin and J.H. Lipps (eds), *Evolutionary Paleobiology*, University of Chicago Press, Chicago: 185–210.

Benton, M.J., 1996b, Testing the roles of competition and expansion in tetrapod evolution, *Proceedings of the Royal Society, London*, **B263**: 641–646.

Benton, M.J., 1997a, *Vertebrate Palaeontology* (2nd edn), Chapman and Hall, London: 452 pp.

Benton, M.J., 1997b, Models for the diversification of life, *Trends in Ecology and Evolution*, **12**: 490–495.

Benton, M.J., 1998, The history of life: large databases in palaeontology. In D.A.T. Harper (ed.), *Statistical Methods in Palaeobiology*, Wiley, Chichester.

Benton, M.J. and Harper, D.A.T., 1997, *Basic Palaeontology*, Addison-Wesley, London: 325 pp.

Benton, M.J. and Hitchin, R., 1996, Testing the quality of the fossil record by groups and by major habitats, *Historical Biology*, **12**: 111–157.

Benton, M.J. and Hitchin, R., 1997, Congruence between phylogenetic and stratigraphic data on the history of life, *Proceedings of the Royal Society, London*, **B264**: 885–890

Benton, M.J. and Simms, M.J., 1995, Testing the marine and continental fossil records, *Geology*, **23**: 601–604.

Benton, M.J. and Storrs, G.W., 1994, Testing the quality of the fossil record: paleontological knowledge is improving, *Geology*, **22**: 111–114.

Benton, M.J. and Storrs, G.W., 1996, Diversity in the past: comparing cladistic phylogenies and stratigraphy. *In* M.E. Hochberg, J. Clobert and R. Barbault (eds), *Aspects of the Genesis and Maintenance of Biological Diversity*, Oxford University Press, Oxford: 19–40.

Briggs, D.E.G. and Crowther, P.R., 1990, *Palaeobiology: A Synthesis*, Blackwell Scientific, Oxford: 583 pp.

Carroll, R.L., 1977, Patterns of amphibian evolution: an extended example of the incompleteness of the fossil record. *In* A. Hallam (ed.), *Patterns of Evolution as Illustrated by the Fossil Record*, Elsevier, Amsterdam: 405–437.

Carroll, R.L., 1987, *Vertebrate Paleontology and Evolution*, W.H. Freeman, San Francisco: 698 pp.

Chatterjee, S., 1995, The Triassic bird *Protoavis*, *Archaeopteryx*, **13**: 15–31.

Clarkson, E.N.K., 1993, *Invertebrate Palaeontology and Evolution* (3rd edn), Chapman and Hall, London: 434 pp.

Courtillot, V. and Gaudemer, Y., 1996, Effects of mass extinctions on biodiversity, *Nature*, **381**: 146–148.

Cowen, R., 1990, *History of Life*, Blackwell Scientific, Boston: 470 pp.

Donovan, S.K., 1991, *The Processes of Fossilization*, Belhaven Press, London: 303 pp.

Flessa, K.W., 1990, The 'facts' of mass extinctions, *Geological Society of America Special Paper*, **247**: 1–7.

Fürsich, F.T., 1990, Fossil concentrations and life and death assemblages. *In* D.E.G. Briggs and P.R. Crowther (eds), *Palaeobiology; A Synthesis*, Blackwell Scientific, Oxford: 235–239.

Gauthier, J., Kluge, A.G. and Rowe, T., 1988, Amniote phylogeny and the importance of fossils, *Cladistics*, **4**: 105–209.

Harland, W.B., Holland, C.H., House, M.R., Hughes, N.F., Reynolds, A.B., Rudwick, M.J.S., Satterthwaite, G.E., Tarlo, L.B.H. and Willey, E.C., 1967, *The Fossil Record; A Symposium with Documentation*, Geological Society of London, London: 827 pp.

Hoffman, A., 1985, Island biogeography and palaeobiology: in search for evolutionary equilibria, *Biological Reviews*, **60**: 455–471.

Huelsenbeck, J.P., 1994, Comparing the stratigraphic record to estimates of phylogeny, *Paleobiology*, **20**: 470–483.

Jablonski, D., 1991, Extinctions: a paleontological perspective, *Science*, **253**: 754–757.

Kidwell, S.M., 1986, Models of fossil concentrations: paleobiologic implications, *Paleobiology*, **12**: 6–24.

Kidwell, S.M. and Brenchley, P.J., 1996, Evolution of the fossil record: thickness trends in marine skeletal accumulations and their implications. *In* D. Jablonski, D.H. Erwin and J.H. Lipps (eds), *Evolutionary Palaeobiology*, University of Chicago Press, Chicago: 290–336.

MacArthur, R.H. and Wilson, E.O., 1967, *The Theory of Island Biogeography*, Princeton University Press, Princeton, New Jersey: 203 pp.

Maxwell, W.D. and Benton, M.J., 1990, Historical tests of the absolute completeness of the fossil record of tetrapods, *Paleobiology*, **16**: 322–335.

Norell, M.A. and Novacek, M.J., 1992a, The fossil record and evolution: comparing cladistic and paleontologic evidence for vertebrate history, *Science*, **255**: 1690–1693.

Norell, M.A. and Novacek, M.J., 1992b, Congruence between superpositional and phylogenetic patterns: comparing cladistic patterns with fossil records, *Cladistics*, **8**: 319–337.

Paul, C.R.C., 1992, The recognition of ancestors, *Historical Biology*, **6**: 239–250.

Platnick, N.I., 1979, Philosophy and the transformation of cladistics, *Systematic Zoology*, **28**: 537–546.

Raup, D.M., 1979, Biases in the fossil record of species and genera, *Bulletin of the Carnegie Museum of Natural History*, **13**: 85–91.

Retallack, G., 1984, Completeness of the rock and fossil record: some estimates using fossil soils, *Paleobiology*, **10**: 59–78.

Rieppel, O., 1984, The problem of extinction, *Zeitschrift für Zoologischer Systematik und Evolutionsforschung*, **22**: 81–85.

Romer, A.S., 1966, *Vertebrate Paleontology* (3rd edn), University of Chicago Press, Chicago: 468 pp.

Rosenzweig, M.L., 1975, On continental steady states of species diversity. *In* M.L. Cody and J.M. Diamond (eds), *The Ecology of Species Communities*, Belknap Press of Harvard University, Cambridge, Massachusetts: 121–140.

Rosenzweig, M.L., 1995, *Species Diversity in Space and Time*, Cambridge University Press, Cambridge: 436 pp.

Sadler, P.M., 1981, Sediment accumulation rates and the completeness of stratigraphic sections, *Journal of Geology*, **89**: 569–584.

Sansom, I.J., Smith, M.M. and Smith, M.P., 1996, Scales of thelodont and shark-like fishes from the Ordovician of Colorado, *Nature*, **379**: 628–630.

Sepkoski, J.J., Jr, 1978, A kinetic model of Phanerozoic taxonomic diversity: I. Analysis of marine orders, *Paleobiology*, **4**: 223–251.

Sepkoski, J.J., Jr, 1979, A kinetic model of Phanerozoic taxonomic diversity: II. Early Phanerozoic families and multiple equilibria, *Paleobiology*, **5**: 222–251.

Sepkoski, J.J., Jr, 1984, A kinetic model of Phanerozoic taxonomic diversity: III. Post-Paleozoic families and mass extinctions, *Paleobiology*, **10**: 246–267.

Sepkoski, J.J., Jr, 1993, Ten years in the library: how changes in taxonomic databases affect perception of macroevolutionary pattern, *Paleobiology*, **19**: 43–51.

Sepkoski, J.J., Jr, 1996, Competition in macroevolution: the double wedge revisited. *In* D. Jablonski, D.H. Erwin and J.H. Lipps (eds), *Evolutionary Paleobiology*, University of Chicago Press, Chicago: 211–255.

Skelton, P.W. (ed.), 1993, *Evolution: A Biological and Palaeontological Approach*, Addison-Wesley, Wokingham: 1064 pp.

Stauffer, R.C., 1975, *Charles Darwin's Natural Selection: Being the Second Part of his Big Species Book Written from 1856–1858*, Cambridge University Press, New York: 694 pp.

Valentine, J.W., 1969, Patterns of taxonomic and ecological structure of the shelf benthos during Phanerozoic time, *Palaeontology*, **12**: 684–709.

Wagner, P.J., 1995, Stratigraphic tests of cladistic hypotheses, *Paleobiology*, **21**: 153–178.

Walker, T.D. and Valentine, J.W., 1984, Equilibrium models of evolutionary species diversity and the number of empty niches, *American Naturalist*, **124**: 887–899.

Subject Index

(*Note*: Names of stratigraphic intervals, etc., that are listed in figures and tables are not included in this index; however, those mentioned in captions are included.)

Systematic Index

(*Note*: Names of taxa that are listed in figures and tables are not included in this index; however, those mentioned in captions are included.)